Introductory Physics

A Problem-Solving Approach

Introductory Physics

A Problem-Solving Approach

Jesse David Wall
City College of San Francisco

D.C. HEATH AND COMPANY
Lexington, Massachusetts   Toronto

Illustrated by Paul Hewitt.

This book is dedicated to
the student who will take the time to do the problems.

In practical matters the end is not the mere speculative knowledge
of what is to be done, but rather the doing of it.
                                                    Aristotle, 362 B.C.

Preface

▶ General Objectives

Introductory Physics: A Problem-Solving Approach is written for people who have never studied physics before. It is designed to provide background for those who would like to take additional courses in physics or related sciences and for those who are curious about the physical world around them. A principal feature of this book is that it opens the door to further studies in physics for those students who wish to go further. It thus maximizes options. *Introductory Physics* is sympathetic to the needs of students who are reluctant to jump into the regular general college physics course and compete with students who have already had two semesters of high-school physics. They are able to approach physics without the usual measure of fear.

Physics is a large body of knowledge, and some concepts are best mastered at several levels. Concepts related to mechanical energy, momentum, and electrical voltage, as examples, are quite sophisticated, and most people need to think about them for awhile before they can claim a reasonable understanding of these ideas. This book provides a first exposure to the major ideas of classical physics and illustrates how they are applied to various physical situations.

The approach needed for a college course of this type differs in several respects from that found most useful on the high-school level. The most obvious difference is a question of pace. The material must be highly condensed and carefully selected since the college student is expected to cover it in a shorter time. The selection of topics truly essential for a first exposure has evolved only after years of experience with the course and hours of meetings with colleagues. All of that good material found in the usual high-school physics book has been boiled down, then boiled down again, so that only the best and most essential remains.

Another difference in approach is reading level. The author hopes his readers will not find difficult passages that are hard to understand, but he is even more fearful that his collegiate readers will find dull passages and start to count the pages to the end of the chapter. For that reason, greater attention has been given to making the material forceful and interesting to read than to making it easy. Both pace and reading level have therefore been adjusted to the needs of college students.

▶ To the Student

In using this book, you will gain insight into the world around you and the kind of background you will need to enjoy taking additional courses in physics. Physics is fascinating. You will find that you actually like physics if you take the time to master the skills presented in this book. These skills are simply those that allow you to understand scientific theory more easily and apply that theory to practice by carefully observing the world in which you live.

In your own way, you already know almost all the physical principles dealt with in this book. If you didn't know them, you wouldn't be able to tell magic from ordinary reality. When you watch a magician, you know that he is tricking you in some way when he pretends to do things that violate physical principles. This text will convince you that you really do understand physical principles and will show you how those principles can be expressed mathematically. You will learn to see through an algebraic expression as a mere statement of something you already know.

Applying theory to practice involves being able to work mathematical problems. You will find the rules of algebra to be powerful tools for reasoning. You may find that you actually enjoy working physics problems when you see how the puzzles they represent yield to the rules of logic you have already learned. The problems you will be asked to do amount to applications of physical theory to concrete situations. You will develop skill in observing which physical principles are involved in various situations and will learn how to manipulate the expression of those principles to give a quantitative answer. Your experience will grow as you learn the rules of problem solving, and this will allow you to work problems of increasing difficulty.

Problem Sets

The problem section at the end of each chapter has been set up in a programmed format to facilitate your mastery of the problems. You will find, first, a set of *Example Problems* selected to illustrate each of the physical principles covered in the chapter. A statement of the problem is provided on one side of the page together with a work space and a preliminary discussion. On the back of this page you will find a sample solution and a postmortem discussion so that you can see how the author would solve the problem. The idea is to give you a chance to work the problem without the answer staring you in the face.

If you can work the problem, then turn the page to compare your solution with the sample solution. It is likely that some difference will exist between your solution and the sample solution since there are usually several ways to solve a physics problem. Reading through the sample solution therefore gives you a chance to see the problem you just solved in perhaps a slightly different light. You need not think of the sample solution as better than your own, as long as you agree with the final result, but you should be able to see parallels between the two approaches.

If you are unable to work the problem, you should make your best attempt and then turn the page and study the sample solution so that you will be able to work the problem on your next pass through the Problem Set. You should then continue on to the next problem until you have finished all the examples. At that point, you should return to the Example Problems and make certain that you are now able to work each example without turning the page.

After you have completely mastered the Example Problems, go on to the next set, called *Essential Problems*. You will find that there are exactly the same number of these as there were Example Problems and that each Example Problem is matched to an Essential Problem that is to be solved in much the same way. The Essential Problems involve a different application of the same physical principle, and there may be a new twist here and there, but you should be able to breeze right through the Essential Problems if you fully understood the examples. If you happen to get stuck on the fourth Essential Problem, you had better go back and check the fourth Example Problem and figure out what you missed.

If you find yourself spending an outrageous amount of time on the Essential Problems, the chances are that you are not giving enough time to the Examples. Time invested in fully understanding the examples will save time in the long run.

Finally, under the heading *More Interesting Problems,* you will find problems that are more interesting, either in the sense that they are a bit more difficult or in the sense that they illustrate still other applications of a physical principle. In either sense, these problems are not worked in exactly the same way as an example. They allow you to try your hand at solving problems without being told exactly how. Often these problems are harder only because they don't call for all the steps you need to get at the answer. You will have learned enough, hopefully, from the examples and essential problems to select the proper intermediate steps you will need.

The answers to all problems, not just even- or odd-numbered problems, are given at the end of each chapter. All the answers are given to encourage you to work all the problems. Many of the problems represent important applications that would be treated as discussion in a fatter physics book. All the problems are stated in concrete terms, nameless objects and bodies being referred to as automobiles and bottles of liquid refreshment, and so on. You may find the realism (or surrealism, as the case may be) more interesting and entertaining than the nameless object style of problem statement, but you will need to work a number of problems stated in concrete terms to see that the methods apply to a whole class of problems. The result will be that you are better able to relate your problem-solving skill to real life situations.

The problems mimic real life situations in other ways. Some problems give you more information than you actually need. More information isn't given just to fool you, but to give you exercise in abstracting only that information you need to know. Some data are given to a greater number of significant figures than will affect the answer. This is just another form of redundant information that frequently shows up in real life, but which you can quickly learn to disregard.

You will find the answers in this text to only that accuracy justified by the data, except that an extra significant figure may appear in an intermediate solution to avoid a round-off error in the final answer. In determining the number of significant figures in the data, the widely accepted convention is adopted here of assuming that **zeros used to place the decimal point are not significant.** For example, the statement that 45,000 people live in a town would, in this text, mean forty-five thousand people give or take a few hundred. Do not be too concerned if your answer disagrees with the book's answer by as much as one significant figure; but if you are off by more than that, go back to see whether you have made a conceptual error. The use of the proper number of significant figures in doing calculations is a small matter, particularly in these days of the digital calculator, but sometimes all of those extra insignificant digits can stand in the way of your seeing what is really going on.

You will find yourself using a good amount of algebra and trigonometry in solving physics problems. Please don't let that discourage you, even if you didn't particularly like mathematics or found it difficult. Many people find that algebra is easier and more fun to use than it was to learn. If you feel weak in algebra or if you are only now taking trigonometry, you will find that this book has been arranged for your needs. Optics is taken up first, for example, because it requires little algebra or trigonometry. You can learn a good amount of physics even if you are struggling with algebra. The only trigonometry needed is the ability to look up the sine and tangent of an angle in a trig table. What is more, optics is not essential to the later development of the course. Nothing you might miss while you are practicing your math will be needed until you study optics again in a subsequent course. Fluid density and pressure are taken up next, delaying the need for trigonometry until as late in the course as possible. At about Chapter 4, however, the course starts to build. The chapter on Newton's laws of motion, Chapter 6, requires an understanding of the previous two chapters on force and acceleration. Everything else depends on an understanding of Chapter 6. There are just a few mathematical skills you must acquire before the course starts to build, but if you feel weak in mathematics, you had better start to work now. You will find appendices at the end of the book to help you.

▶ To the Instructor

Rush right along; don't get bogged down. The point of this book is to serve as a basic introduction while covering a broad range of topics. The most tempting error for those accustomed to teaching general college physics is to try to attain the same depth in this course. This often results in covering only the first half of the book, leaving the student with little to learn in the first half of general college physics and with no background for the second half of the course. Several features of this book will allow you to streamline your course if you take advantage of them. The problems are the simplest types that still illustrate the concept being taught. The only picture-frame force problems used, for example, are those in which the upward angles are equal. Topics which are not critical to the central development of the students' background have been omitted even though they might be simple enough in themselves. The fact that the index of refraction of a medium equals c/v, for example, is so nice and neat that it was omitted from the present development only after considerable pain and suffering. Some instructors may find it impossible to treat force equilibrium problems without also mentioning rotational equilibrium. If you find yourself incorporating additional topics that seem to naturally fit with those presented in the book, be careful to either note them as asides or be prepared to sacrifice other material. As it is, you will need to rush to cover the material presented, and you may not get to modern physics at that.

Optics and fluids are discussed out of their usual order to help the student who is weak in algebra and trigonometry. This material is covered first because it provides a good algebra review without presenting essential information needed later in the course. The need for trigonometry is thereby put off until Chapter 4, an advantage to those students who are taking trigonometry concurrently with this course. The main difficulty with using fluid pressure as an introduction to the idea of force, rather than as an application of that idea, is that the relationship between mass and weight has not been fully covered before it is needed for talking about density. The relationship between mass and weight is therefore introduced in Chapter 1 as an example of a proportional model. The early introduction of this important relationship sets the stage for its more complete development in Chapter 6.

The problems in this text are stated in concrete terms as applications of physical principles to lifelike situations. This style of problem statement has at least two advantages at this level of physics instruction. First, students who are making the transition between Piaget's concrete and formal levels of learning are far more comfortable with tangible situations. Even students who have achieved formal levels of thought in other subjects seem to enjoy their first physics problems more if they are stated in realistic, or even surrealistic, terms. Second, many of the problems lend themselves to incorporation into the lecture, either as a story or as a demonstration. The position of the virtual image in a plane mirror, for example, is far more interesting to the student doing Example Problem 2-1 if that student has actually seen the illusion of a ghostly candle burning under water produced in this easily performed lecture demonstration. The Atwood machine is a wonderful physics problem and an easy demonstration, but it is even more fun for students to work after they have heard Gerald Hoffnung's bricklayer story (available on BBC Records, REB 87M). With a little thought and planning, the problems stated in concrete terms can be made to complement the lecture.

A list of appropriate demonstrations is available from the author along with his lecture notes. Write to him at City College of San Francisco, Physics Department, San Francisco, California 94112.

Insist that your students solve problems in a suitable format. Some students will follow the lead of the sample solutions in the book because it is obviously a good idea to form good habits early. Others will use algebra and show their work only if their grade depends on it. The problems become more difficult in the last part of the book and require proper use of pencil and paper as tools for reasoning. By that time, the students should have acquired the knack of working problems with algebra before inserting data and calculating the answer—using the proper units of course.

If most of your students are going to take general college physics or other science courses, and thus have an opportunity to further develop their problem-solving skills at a later time, you may wish to limit the level of skill they need to be responsible for in this course. One way to limit the course is to use homework problems for all exams—word for word with only the numbers changed. This is a good way to take the fear out of physics. Your students know exactly what is expected of them. They need to learn how to work all the physics problems in the text, but that is all that they need to know. They can get through their first course in physics on pure diligence. You might think that this procedure would ruin your grade distribution, but it turns out that diligence falls on a bell-shaped curve like everything else. The author has been limiting all exams but the final to homework problems (all the problems in the book) for five of the seven years he has been teaching this introductory level course. The results vary only slightly from semester to semester. A reasonable distribution always results, and midterm exams based entirely on homework problems give results that correlate well with the final exam based on entirely new problems that the students haven't seen before. If the students can work the problems in the book, it turns out that they can work similar problems that aren't in the text.

You may find that this entry level physics course becomes a favorite course to teach. The students at this level are more interested in mastering the subject matter than in the grade they get. They know they need the skills to do well in subsequent courses, so they are easy to motivate. Even so, you can devote more than the usual amount of your effort to nurturing the students' appreciation and enjoyment of physics. They are going to need all of the enjoyment that can muster when they take future physics courses. Turning your students on to something you like is always fun, and this is a course where fun is a primary objective.

Acknowledgments

I would like to express special gratitude to Paul Hewitt for inspiring me and helping me to develop my pedagogical style as well as for doing the drawings for this book. I would also like to thank artistic Mary Jew for the beautiful job she did in hand lettering the example solutions. Since this book emphasizes problem solving, students will spend many hours enjoying Mary's handwriting.

Several students provided much-needed inspiration, criticism, and help. Martha Stasinos was particularly influential in helping me develop an adequate writing style by extensively copyediting the second syllabus edition. Mary Jo Callahan and Julie Racz also provided much encouragement and correction in the preparation of the third manuscript. Sharon Akutagawa extensively copyedited that manuscript and gave valuable style criticism and suggestions, which helped in the final rewrite. There are many other students, too numerous to list, whose comments and suggestions are reflected in this book and to whom I am very grateful and express thanks.

I would like to thank the many reviewers who looked at the manuscript and gave helpful suggestions—in particular, Mario Iona and Jackie Spears, who went through the third manuscript with a fine tooth comb and found many weak places that required revision. The responsibility for remaining errors, of course, rests with me.

Thanks, also, to Lewis Epstein for help on the last chapter and to my other friends in the Physics Department at City College for their help on this book.

I also thank my sister-in-law, Joyce Cauthen, and my wife, Ellen Wall, for help in getting started on this project. To my wife I owe even greater appreciation for encouraging me to get finished. It would never have been possible without her.

Contents

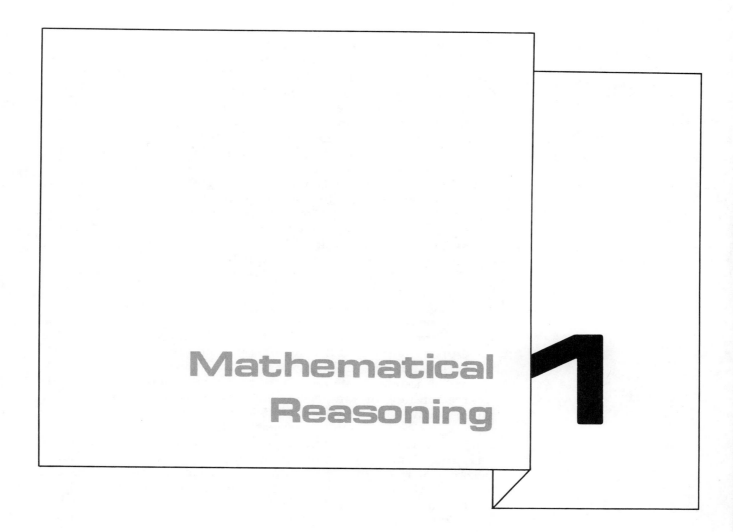

Mathematical Reasoning 1

People starting to study physics are often intimidated by the mathematical thinking involved. Actually, the reasoning used in introductory physics is far simpler than the complex and sophisticated reasoning used in everyday social interaction.

▶ Models

The "scientific method" resembles the kind of logic people use in making friends. A scientist constructing a hypothesis is building a mental model of reality in very much the same way you form an opinion about a person you have just met. Your judgment of that person is a mental model based on a large number of observations of that individual integrated with a whole lifetime of past experiences with other people. Synthesizing a model from individual facts is known as "inductive reasoning." Scientists do it, but so does everyone else.

A hypothesis, once formed, is tested by seeing if its application produces correct results. Going from general information to specific facts is called "deductive logic." You use this type of reasoning when you try to anticipate what a person will do in a given situation. If

FIGURE 1-1 *The act of thinking involves the making and using of mental models.*

1

FIGURE 1-2 *You smile and wave.*

your new friend acts as expected, your model of that person is confirmed. You smile and wave; your friend smiles and waves back. Nice person. If this new acquaintance grunts instead and makes an obscene gesture, you naturally revise your model of this friend. This depends on the facts, of course. Maybe you just cut him out of a parking space. All of these complicated factors must be taken into account if your revised mental model is to be any good.

The more you know about a person, the more sophisticated your model becomes. No matter how sophisticated, however, a mental model is never more than an approximation of reality. People sometimes lose sight of this fact and confuse their models with reality itself. In other people, this phenomenon is called prejudice. The model may represent a person or a whole group of people, and facts inconsistent with this model are largely ignored or accommodated as "special cases" without revising the basic model. (Prejudice, of course, occurs only in "other" people.)

This same tendency to forget the inherently approximate nature of mental models has existed in the physical sciences. The history of science has been a continual revolution, in which physical laws continue to be discovered and old models of the universe give way to new theories. From time to time it might seem that we have such a good understanding of the universe that no further revision of our theories is possible and that we can start ignoring minor inconsistencies or accommodate them as special cases. Prejudice thus tends to creep in our thinking about physical, as well as social, matters whenever we forget the limitations of our mental models.

The "scientific method," at least as it is applied to the physical sciences, seems to be a more formal way of thinking than the everyday kind of thinking used to get along in life. Students should not become alarmed by this, however, because "formal" means "simple" in this case. The inductive logic used in this book will be, for the most part, based on the direct proportion. The deductive logic will be nothing more than the rules of algebra that you should be somewhat familiar with. If this reasoning seems mathematical, it is only because it is simple enough to be reduced to a few symbols on a piece of paper. In the physical situations we shall deal with, it will be both helpful and easy to see through the mathematics as nothing more than simple thoughts. Compared to the vastly more complex thinking necessary just to survive in society, the simple thinking used here should be relatively easy to master.

This chapter will first develop the idea of a direct proportion so that you can understand the inductive logic used in this book to construct mental models. Then it briefly goes over the rules of algebra and computation you will need to manipulate these mental models and apply them to specific situations. If at the end of this chapter you feel yourself in need of a more thorough algebra review, you will find one at the end of the book in Appendix I.

▶ Proportions

A cause-and-effect relationship in physical situations is expressed mathematically as a **functional relationship.** One thing is said to be a function of another thing when a change in one is related to a change in the other. The speed of a train might be thought of as a

FIGURE 1-3 *Steam locomotive and passenger train resistance. (Data taken from Lewis Sillcox, Mastering Momentum (Simmons-Boardman Publishing Co., 1941), p. 11.)*

function of various things: the thrust of the locomotive, the load and number of cars, and the grade of the track would all have an effect on the train's speed. Frequently the relationship between variables is quite complicated if we try to take everything into account. In a surprising number of cases, however, useful results can be obtained from a simplified model based on the most elementary of all functional relationships—the **direct proportion.**

The resistance of a train, for instance, increases with speed. The actual relationship, from the experience of people who have taken the trouble to measure such things, is shown in the graph in Figure 1-3. If we were to say that the resistance, R, is directly **proportional** to the speed, s, expressed mathematically with proportionality sign $\propto$, then

$$R \propto s$$

we would be saying that equal increases in R are brought about by equal increases in s and that they are both equal to zero at the same time. On a graph, this would be a straight line going through zero. A close look at the data shows that this is not exactly what happens. The actual function is not a straight line but a gentle curve, a curve that doesn't even quite go through zero. In an actual case, it takes some force to overcome what is called static friction to get a train rolling from rest. If we were to say that resistance is proportional to speed, we would be ignoring these little difficulties. We would be ignoring static friction and the fact that the line is curved. That might be good enough, however, depending on how accurate we need to be in a given situation.

Suppose, for example, you only need a rough estimate of how fast a locomotive will pull a certain train. For practical purposes, the straight-line proportionality approximation is probably good enough. If you need a closer estimate of the train's speed, you can always use a more sophisticated approximation to the true curve. You could still use the idea of a proportion if you just restrict the range over which the proportionality is to hold true. Choose a small segment of the curve in the area of the graph in which you are interested and approximate it with a straight line. If the straight line doesn't go through zero, the relationship is properly called a linear relationship. The idea is the same as a proportionality except that

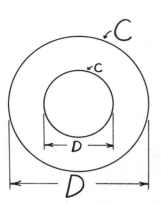

FIGURE 1-4 *The circle.*

you add in a constant. We will need this trick when we deal with absolute temperatures and pressures, but for most of our work the straight-line proportionality function passing through zero will give us a good enough approximation of reality.

The proportionality is a little easier to manipulate by the rules of algebra, which we have come to know and love, if we replace the proportionality sign, $\propto$, with an equality sign, $=$, and a proportionality constant, K, where K is some number.

$$R = Ks$$

One of the best known proportionality constants is represented by the Greek letter π (pi). The ancient Greeks noticed that the circumference of a circle, C, is proportional to its diameter, D:

$$C \propto D$$

That just means that the bigger circles will have bigger circumferences C, but the ratio of the larger circumference to the larger diameter D will be the same as for the smaller circle. π is simply defined as the ratio of the circumference to the diameter of a circle.

$$\pi \equiv \frac{C}{D} = \frac{C}{D}$$

(The three-line equality sign, $\equiv$, is used in this book to mean "is defined as," which says just a little more than "is equal to.") Thus, the well-known formula for the circumference of a circle

$$C = \pi D$$

may be looked upon as a simple statement of proportionality, with the equality sign, $=$, together with the proportionality constant π replacing the proportionality sign, $\propto$.

The use of proportionality constants has become as much a part of everyday thought in modern society as the stereotype, a mental trick used frequently for understanding society. Consider Archie, a cartoon teenager—one of the oldest teenagers around—in comic books for the past 30 years and more recently on Saturday morning television. Archie is a stylized stereotype whose understanding of reality might well be used to represent that of the ordinary person. Assume he has discovered from considerable experimental evidence that he must put gasoline in his jalopy if he is to go anywhere. He has further noticed that the amount of gasoline he needs is proportional to the distance he goes. Archie only puts in the amount of gasoline he actually needs. He is not wealthy enough to buy gasoline by the tankful.

The proportionality between the distance he can go and the gasoline he needs is a mathematical model, even if Archie does not think of it as such. The model can be expressed on paper as a proportionality between distance, d, and amount of gasoline, G:

$$d \propto G$$

or

$$d = wG$$

FIGURE 1-5 *The concept of gasoline mileage is a mathematical model based on the direct proportion. (© 1976, Archie Comic Publications, Inc.)*

The proportionality constant w is called a "whatkind" by Archie's peer group. Jughead might ask, "Whatkind of gas mileage do you get, Arch?" Archie is always careful to ascribe the proper units to the proportionality constant whenever he gives its value, even in ordinary conversation. He says, "Oh, I get about 29 miles per gallon."

Archie's model has limitations, as do all mental models, but he recognizes the limitations of his model. He specifies that he gets 29 miles per gallon on the open road, while he only gets 23 miles per gallon in town. He has then improved his model by specifying that the whatkind is not really constant but is a function of road conditions and the speed he drives. His model could be further improved by accounting for the pressure of his foot on the gas pedal, the kind of gasoline, tire inflation, the tune of his engine, and even the weather. Accounting for all of these variables would be difficult and unnecessary for most purposes. For most purposes, the simplest possible model based on a straight proportionality gives usable results, but Archie can easily use it in an even more sophisticated form if he feels so inclined.

The concept of gasoline mileage is but one of many examples where proportional thinking enters into a person's daily life. This same type of reasoning will serve us through most of this course. Some people think of physics as mathematical and therefore difficult. Physics truly is mathematical but only because the rules of deductive logic that apply are so simple that they can be clearly spelled out. These rules of logic are collectively called algebra.

► Algebra

Once an algebraic model of reality is constructed, it will yield the solution for any one variable by a series of algebraic operations. These mathematical manipulations have become a part of the thinking of ordinary people in our society. Archie is able to give you the value of his "whatkind," w, only because he is able to solve his basic model for that variable. That is, he is able to say that distance equals whatkind times gasoline. The basic proportional model

distance = (whatkind) (amount of gasoline)

can be expressed in terms of algebraic symbols as

$$d = wG$$

Archie is able to tell you without hesitating that this is equivalent to

$$w = \frac{d}{G}$$

Most people seem to know that it is "legal" to move a variable from the numerator on one side of an equation to the denominator on the other side and to switch sides of an equation.

All of the manipulations considered acceptable in algebra are based on just a few premises. One is that you can always replace something with something else of equal quantity. That is why it is okay to switch sides of an equation. Another premise is that you can do almost anything you wish to one side of an equation as long as you do it to the other. It certainly seems reasonable that if you have two things that are equal to begin with, you should come out with the same result if you do exactly the same thing to both of them.

FIGURE 1-6 *Algebra is just simple logic.*

Thus, we can start with the gasoline mileage proportionality model

$$d = wG$$

and divide both sides of the equation by the volume of gasoline, G, to get

$$\frac{d}{G} = \frac{wG}{G}$$

A third premise is that one times anything is that thing itself. We can multiply one side of the equation by unity (one) without disturbing the equality. Since the fraction $\left(\frac{G}{G}\right)$ is equal to unity, the expression

$$\frac{d}{G} = w\left(\frac{G}{G}\right)$$

is equivalent to

$$\frac{d}{G} = w(1)$$

and we can ignore the one.

$$\frac{d}{G} = w$$

Sometimes this is referred to as "canceling." Such language tends to obscure the simple logic of dividing something by itself and ignoring the unity, but it is in common use because it is easier to say.

We have just seen that the operation of moving G from the numerator on the right side of the equation to the denominator on the left is acceptable because it is the combination of two basic operations that are obviously legitimate. Several other combination operations are useful shortcuts and are reviewed in the appendix on algebra, but there is one more operation worthy of special note.

It is acceptable to move something from the denominator of the denominator into the numerator. Suppose after several manipulations we wind up with G in the denominator of the denominator. We can see that this is the same thing as having G in the numerator

$$\frac{1}{\frac{1}{G}} = G$$

if we multiply the left-hand side of the equation by $\frac{G}{G}$, which is the same as unity.

$$\frac{1}{\frac{1}{G}}\left(\frac{G}{G}\right) = \frac{G}{\left(\frac{G}{G}\right)}$$

Then G is in both the numerator and the denominator of the denominator and can be ignored. This rule, that **the denominator of the**

denominator is really the numerator, will be especially useful in dealing with units.

► Units

The best way to solve a simple physics problem is to state the "equation," or mathematical model, in its general form, solve it algebraically for the unknown variable, and then substitute the given data explicitly as stated in the problem before doing any of the arithmetic. It is important to note that the data consist of both numbers and units, and the same operations performed on the numbers can also be performed on the units.

Suppose, for example, that Archie needs to know how much gasoline will be required to make a round trip of 123 miles, assuming his whatkind is 29 miles per gallon. Archie would probably do everything in his head, but there are advantages to working the whole thing out on paper: one is you can convince yourself that you know what you are doing. Another is you can do harder problems.

First we state the formula:

$$d = wG$$

Although we call this a formula, we should recognize that it is a statement of our model of reality, in which distance traveled is proportional to the amount of gasoline used and w is the proportionality constant.

Next we solve for the unknown, which is gasoline consumed G:

$$G = \frac{d}{w}$$

(This is, of course, a combination operation, in which w is moved from the numerator on one side to the denominator on the other and the sides of the equation are switched.)

Then we insert data, taking care to include both numbers and units:

$$G = \frac{123 \text{ miles}}{\dfrac{29 \text{ miles}}{\text{gallon}}}$$

Gallons in the denominator of the denominator are, according to the rules, the same thing as gallons in the numerator:

$$G = \frac{123}{29} \cdot \frac{\text{miles} \cdot \text{gallon}}{\text{miles}}$$

Performing the same operations on the units as on the numbers gives us confidence that our answer is correct. G should, in fact, be expressed in gallons:

$$G = 4.2 \text{ gallons}$$

Many errors are spotted by the failure of the units to work out right. People should *never* leave out the units and thereby deprive themselves of the joy of seeing everything fall into place.

FIGURE 1-7 *A cartoon is a two-dimensional mental model of a three-dimensional thing.*

Units may act as a tool for reasoning as well as for checking. It frequently happens that the needed operations can be determined by figuring out what changes of units are necessary. The mathematical tool that does this for us is known as the **conversion factor.**

▶ Conversion Factors

Suppose that, because of inflation, they start selling gasoline by the quart rather than by the gallon. Archie would certainly be able to convert from gallons to quarts. If he knew that he needed 4.2 gallons of gasoline, he would be able to tell us that he would need four times that many quarts. There is, however, a formal algebraic way of handling the units based on the premise that one times anything is that thing itself.

$$G = 4.2 \text{ gallons}(1)$$

Multiplying by unity doesn't change the equality. Even if the thing inside the parenthesis is a fraction, it won't change the equality as long as the fraction is equal to unity.

$$G = 4.2 \text{ gallons} \left(\frac{4 \text{ quarts}}{1 \text{ gallon}} \right)$$

The fraction we elect to use has quarts in the numerator and gallons in the denominator, since we are trying to get rid of gallons in the numerator. The whole fraction is equal to one since the numerator equals the denominator. Multiplying by such a fraction, called a **conversion factor,** has the effect of changing our units without disturbing the equality. This mathematical trick helps us keep track of whether we should be multiplying or dividing to get our answer. This is a major asset in difficult problems, where a number of such operations are needed.

Some problems that could be worked as proportions can also be handled with conversion factors. If the price of gasoline is one dollar per gallon, Archie would immediately be able to tell you that he needs $4.20 worth of gasoline. Even numbers are nice; odd numbers are a bit more difficult. Suppose Archie discovers a station that sells gasoline for the bargain price of 93 cents per gallon. If we could visualize the thought process he would use, it would probably look very much like our conversion factor problem. Perhaps he would reason that if 93 cents will buy him one gallon, 4.2 gallons will require 4.2 times as much money. We can duplicate this reasoning with conversion factors, treating money as a measure of gasoline—another mental model. From this point of view, we are converting money into gasoline. It may not literally be true that we are changing pieces of paper into a liquid, but it is common for many people to make that kind of conversion in their heads when they go into a gas station.

$$G = (4.2 \text{ gallons}) \left(\frac{93 \text{ cents}}{1 \text{ gallon}} \right)$$

We use cents in this case to stand for "cents worth of gasoline." In that restricted sense it is proper to say that the numerator is equal to the denominator. We can then tack on another conversion factor to come out with dollars:

$$G = (4.2 \text{ gallons}) \left(\frac{93 \text{ cents}}{1 \text{ gallon}} \right) \left(\frac{1 \text{ dollar}}{100 \text{ cents}} \right)$$

$$= 3.91 \text{ dollars}$$

Some seemingly difficult problems can be solved by just stringing together enough conversion factors until you come out with the right units.

► Think Metric

Perhaps you find the idea of buying gasoline by the quart instead of the gallon to be a bit remote, but that is nearly what the world is coming to. In fact, all the world except the United States has already come to the point of selling gasoline by something which is very close to a quart. Gasoline is sold by the liter (LEE-ter). Instead of four to a gallon, as for the quart, there are 3.8 liters in a gallon. It might seem silly for us to make such a change, but the change is inevitable. All the world is using the metric system of units and the United States is in the process of following suit.

The metric system has a number of advantages over the crazy-quilt system of British units. One of the nicest things about it is that most conversion factors are in powers of ten. You can convert from one unit to another by just moving the decimal point. The price of spice bought by the kilogram and given away by the gram can be easily computed, since there are an even thousand grams in a kilogram. Converting from pounds to ounces, on the other hand, means dividing or multiplying by 16.

Multiples and parts of basic metric units have Latin prefixes. There are 100 **centimeters** (CEN-ti-meters; cm) in a meter just as there are 100 cents in a dollar. There are 1,000 **millimeters** (MILL-i-meters; mm) in a meter just as there are 1,000 tax mills in a dollar. Once you know that "centi" stands for one-hundredth and that "milli" stands for one-thousandth, you can apply these same prefixes to other units. A **millisecond** is one-thousandth of a second. You thus have a way of extending your knowledge. Even if you don't fully understand the meaning of a volt, you can easily guess that a **millivolt** is one-thousandth of whatever a volt is.

The basic unit of length in the metric system is the **meter** (m), which is a little longer than a yard. The meter was originally defined as one ten-millionth of the distance along the surface of the earth from one pole to the equator. That distance turns out to be 39.37 inches. In fact, this was the legal definition of the inch in the United States for nearly a century. The meter was defined in terms of a standard bar, a replica of one kept in France, and the inch was defined as something that it took 39.37 of to make up a meter. The meter is still the basic standard of length in the United States, but the legal definition of the inch was changed slightly in 1959. Now 1 inch is defined as exactly 2.54 centimeters. We shall see in the examples that the difference only appears in the seventh significant figure.

CHECK QUESTIONS

1. If, as was true in the older definition of the inch, we could say that 39.37 inches equals 1 meter, how big would a 10-centimeter object be in inches?

   Answer: 3.937 in.

2. How big would a 1-centimeter object be in inches?

   Answer: 0.3937 in.

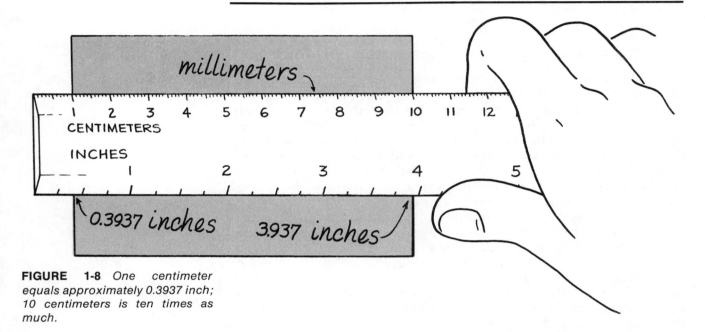

FIGURE 1-8 *One centimeter equals approximately 0.3937 inch; 10 centimeters is ten times as much.*

Since there are 1,000 millimeters in a meter and 100 centimeters in the same distance, there are 10 millimeters in a centimeter. The millimeters are the tiny marks on a meter stick between the numbered centimeter marks.

Highway distances, which we measure in miles in the United States, are measured in **kilometers** (Kil-OM-eters; km) in the rest of the world. A kilometer is equal to 1,000 meters or 0.621 miles. Going the other way, a mile is equal to 1.61 kilometers, 1.61 being the reciprocal of 0.621. The United States government has embarked upon an information campaign to implant a conversion factor from miles to kilometers in the minds of the public. Large signs have been posted on interstate highways 100 miles from major cities saying, for example, *St. Louis—100 miles—161 kilometers.* This is a pleasant and painless way of giving you a conversion factor saying that 1 mile equals 1.61 kilometers.

▶ Volume and Mass

The measure of volume in the metric system is the cubic meter. A block 1 meter on a side is a little bigger than a cubic yard. Building contractors buy concrete in cubic yards in the United States and in

FIGURE 1-9 *One mile equals 1.61 kilometers.*

cubic meters every place else. This unit is too large, however, for dealing in commodities like gasoline and milk. The unit in general use for measuring volume is the **liter.** A liter is the volume of a square box 10 centimeters on a side. A stein of beer in Germany holds 1 liter, just about the right amount.

A smaller unit of volume in common use is the **milliliter** (ml), which is one-thousandth of a liter. This unit used to be called the cubic centimeter (or cc) because it is the volume of a cube 1 centimeter on a side. The cubic centimeter is the same as the milliliter. There are 1,000 cubic centimeters in a liter.

$$1 \text{ liter} = (10 \text{ cm})^3 = 10^3 (\text{cm})^3 = 10^3 \text{cm}^3$$

The milliliter is in common use in medicine because it is a convenient measure for drugs.

The original unit of mass was based on a cubic centimeter of pure water under standard conditions, which was defined as having a mass of one **gram.** This definition remained until balances were developed to such a high degree of sensitivity that they could actually sense a difference between identically prepared samples of water due to differences in naturally occurring abundances of heavy water. At that point, the French machined a block of platinum and iridium alloy so that it had a mass as close as possible to that of a liter of pure water. This primary standard of mass is carefully stored in a vault under double bell jars and taken out only to compare with secondary standards. Because of the original definition of the gram, 1,000 cubic centimeters, or 1 liter, of pure water has the mass very close to 1 kilogram.

▶ Weight

A kilogram (kg) of mass has a weight of 2.2045 pounds at a certain place on the surface of the earth.[1] The equivalence of 2.2 pounds (approximately) to the kilogram is frequently used as a conversion factor to go from British to metric units. There is, however, a difference between **mass** and **weight.** The mass of an object is a property of the object alone and is the same on the earth, on the moon, or anywhere in the universe. The weight of an object is a force resulting from the gravitational attraction between the object and the earth, the moon, or some other body. Weight is therefore not a property of the object alone but depends upon both the mass of the object and the place it happens to be. The weight, wt, of an object at a given location is, however, directly proportional to its mass, m:

$$\text{wt} \propto m$$

We can write this in terms of a proportionality constant, which turns out to be the acceleration due to gravity, g.

$$\text{wt} = gm$$

The g is usually written after the m, however, because this relation-

---

[1] That would be a place where the acceleration due to gravity would be 32.174 ft/sec². (Sea level at about 45 °N latitude.) This acceleration due to gravity would include the gravitational attraction of the earth as well as the slight balancing effect of the earth spinning on its axis.

ship turns out to be a special case of Newton's second law of motion, as will be explained in a later chapter.

$$\text{wt} = mg$$

Meanwhile, we can think of this relationship as a simple proportionality.

We are accustomed to using the weight of a substance as a measure of its mass because we have been confined to living in a thin shell in which the acceleration due to gravity, g, is relatively constant. As we move out of our shell, or as we begin to deal with finer measurements (to the third significant figure), we need to become aware of the distinction between mass and weight.[2] A common folk-saying states, "A pint is a pound the world 'round." But a kilogram is a kilogram even on the moon.

Consider the weight of a 1-kilogram mass. Weight depends not only on the mass but on the acceleration due to gravity which, to two significant figures, equals 9.8 meters/second² (m/s^2) on the surface of the earth. The relationship

$$\text{wt} = mg$$

in this case is

$$\text{wt} = (1 \text{ kg})(9.8 \text{ m/s}^2)$$
$$= 9.8 \text{ kg} \cdot \text{m/s}^2$$

There is a special name for this particular combination of units. It is called the **newton,** abbreviated N.

$$1 \text{ N} \equiv 1 \text{ kg} \cdot \text{m/s}^2$$

The weight of a 1-kilogram mass is

$$\text{wt} = 9.8 \text{ N}$$

This is true only on the surface of the earth. On the moon's surface, g is one-sixth as great and the weight there would be one-sixth as many newtons.

Strictly speaking, we cannot convert pounds to kilograms. The pound is a unit of force; we can only convert it to newtons, another unit of force. Since the newton is the basic unit of force in the metric system, you might be interested to see how big it is in pounds. Since 1 kilogram weighs 2.2046 pounds in British units and 9.8 newtons in metric units,

$$9.8 \text{ N} = 2.2 \text{ lb}$$

This is just a little less than 10 newtons. One newton is thus just a little less than a quarter of a pound.

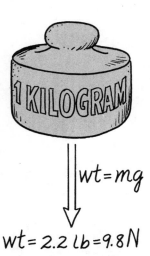

$$wt = mg$$

$$wt = 2.2 \, lb = 9.8 \, N$$

FIGURE 1-10 *The weight of an object is the product of its mass and the acceleration due to gravity. On the surface of the earth, a 1-kilogram object weighs 2.2 pounds, or 9.8 newtons.*

[2] Values for g at different locations can be found in handbooks. A few examples are as follows:

Location	m/s^2
North Pole	9.832
Equator	9.780
New York City	9.802
San Francisco	9.799
Pikes Peak	9.789

Thus, $g = 9.8$ m/s² is only correct to two significant figures.

$$1 \text{ N} = 0.22 \text{ lb}$$

A quarter-pound stick of butter weighs about 10 per cent more than a newton. A stick of butter with one slice removed weighs a newton. The slice of butter should be about the size of a fig cookie. That, of course, is where the fig cookie got its name.[3]

CHECK QUESTIONS

1. How much does a 2.0-kg block weigh on the surface of the earth in (a) newtons and (b) pounds?

    Answer: (a) 19.6 N, (b) 4.4 lb

2. How many newtons does 1 lb of butter weigh?

    Answer: 4.5 N

▶ Calculations

Often people taking their first physics course are inclined simply to do all computations by hand. This is an inefficient use of their time. Arithmetic is the least important and most tedious part of a physics problem, and yet it can be the most time-consuming. Most of the mental models we shall use in this course are based on the proportion. Thus, most of our computations will be multiplication and division—the most difficult arithmetic to do by hand. This course, therefore, calls for the use of a calculator. The best and least expensive calculator for this course is a slide rule. Since a slide rule requires a bit of study to master, however, many students will prefer using an inexpensive digital calculator.

Some people believe that a calculator gives them greater accuracy than a slide rule. They see eight numbers light up on the display and they think that all of those numbers have meaning. But in most of the problems we shall look at, only the first two digits will have any significance. There is hardly a problem in this book for which the numbers past the third digit will have any meaning except insofar as they place the decimal point. Those people who prefer using a digital calculator to solve physics problems need to learn how many of those numbers have significance. People using a slide rule will probably have less trouble with the rules of significant figures.

▶ Significant Figures and Power-of-Ten Notation

Numbers are frequently used to describe not only what we know but how accurately we know it. If Archie says that he has traveled a distance of 129 miles, we assume he knows that distance to the closest mile. If he were to say that he had traveled a distance of 129.0

[3] "Learn from science that you must doubt the experts . . . science is the belief in the ignorance of experts." Richard P. Feynman, "What Is Science," *The Physics Teacher,* (September 1969) in *Teaching Introductory Physics* (American Association of Physics Teachers), p. 333.

miles, we would assume that he knew the distance to the closest tenth of a mile. According to the generally accepted rules of describing such things as distance, 129 miles is not the same thing as 129.0 miles. A distance of 129 miles means anything between 128.5 miles and 129.5 miles, whereas 129.0 miles could refer to an actual distance of anything from 128.95 to 129.05 miles. If Archie were to describe a distance as 129.0 miles when he did not in fact know it to the closest tenth of a mile, he would be claiming more knowledge than he actually possessed. The use of too many figures to describe a measurable quantity is a qualitative error in thinking.

There is some ambiguity in the way most people express large numbers. It is commonly said, for example, that the speed of light is 186,000 miles per second. There is little room for misunderstanding in the first three digits, but one might question whether the last three zeros are really zero values or whether they merely place the decimal point. The convention used by most people, and the one adopted by this book, is that zeros at the end of a large number are there only to place the decimal point. In fact, the speed of light is 186,282 miles per second.

The ambiguity in the significance of the final zeros can be resolved by finding another way to place the decimal point. The way that has been adopted in most scientific literature is to express the number as a product of a number between 1 and 10 times 10 raised to the appropriate power. For example, 186,000 miles per second, give or take 500 miles per second, would be expressed as follows:

$$(186{,}000 \pm 500)\,\text{mi/s} = 1.86 \times 10^5\ \text{mi/s}$$

If you wanted to say that a spaceship is traveling at 282 miles per second less than the speed of light, you might express its speed as 1.86000×10^5 miles per second. There would then be no doubt that the zeros at the end are zero values, since they would serve no purpose other than to indicate precision of knowledge.

The number of digits necessary to express a measurement in this power-of-ten notation (sometimes called scientific notation) is called the number of **significant figures.** The precision of the data, as indicated by the number of significant figures with which the data are stated, tells you how careful you need to be in your computations. Zeros used to place the decimal point are not counted in determining the number of significant figures.

CHECK QUESTIONS

1. The equatorial radius of the earth is 3,964 miles. Write this in power-of-ten notation and state the number of significant figures.

    Answer: 3.964×10^3 mi; four

2. The valve tappets on the engine of a Triumph TR-3 sports car should be set for a clearance of 0.012 inch. Express this in power-of-ten notation and state the number of significant figures.

    Answer: 1.2×10^{-2} in.; two

Noting the number of significant figures in the data can save you time and effort in making calculations. Answers need never be more accurate than justified by the data. For operations involving multiplication and division, the rule of thumb is that an answer should be rounded off to the same number of significant figures as the least accurate data. To see why this rule is reasonable, let us first consider an example worked out on the slide rule. We will then consider another example worked out on a digital calculator.

Suppose we wanted to figure out how long it would take to stroll around the perimeter of the earth, a distance of 2.5×10^4 miles (25,000 mi), traveling at a nice steady walking speed of 5 miles per hour. Of course, we understand the limited accuracy of this speed, since it is difficult to keep such a pace exactly constant, especially on water. Let us therefore agree that by 5 miles per hour we mean anything between 4.5 and 5.5 miles per hour. The general solution to the problem of finding the time necessary to get somewhere can be found from the definition of velocity:

$$v \equiv \frac{d}{t}$$

Multiplying both sides by time and dividing both sides by velocity gives us the general result

$$t = \frac{d}{v}$$

Inserting the data for this particular problem,

$$t = \frac{2.5 \times 10^4 \text{ mi}}{5 \text{ mi/hr}} = 5 \times 10^3 \text{ hr}$$

This answer is properly reported to only one significant figure, since we really don't know what the next digit should be. The situation can be easily visualized in terms of a slide rule set up to calculate our answer. To divide 2.5 in the numerator by 5 in the denominator, we put the 5 on the B scale under the 2.5 on the A scale. We could afford to be reasonably sloppy in lining up the 5 with the 2.5, since 5 means anything from 4.5 to 5.5—quite a distance on our slide rule scale. The permissible error in the answer can be visualized by seeing how far the index on the B scale would move if we were to scoot that scale back and forth between the limits of accuracy associated with the denominator.

The rule of thumb that states that an answer should not be reported to more significant figures than the least accurate data is not

FIGURE 1-11 *The error in the data can be visualized as an error bracket on a slide rule scale. The error will propagate as the same size error bracket in the result.*

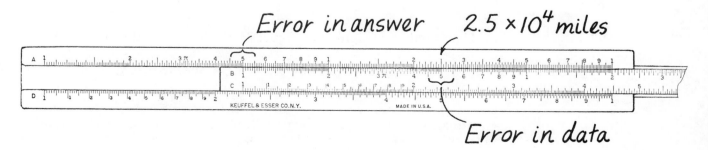

Error in answer 2.5×10^4 miles

Error in data

always strictly true. The error that is propagated by multiplication and division is a certain percentage of the data involved and may be properly visualized in terms of an error bracket on a slide rule. The size of this error bracket depends upon the end of the slide rule you are looking at. The spacing between the numbers changes from one end of the scale to the other just as the percentage error, which is necessarily passed on to the answer, gets smaller for larger numbers. Slide rules have an advantage over calculators for doing physics problems in that they permit you to see literally the inaccuracy you are dealing with.

Since the size of the error associated with the data depends on the size of the numbers you are dealing with, most people like to keep one extra figure in their calculations to hedge against a round-off error, which might creep into the answer as a result of rounding off too soon. No benefit results, however, in keeping more than one extra (insignificant) figure in the calculations, as you can see by considering the following example worked out on a digital calculator.

The distance around the earth, given in the previous example as 2.5×10^4 miles, was calculated from the radius as given in reference books. Actually, two radii are listed since the earth isn't exactly round. The books state that the equatorial radius is 3,964 miles, the polar radius 3,950. Thus, we would have two different distances, depending on which way we decided to walk. The polar route is a bit shorter, but we would stay warmer if we walked the equatorial route. The formula for the circumference of a circle approximates the distances around the earth by these two paths:

$$d = 2\pi r$$

The value of π is known to thousands of significant figures, but we might use $\pi = 3.1415926$, since only eight figures will fit into our calculator. You might argue that we need use no more than five significant figures of π, since each of the radii are given to only four figures. Let's test that theory using the equatorial radius as an example. Using all eight figures of accuracy for π, we get the equatorial circumference

$$d_e = 2(3.1415926)(3,964 \text{ mi}) = 24,906.546 \text{ mi}$$

Using only five significant figures for π, the value for the same distance turns out to be

$$d_e = 2(3.1416)(3,964 \text{ mi}) = 24,906.604 \text{ mi}$$

Thus, the equatorial circumference appears to be 24,907 miles whether you use five or eight digits of π. The same result is found to hold for the polar radii. Using eight significant figures for π, the polar circumference turns out to be

$$d_p = 2(3.1415926)(3,950 \text{ mi}) = 24,818.581 \text{ mi}$$

Using only five significant figures for π, the same distance turns out to be

$$d_p = 2(3.1416)(3,950 \text{ mi}) = 24,818.640 \text{ mi}$$

Here again, the distance comes out essentially the same, 24,819 miles, whether we use five or eight digits of π. The lesser accuracy affects only the figures past the first six, and these weren't significant anyway. As a rule, it is a waste of time to use data that are accurate to more than one significant figure than your least accurate data. Don't be afraid to round off data to this extent if it helps the arithmetic.

The difference between the polar circumference, 24,819 miles, and the equatorial circumference, 24,907 miles, loses its importance if we are only trying to decide which way will take us longer to walk. Both distances are the same up to the third significant figure, and we only know our velocity to one figure. Two significant figures are more than we need for the distance, since the inaccuracy with which we know our velocity masks the difference we could conjure up over the route we should take. The distance we plan to walk is 25,000 miles either way.

Keeping track of the accuracy of data can save you time and effort in working a problem. There is no point in being really careful with part of your data if the rest of your information is sloppy. You can be intentionally sloppy in your calculation if you know that it won't affect your answer. In the preceding example, there is no point in keeping track of the distance to more than two significant figures if the answer is limited to one figure of accuracy by the inaccuracy of the velocity. This is not even the most extreme example. Some data are known to zero significant figures and can still produce interesting results.

Calculations based on data known to zero significant figures are called **order of magnitude** calculations. The order of magnitude refers to the exponent of the power of ten you need to express that data in power-of-ten notation. Sometimes that is all you know. For example, it happens that 1 liter of air contains, to one significant figure, about 3×10^{22} molecules:

$$1 \text{ liter} = 3 \times 10^{22} \text{ molecules}$$

We could therefore say that a breath of air contains on the order of 10^{22} molecules:

$$1 \text{ breath} = 10^{22} \text{ molecules}$$

A breath of air occupies something like a liter in volume, but its exact size depends on whether it is a little puff or a great big lungful of air. The exact number multiplied by 10^{22} depends on the size of the breath, but any breath of air will be something times 10^{22} molecules. Thus, we don't even know these data to one significant figure, but we do know the order of magnitude of the number we are talking about.

Ten raised to the twenty-second power is one big bunch of molecules—that is, ten thousand million million million molecules. In the whole atmosphere of the earth, there are only something like 10^{44} molecules:

$$1 \text{ atmosphere of the earth} = 10^{44} \text{ molecules}$$

This does not mean, of course, that there are only two breaths of air in the atmosphere of the earth. 10^{44} is not twice as big as 10^{22}; it is 10^{22} squared. This leads us to conclude that there are about the same number of breaths of air in the atmosphere of the earth as there are molecules of a single breath:

$$1 \text{ atmosphere of the earth} = (10^{22})(10^{22}) \text{ molecules}$$
$$= (10^{22})(\text{breaths of air})$$

FIGURE 1-12 *A breath of air is about a liter in volume.*

You can therefore gain an appreciation for the smallness of one molecule by looking up into the sky and thinking about how many breaths of air are out there covering the whole earth. One molecule bears about the same relationship to a single breath as that breath bears to the whole atmosphere. You could take the molecules in one single breath of air, and there would be just about enough of them to put one molecule in every breath that's out there. You don't even need to go to all that trouble: just release a breath of air and wait for a long time. "If we assume that the last breath of, say, Julius Caesar has by now become thoroughly scattered through the atmosphere, then the chances are that each of us inhales one molecule of it with each breath we take."[4]

CHECK QUESTIONS

1. You are going to visit a friend in a distant city. You know that the distance from your driveway to your friend's front door is 235.23 miles, and you plan to maintain an average speed of 55 mph. To how many significant figures of accuracy do you know the (a) distance and (b) the speed?

    Answer: (a) 5, (b) 2

2. In calculating the time necessary to travel 235.23 miles at an average speed of 55 mph, you divide 235.23 by 55 on a calculator to get the answer in hours. The calculator says 4.276909091. You then multiply by 60 to convert this answer to minutes. The calculator says 256.6145455. (a) How many hours will it take you to travel the given distance at the given speed? (b) Express this time in minutes to the proper number of significant figures.

    Answer: (a) 4.3 hours, (b) 2.6×10^2 minutes

▶ Other Proportional Reasoning

Many of the relationships in nature can be approximated by mental models based on the **direct proportion,** such as the gasoline consumed and the distance traveled in an automobile. We have seen how this kind of model can be expressed either in terms of a proportionality sign

$$d \propto G$$

or in terms of a proportionality constant, which, as it turns out, is defined as the ratio of these two variables:

$$\frac{d}{G} = \text{constant}$$

[4] Sir James Jeans, *Kinetic Theory of Gasses* (New York: Cambridge University Press, 1940), p. 32.

The name of this constant and the variable we elect to put in the numerator of its definition are arbitrary.

So far, we have dealt only with examples of the direct proportion. There is another class of proportional relationships in which one variable may grow larger as the other grows smaller, but equal changes in one still bring about proportionally equal changes in the other. This is called the **inverse proportion.** This kind of proportionality is actually a direct proportion between one variable and the reciprocal of the other. The pressure in a gas, for example, decreases as the volume increases. The pressure is found to become half as great each time the volume doubles (as long as the temperature is held constant). Thus, the pressure is proportional to the reciprocal of the volume:

$$P \propto \frac{1}{V}$$

A proportionality constant and an equality sign can be used to replace the proportionality sign:

$$P = (\text{constant}) \frac{1}{V}$$

It is easy to see that this proportionality constant is therefore equal to the product of the two variables:

$$PV = (\text{constant})$$

This proportionality constant may in fact have a value that depends on other factors. In the example we are discussing, the value of the constant depends on the temperature of the gas and the number of molecules in the container. When these variables are taken into account, this mental model is called the **ideal gas law.** It is discussed in Chapter 3.

Still other relationships in nature are best approximated by a model in which a variable is squared, cubed, or taken to some other exponential power. The area of a circle, for example, is proportional to the square of its radius. The volume of a sphere, on the other hand, is proportional to the cube of its radius. These might seem like mathematical things, circles and spheres, but let us relate them to area and volume in an imaginary physical situation.

Suppose while you are studying, you glance down at the floor and notice an ant tugging at your shoe. This is a large ant. Due to some science fiction process the ant has grown to a thousand times its ordinary size. You wonder if this is cause for alarm. You wish to decide if this ant, about the size of a bulldog, is a thousand times as strong as an ordinary ant.

While the monster is eyeing you and waving its feelers about, you proceed to construct a model by which you can judge the relationship between strength and size. The strength of the ant's bones and muscles is proportional to the cross-sectional area. Strength results from individual molecules hanging together across an imaginary cross-section taken through a muscle or bone. The greater the cross-sectional area, the greater the number of molecular bonds. The cross-sectional area is proportional to the square of linear dimension, l. As the ant gets bigger, its muscles and bones increase in both width and thickness. If the ant is still made of the same sub-

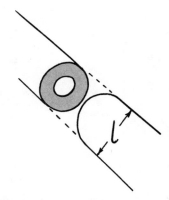

FIGURE 1-13 *The strength of an ant bone is proportional to the cross-sectional area.*

stance as before, the number of molecular bonds, and hence the strength, s, is proportional to the square of linear dimension, l.

$$s \propto l^2$$

Instead of being a thousand times as strong, our monster is a thousand *squared,* or a million, times as strong.

This does not mean that someone could breed ants for size to produce our next superweapon. There is another model we should consider before becoming alarmed.

Weight should be proportional to the total number of molecules, not in a cross-sectional area, but in the entire ant's *volume.* As the ant grows larger, it increases not only in thickness and height, but in length as well. Thus, weight wt is proportional to the *cube* of linear dimension, l.

$$\text{wt} \propto l^3$$

If the ant is a thousand times as big, it is a thousand cubed, or a billion, times as heavy. The ratio of strength to weight is inversely proportional to linear dimension:

$$\frac{s}{\text{wt}} \propto \frac{l^2}{l^3} = \frac{1}{l}$$

Although there are a million times as many molecular bonds in any cross-section of bone and muscle, each bond must support a thousand times as much weight. The ant would collapse of its own weight into a pool of ant juice.

Physics is wonderful.

The problems in this chapter are related to the kind of mathematical reasoning that will be used in the rest of this book. Most of the mathematical models we will use are based on the simple direct proportion. The variable x is said to be directly proportional to y if equal increases in one variable are accompanied by equal increases in the other:

$$x = Cy$$

The proportionality constant C allows the proportion to be expressed with an equality sign rather than a proportionality sign.

Data in a problem consist of both numbers and units. Units should always be inserted in the general solution of a problem along with the numbers as a check on the algebra used to obtain the general solution.

Units may also serve as a tool for reasoning. When you know that the units you have are not the units you want, you can frequently correct the situation with one or more conversion factors. A conversion factor is just a fraction whose numerator is equal to the denominator but is expressed in different units so that the fraction equals unity (1). The units of the numerator and the denominator are not the same, however, but are chosen to cancel the units you are trying to get rid of and replace them with the ones you want.

The easiest system of units to use is the metric system where larger and smaller units for the same quantity are related to each other by even powers of ten. For example, the basic unit of length is the meter; the millimeter, centimeter, and kilometer are, respectively, a thousandth, a hundredth, and a thousand meters.

Millimeter	1 mm = 10^{-3} m		.001 meter
Centimeter	1 cm = 10^{-2} m		.01 meter
Kilometer	1 km = 10^{3} m	1,000	meters

The same Latin prefaces are used for other quantities. A milliliter, for example, is the volume of a cube 1 centimeter on a side and is one thousandth (10^{-3}) of a liter, a volume defined by 10 centimeters on a side. A kilogram, originally defined as the mass of 1 liter of pure water, is one thousand times as big as the gram.

| Milliliter | 1 ml = 10^{-3} liter | .001 liter |
| Kilogram | 1 kg = 10^{3} g | 1,000 grams |

While we are converting to the metric system from the English system, it will be useful to know a few common conversion factors by memory. Some good ones are:

$$1 \text{ inch} = 2.54 \text{ cm} \quad \text{(exactly)}$$
$$1 \text{ mile} = 1.61 \text{ km}$$
$$1 \text{ gallon} = 3.78 \text{ liter}$$

The relationship between mass and weight is introduced in this first chapter because of its great importance. Mass is an inherent property of an object and would be the same anyplace in the universe. Weight, on the other hand, is a force that is proportional to mass but depends on a gravitational field. The proportionality constant g is the acceleration due to gravity:

$$\text{wt} = mg$$

where $g = 9.8$ m/s^2 on the earth's surface. We will later think of this relationship as a special case of Newton's second law. At that point, we will be thinking of mass as the proportionality constant. The acceleration due to gravity will be a special case of acceleration.

The results of calculations should never be reported to more significant figures than justified by the data. Generally, an answer should not be reported to more significant figures than the least accurate data. Sometimes you may wish to keep one more significant figure in an intermediate answer to avoid round-off error in the final result, but no more than one extra (insignificant) figure is ever useful. Keeping too many significant figures is incorrect, because they imply that you know more than you actually do.

EXAMPLE PROBLEMS

1-1 A student trying to optimize his homework grade notes that he earns a higher grade when he uses more paper.

Given: He gets only 2 points out of 10 possible when he crowds everything on to only 1 page. As he gets more generous with himself, he gets $4\frac{1}{4}$ points for 2 pages, $6\frac{1}{2}$ points for 3 pages, and $8\frac{1}{2}$ points for 4 pages. He hypothesizes that the relationship between grade G and pages P is proportional.

Find: (a) State the hypothetical proportionality as an equality using a proportionality constant C.
(b) Solve this equality for the constant and test its value on the data.

Your Solution

Discussion

This proportional model assumes that factors other than pages are not important and is therefore only a first approximation to reality. This particular proportional model is surely a gross oversimplification of reality—assuming the instructor is taking time to look at the work that is on the paper. Still, there is a relationship. Pencil and paper are used in this course as tools for reasoning; a proper and generous use of paper frequently helps the grade.

Sample Solution 1-1

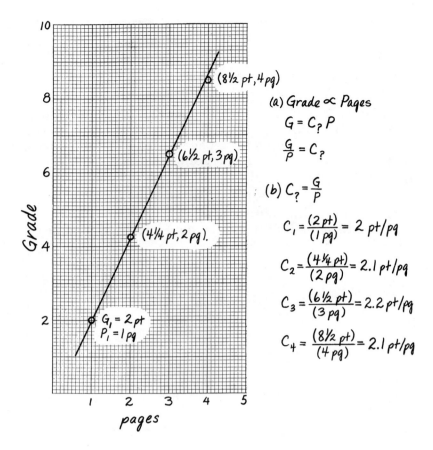

(a) Grade $\propto$ Pages

$$G = C_? P$$

$$\frac{G}{P} = C_?$$

(b) $C_? = \frac{G}{P}$

$$C_1 = \frac{(2\,pt)}{(1\,pg)} = 2\,pt/pg$$

$$C_2 = \frac{(4\frac{1}{4}\,pt)}{(2\,pg)} = 2.1\,pt/pg$$

$$C_3 = \frac{(6\frac{1}{2}\,pt)}{(3\,pg)} = 2.2\,pt/pg$$

$$C_4 = \frac{(8\frac{1}{2}\,pt)}{(4\,pg)} = 2.1\,pt/pg$$

Graph labels: $(8\frac{1}{2}\,pt, 4\,pg)$, $(6\frac{1}{2}\,pt, 3\,pg)$, $(4\frac{1}{4}\,pt, 2\,pg)$, $G_1 = 2\,pt$, $P_1 = 1\,pg$. Axes: Grade (vertical), pages (horizontal).

Discussion

Note that both units and numbers are inserted as data and that units remain in the final answer. The grade is not 2 or $4\frac{1}{2}$ but 2 points and 4 points. The difference between apples and oranges, so to speak, is frequently a matter of importance in physics problems.

The question mark subscript on $C_?$ is used just to emphasize the fact that it is an unknown, which is later to be solved for in the problem. Your solution would not necessarily use this notation.

This sample solution also contains a graph that would not necessarily be in your solution. The problem does not call for a graph, but a picture or a graph often helps understanding in solving physics problems.

According to the data, the relationship between points and paper does seem to be nearly proportional, at least to about two significant figures. The graph produces points that lie nearly on a straight line, and the proportionality constant is nearly the same for each set of data points. This proportionality begins to fail, however, in about the second or third significant figure. The first data point produces a value of C that is about 5% less than the average of 2.1 pt/pg, and the third data point produces a value that is about 5% greater.

1-2   As we convert to the metric system, most people will come to know that 1 inch is equal to 2.54 centimeters and that 1 mile is equal to 1.61 kilometers. These common length conversions are based on precise legal definitions.

Given: Prior to 1959, an inch was legally defined by the equality 39.37 inches = 1 meter. A mile is defined by 1 mile = 5280 feet, just as 1 foot = 12 inches exactly.

Find: (a) Use the old legal definition of an inch to show that an inch equals about 2.54 centimeters.
      (b) Use the old legal definition of a mile to show that it is equal to about 1.61 kilometers.

Your Solution

Discussion

Many problems that could be solved by a proportional model can also be solved with a conversion factor. Using this approach, we start with an equality and multiply by one or more conversion factors, which are fractions equal to unity (1) and therefore do not disturb the equality. This problem is an exercise in the use of conversion factors and allows you to prove for yourself that the length conversions in common use are in accordance with the legal definitions.

Sample Solution 1-2

(a) $1 \text{ in} = 1 \text{ in} \left(\dfrac{1 \text{ m}}{39.37 \text{ in}} \right)$

$\quad = \dfrac{1}{39.37} \text{ m} \left(\dfrac{100 \text{ cm}}{1 \text{ m}} \right)$

We are conversion factors. Our numerators equal our denominators.

$\quad = \dfrac{100}{39.37} \text{ cm}$

$\quad = \boxed{2.540005800102 \text{ cm}}$

We are conversion factors also. We don't disturb the equality because we equal UNITY!

(b) $1 \text{ mi} = 1 \text{ mi} \left(\dfrac{5{,}280 \text{ ft}}{1 \text{ mi}} \right) \left(\dfrac{12 \text{ in}}{1 \text{ ft}} \right) \left(\dfrac{1 \text{ m}}{39.37 \text{ in}} \right)$

$\quad = \dfrac{5{,}280 \times 12}{39.37} \text{ m} \left(\dfrac{1 \text{ km}}{10^3 \text{ m}} \right)$

UNITY for all us conversion factors!

$\quad = \boxed{1.609347218694437 \text{ km}}$

Discussion

The units in the conversion factors are selected to cancel the units you are trying to get rid of. If they don't leave you with the units you want, just use another conversion factor. You don't need to do the arithmetic at each step, however, as it is usually easier to string up a bunch of conversion factors and do the arithmetic all at once as shown in part (b) of the sample solution.

There is no reason for your solution to carry out the computation to more than three significant figures, slide rule accuracy, as the problem does not call for anything more. It is done to 13 significant figures in the sample solution just for fun so that you can see what the numbers are that follow the commonly used conversion factors. This is one of the rare situations where it is not wrong to crank out the computation as far as you might like. One meter is exactly equal to 39.37 inches according to the old legal definition, to as many significant figures as you like, just as 1 foot is exactly equal to 12.00000. . . inches by definition.

You may find it interesting that 2.54 centimeters to the inch is good to at least five, and almost six, significant figures.

The older legal definition of the inch, that 39.37 inches equals 1 meter, led to the approximation that 1 inch is 2.54 centimeters. This approximation was good to six significant figures, as shown. The new definition of the inch just made this approximation exact. Now 1 meter is only approximately equal to 39.37 inches, but the approximation differs from the exact figure only in the seventh significant figure.

1-3 A physics instructor climbs on a scale in class to demonstrate the fact that weight is proportional to mass, the proportionality constant being g, the acceleration due to gravity.

Given: The physics instructor claims to have a mass of 90 kilograms. His weight, before going on vacation, is indicated by the scale to be 882 newtons. Upon returning from vacation, however, he finds that the gravitational attraction of the earth for his body has mysteriously increased to 1,000 newtons.

Find: (a) What value of g is indicated by the instructor's claim and his weight before he left for vacation?
(b) What is his mass after returning, assuming g has not increased?

Your Solution

Discussion

Looking upon the acceleration due to gravity g as a measure of the earth's gravitational field, it seems reasonable to use it as the proportionality constant between mass and the force of gravity called weight. This problem serves as an introduction to the relationship between mass and weight, a relationship for which we will find much use. We will later see it as a special case of Newton's second law.

Sample Solution 1-3

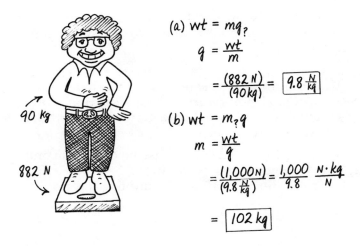

(a) $wt = mg$?

$$g = \frac{wt}{m}$$

$$= \frac{(882\,N)}{(90\,kg)} = \boxed{9.8\,\tfrac{N}{kg}}$$

(b) $wt = m_?g$

$$m = \frac{wt}{g}$$

$$= \frac{(1,000\,N)}{(9.8\,\tfrac{N}{kg})} = \frac{1,000}{9.8}\,\frac{N \cdot kg}{N}$$

$$= \boxed{102\,kg}$$

Discussion

It is good form to start a physics problem by explicitly stating the appropriate mathematical model. It is like saying, "Oh, this is one of those wt = *mg* problems." This shows where you are starting from (which might well be good for part credit if you later make some stupid algebra mistake) at the same time that it reduces the chances that you will make an algebra error by trying to do too much in your head.

The model is then solved algebraically to get the general solution to the problem before inserting data, both numbers and units, to get the particular solution or answer. Recording your thoughts on paper in this way allows you or someone else to find your error should your answer be incorrect.

1-4 A piece of typing paper can be measured with a meter stick to get its length and width to four significant figures by estimating to tenths of a millimeter, but its thickness can only be measured to two significant figures even by using a micrometer caliper, which can be read to a thousandth of a millimeter.

Given: A piece of typing paper is measured to be 28.87 centimeters long, 21.61 centimeters wide, and 1.5×10^{-3} centimeter thick.

Find: (a) Calculate the area of the paper to the proper number of significant figures.
(b) Calculate the volume of the piece of paper to the proper number of significant figures.

Your Solution

Discussion

This problem illustrates the usefulness of the rules of significant figures in limiting the accuracy with which you need to do computations. It also points out the difference between precision and accuracy. The micrometer is much more precise than an ordinary meter stick in measuring tiny distances, but it may be even less accurate if the thing being measured is also very tiny. Accuracy is relative to that which is being measured.

Sample Solution 1-4

(a) Area = Length × Width

   = (28.87 cm) (21.61 cm)

   = $\boxed{623.9 \ cm^2}$

(b) Volume = Area × Thickness

   = (623.9 cm²) (1.5 × 10⁻³ cm)

   = $\boxed{9.4 \times 10^{-1} \ cm^3}$

← Workpiece being measured

← Spindle runs on a screw having a pitch of one or two turns per millimeter.

← Sleeve has a main scale reading directly in millimeters.

Thimble turns with spindle on same screw. Has a micrometer scale which divides each main scale division into 100 parts in terms of parts of a turn of the screw.

Discussion

The area is part (a) should be computed to four significant figures, because both length and width are known to that accuracy. Long-hand multiplication or a calculator turns out the product of the two numbers as 28.87 × 21.61 = 623.8807, but the last three figures are not significant as can be seen by tacking a little bit on to one or both of the numbers. For example, 28.871 × 21.611 = 623.93118.

The volume in part (b) is limited to two significant figures by the accuracy with which we know the thickness. If we believed everything that a calculator said, we might think that 623.9 × 1.5 = 935.8, but we can't rely on the 5.8 at the end because 623.9 × 1.51 = 942.08. The thickness might just as well be 1.51 × 10⁻³ centimeter as 1.50 × 10⁻³ centimeter for all we know.

The picture of the micrometer caliper is only provided for general information.

1-5 Harry Homeowner discovers that he uses electricity faster in his home the more rooms he keeps lit.

Given: He finds that he uses 3 kilowatt hours (kWh) per day when he is alone and only keeps one room at a time lit. He finds that it takes 9 kWh/day to keep 3 rooms at a time lit and 12 kWh/day when his family keeps 4 rooms at a time lit.

Find: (a) State a hypothetical proportionality between electricity used, E, and the number of rooms, R, using an equality sign and a proportionality constant C.

   (b) Solve this equality for the proportionality constant C and test its value on the data.

1-6 An auto mechanic drops his American-sized wrench into a grease pit. He must either find a slightly oversized metric wrench in his tool box or climb down into the grease pit. He knows that there are 2.54 centimeters to the inch and 12 inches to the foot.

Given: The size of the wrench he dropped is $\frac{9.0}{16}$ inch, and the grease pit is 6.0 feet deep.

Find: (a) What is the smallest size metric wrench, in millimeters, that will fit?

   (b) How deep is the grease pit in meters?

1-7 An astronaut weighs very little on the moon because the proportionality constant between weight, wt, and mass, m, is much smaller on the surface of the moon than it is here on earth. This constant is called the acceleration due to gravity, g.

Given: An astronaut in his space suit with certain equipment has a mass of 175 kilograms and weighs 286 newtons when standing on the surface of the moon. The astronaut without his suit, however, only has a mass of 84 kilograms. (Remember $1\ \text{N} \equiv 1\ \text{kg} \cdot \text{m/s}^2$.)

Find: (a) What value of g is indicated by the astronaut's weight while standing on the surface of the moon?

   (b) What is the weight of the astronaut alone when standing inside his spacecraft without his suit and equipment?

1-8 Land is frequently measured in acres, square feet, or square miles, depending on how much you are talking about. An acre is 43,560 square feet and a mile is 5,280 feet. Farmland was originally laid out in the Midwest in square miles, with a road about 20 feet wide along each boundary. City lots are much smaller.

Given: A parcel of farmland is 1 mile square, give or take 10 feet on each side, whereas a typical house lot in San Francisco is 100 feet long and 25 feet wide, give or take a foot for fences.

Find: (a) Calculate the arable land (the area that can be used for farming) in acres of a square mile to the proper number of significant figures.

   (b) Calculate the area of a San Francisco house lot in acres to the proper number of significant figures.

MORE INTERESTING PROBLEMS

$g = 9.81 \ m/s^2$

1-9 An object in free fall will accelerate uniformly, traveling a certain distance in a certain time regardless of its weight.

Given: A falling object will accelerate at 9.81 meters per second squared. It will have fallen 44.1 meters at the end of 3.00 seconds, at which time it will be traveling at a velocity of 29.4 meters per second.

Find: Convert
(a) 9.81 meters per second squared into feet per second squared,
(b) 29.4 meters per second into miles per hour, and
(c) 44.1 meters into inches.

1-10 The British units for highway speed and gasoline consumption will soon be as obsolete here as they now are in Britain. Someday little children will ask, "Mommy, what is a mile?" You might answer, "1.61 kilometers, my dear." Or the children might ask, "Daddy, what does a gallon mean?" To this, you might answer, "It's 3.78 liters, kid."

Given: A 1967 Chevy will get 15 miles per gallon of gas when driven at 80 miles per hour.

Find: (a) What is this speed in kilometers per hour?
(b) What is this gas consumption in kilometers per liter?

1-11 The force necessary to drag a trunk across a floor is found by experiment to be directly proportional to the trunk's weight. The proportionality constant involved is called the **coefficient of friction.**

Given: A 50-pound trunk requires 20 pounds of force to make it slide. The same trunk, with more stuff in it, so that it weighs 75 pounds, requires 30 pounds of force to make it slide.

Find: (a) Plot a graph of force versus weight.
(b) What is the coefficient of friction in this case?
(c) How much force would be necessary to pull the trunk if it contained lead, so that it weighed 125 pounds?

ANSWERS

1-5   (a) $E = CR$; (b) 3 kWh/(day · room).
1-6   (a) 14.29 mm ≅ 15 mm; (b) 1.8 m.
1-7   (a) 1.63 N/kg = 1.63 m/s²; (b) 137 N ≅ 140 N.
1-8   (a) 640 acres; (b) 0.057 acre.
1-9   (a) 32.2 ft/s²; (b) 65.7 mph or 65.8 mph; (c) 1.74 × 10³ in.
1-10 (a) 129 km ≅ 130 km/h; (b) 6.4 km/liter.
1-11 (b) 0.4; (c) 50 lb.

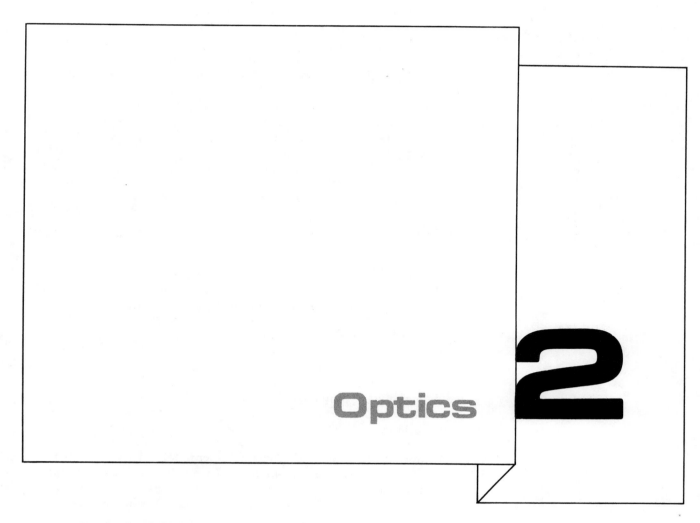

Optics 2

► The Nature of Light

A physics student lies in bed in the mid-afternoon contemplating the scene outside her window. A wild bee takes nectar from a flower. The yellow-brown fuzz of its body is haloed by the afternoon sun. A robin, chirping in a nearby tree, is framed in the student's mirror, forming a moving picture of red, brown, and green.

Visual impressions are some of the most important and pleasant of all the information gathered by our senses, and so people have always pondered the nature of sight. Some of the ancient Greeks thought of light as rays, originating in the eye and flowing out over objects in the same way that fingers of the hand are outstretched to feel. "Why else," they queried, "do you not see a needle on the floor unless you are looking for it?" This theory of light is still portrayed in cartoons, where a dotted line coming from a character's eye is used as a convention to indicate what the character is looking at.

Sir Isaac Newton suggested that light could either be a wave or a particle. While philosophers have argued over these two models for centuries, it now seems that light is really something between a particle and a wave.

The two simple pictures we can form in our heads, waves and particles, are necessary tools for thinking and relating to our everyday experience. A wave is energy spread out over space, as in a water wave. A particle is something localized in space, like a tiny BB pellet. It may be hard to think about something that is spread out like a wave at the same time that it is localized like a particle, but it happens that light is really like that. Reality is more sophisticated than either of the two simple models we can form in our minds.

Even waves and particles are more sophisticated models than we need for the study of lenses and mirrors. In geometrical optics, light is thought of as a **ray,** a straight line along which light is supposed to travel except when it interacts with matter by reflection or refraction. We may thus think of ourselves as reverting to the simple Greek model or to the cartoon character with a dotted line connecting the observed object and the character's eye.

HIGHLIGHT

In 1801, Thomas Young found evidence that light behaves like waves. He discovered that light passing through two slits forms interference patterns that can only be accounted for by waves. This had not been noticed before, because the waves were so short that they appeared to act like particles. Newton's picture of light as a particle was a useful mental tool, but it seemed that it was not really true. Light, it seemed, was really more of a wave.

As a child, Albert Einstein asked one of his teachers what light would look like if you were to move along with it. His teacher thought he was a dumb kid because he wasn't satisfied with the wave picture. Einstein later said that this unanswered question eventually started him thinking about relativity. He was, in fact, given a Nobel prize in 1921, not for his then controversial theory of relativity, but for showing that light is really something of a particle. The wave picture of light was also a useful mental tool, but it turned out that that was not really true either. Light, as it turns out, is really something somewhere between a particle and a wave.

▶ Reflection

Light being reflected from the bee's body and then bounced off the polished surface of the mirror forms the image seen by the student. Some colors are absorbed, others reflected. She sees the light reflected from the surface of tiny hairs as the color and texture of bee fuzz. It is easiest to visualize the relationship between incident and reflected light for the case of a smooth surface. Light coming toward such a surface is reflected so that it leaves the surface at the same angle. This is called the **law of reflection.** The angle of incidence equals the angle of reflection:

$$\angle i = \angle r$$

These angles are usually measured in relation to the **normal,** a line perpendicular to the surface. Light rays bounce off the mirror in the same way that a perfectly elastic ball would bounce off the floor. Assuming the ball's horizontal velocity remains constant, and that only the *direction* of its vertical velocity changes from down to up, it leaves the floor at the same angle that it approached the floor.

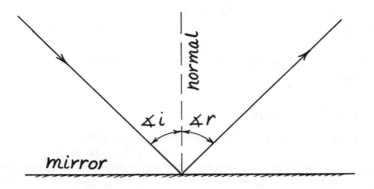

FIGURE 2-1 *The law of reflection states that the angle of incidence equals the angle of reflection.*

CHECK QUESTIONS

Light approaches a mirror along a line making an angle of 53° with the mirror's surface. (a) What is the angle of incidence? (b) At what angle to the normal will the light be reflected?

Answer: (a) 37°, (b) 37°

▶ Images

Every morning you go to the bathroom mirror and look yourself in the eye. The person on the other side of the mirror is called an **image** because he looks just like you. The image does not appear to be located at the surface, but behind the mirror, as though there were someone on the other side of the wall. Light does not really travel behind the mirror, but it reflects from the mirror in such a way that it comes to your eye at the same angle as if it had come from a point behind the mirror. What you see in the mirror is called a **virtual image,** because light does not actually pass through the image position but behaves virtually as if it did. Light from your nose bounces to your eyes along a path similar to the path it would take had it come

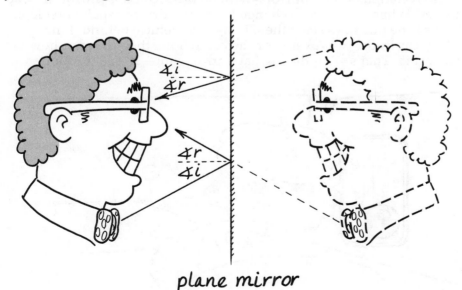

plane mirror

FIGURE 2-2 *The virtual image in a plane mirror looks exactly like the object and is the same distance behind the mirror as the object is in front of the mirror.*

from behind the mirror. From the symmetry of the situation you can see that the distance from your nose to the mirror is the same as the distance from the image to the mirror—as long as the mirror is flat. When the mirror is curved, the object's and image's distances are no longer equal.

▶ Spherical Mirrors

A **convex** mirror may be thought of as a plane mirror that has been bent so that it bulges out in the middle like the surface of a Christmas tree ball. Images formed by such a mirror are smaller than those formed by a plane mirror and, to some people, appear further away. Actually, the images in a convex mirror are closer to the mirror, as we shall see, but their small size seems to fool people. We judge distance in part by size. Things that are far away appear small to our eye, so we tend to believe the converse. In a convex mirror, however, small means close.

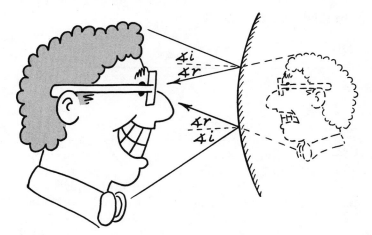

FIGURE 2-3 *Images in a convex mirror are smaller and closer to the mirror than the object.*

Truck drivers use a convex mirror for a wide-angle view. The angle of exposure allows more images, though smaller, to be crowded into the field of view. Truck rear-view mirrors are generally large rectangular flat mirrors with a little convex mirror in one corner. When he wants to change lanes, the driver usually first looks in the little mirror to see if there is another vehicle behind. If he spies a tiny car in the convex mirror, he can tell it's there but not how far away it is. That's what the big flat mirror is for—it is large enough so

FIGURE 2-4 *The rear-view mirror on a truck is frequently a large plane mirror with a small convex mirror in one corner to give a wide-angle view.*

that the driver can see that little car with both eyes and bring his binocular depth perception into play in judging distance. You may have noticed that the rear-view mirror on a truck is about as wide as the driver's face. That is so that he can look at you with both eyes. You may also have noticed that he can tell if there is as little as a few inches to spare.

Convex mirrors are also frequently used in stores for their wide-angle view to spot shoplifters. When you next see one in a store, carefully observe the relative distances between images and objects as well as their relative sizes. A close examination of a store's security systems is always good for a little excitement.

A mirror that bulges in the opposite direction—inward in the middle—is called a **concave** mirror from the same Latin root as *cave*. Since it bends in the opposite direction from a convex mirror, a concave mirror does the opposite thing to the image position and size. While a convex mirror forms an image closer to the mirror and smaller than the object, a concave mirror forms an image that is larger and further away. Here again, the size of the image can fool you about its actual position. While the enlarged size of the image might make it seem close, a careful inspection of the light reflected from the mirror will show that it comes from a point in space actually farther behind the mirror's surface than the object is in front of the mirror, as long as the object doesn't get too far from the mirror.

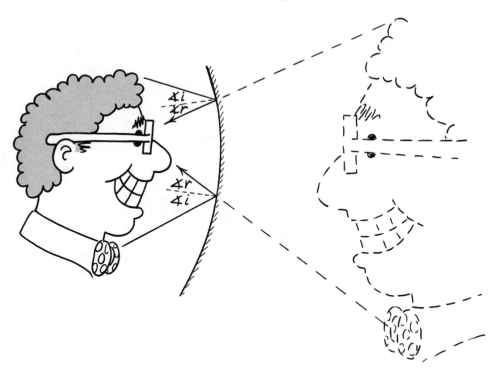

FIGURE 2-5 *Images in a concave mirror are larger and further away from the mirror than the object as long as the latter is within the focal point.*

A concave mirror brings parallel light rays to focus at a place called the **focal point** of the mirror. Figure 2-6 shows how incident light rays, which start out parallel to the axis, are reflected so that they cross at this point, which turns out to be halfway between the mirror and the center of curvature of the mirror. Remember that the angle of incidence is measured to lines drawn normal to the surface. All of the lines normal to the surface pass through the center of curvature. (By definition, all radii of a circle pass through the center and

FIGURE 2-6 *Parallel light is brought to a focus at a point that is halfway to the center of curvature of the mirror.*

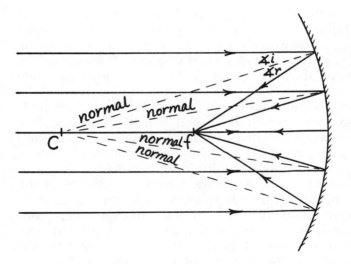

are perpendicular to the circumference.) Incident rays that are further from the axis of the mirror will have larger angles of incidence, and thus larger angles of reflection, and will cross at about the same point as rays closer to the axis.[1]

One of the problems in the use of solar energy has been that the energy from the sun is widely diffused by the time it reaches the earth. Concave mirrors can be used to concentrate the sun's rays and more economically produce electrical energy. A long cylindrical mirror, for example, can be used to focus sunlight on a pipe to heat water and drive a heat engine. Heat engines work more efficiently the higher the temperature of their heat source. This concentration of the sun's rays by mirrors provides higher working temperatures.

CHECK QUESTIONS

You accidently drop and break a round glass Christmas ornament. While cleaning up the glass, you pick up one of the larger pieces and get interested in the concave inner surface. The ball was originally 3.0 inches in diameter. (a) What is the radius of curvature of the concave surface, neglecting the thickness of the glass? (b) What is the focal length of this concave surface?

Answer: (a) 1.5 in., (b) 0.75 in.

Just as parallel light from a far distant object will be brought to focus at this point, light coming from the focal point will be reflected and go out parallel. The law of reflection works the same both ways,

[1] Incident rays that are reasonably close to the axis are called **paraxial rays.** Rays that are a considerable distance from the axis, and that therefore strike at large angles of incidence, actually cross the axis at a point closer to the mirror than the focal point. This smearing of the focal point for larger angles of incidence is called **spherical aberration.** Parabolic mirrors have a shape designed to eliminate spherical aberration, but they only work well for parallel light.

whether bringing parallel light to focus or causing the reverse to happen to light traveling in the opposite direction. The angles of incidence and reflection are interchangeable, since the law of reflection only says that they are equal, without regard to which way the light is traveling. Thus, parallel light is associated with light traveling through the focal point whether the light is coming or going.

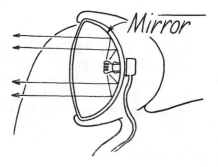

FIGURE 2-7 *A sports car driving light has a concave mirror in the back to produce a parallel beam of light. Unlike a headlight, it has no diffusing lens on its front surface to spread the light in front of the car.*

The ability of a concave mirror to form a parallel beam of light finds even more applications than its ability to bring light into focus. Search lights, flashlights, and automobile headlights all use concave mirrors to form beams of parallel light. A white-hot wire filament, electric arc, or other bright source of light is placed at the focal point. Light coming from this source is reflected into a parallel beam. This beam is so nearly parallel, in fact, an automobile headlamp in addition has a lens with a rippled surface to diffuse the beam and spread it out in front of the car. Sports car enthusiasts frequently add driving lamps to their cars that are like headlights except that they have clear glass in place of the diffusing lens. Such driving lamps must be used judiciously, however, since they produce a "pencil beam," which is so nearly parallel that it can temporarily blind another driver miles away.

▶ Ray Diagrams—Concave Mirrors

An image is a point in space from which light rays appear to come after they have been reflected from a mirror. It is possible to apply the law of reflection to any number of rays to predict the location of the image. Just use a protractor to measure the angle of incidence and construct an equal angle of reflection to determine the direction of a reflected light ray. But, the image can also be located, without a protractor, using three special rays associated with the focal point and center of curvature of the mirror. These three rays do the same thing as all the other rays we could construct, but they are easier to find.

Consider a candle located beyond the center of curvature of a concave mirror. Assume that the candle is right side up with its base resting on the mirror's axis, as shown in Figure 2-8. Of all the light rays impinging on the whole surface of the mirror, the three rays

FIGURE 2-8 *Three rays can be used to locate an image formed by a concave mirror.*

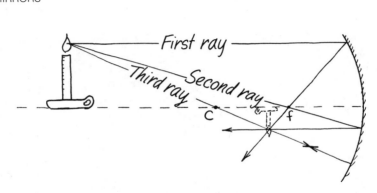

easiest to locate are the following: The first is the one that comes from the tip of the candle and travels parallel to the axis of the mirror. It will be reflected at such an angle as to pass through the focal point. All light coming parallel to the mirror's axis is reflected to the focal point as in the solar energy concentrator mentioned earlier. The second ray comes from the tip of the candle and passes through the focal point on its way to the mirror. This ray strikes the mirror at an angle that will cause it to be reflected out parallel. Just as in the case of the search light, light coming from the focal point is reflected into a parallel beam. The third ray passes through the center of curvature on its way to the mirror. Since this ray strikes the mirror perpendicular to the surface, it travels along a radius of curvature and is reflected back on itself. The image is located where these three rays cross. Actually, two rays could be used to locate the image, but it is a good idea to draw all three when making ray diagrams, to confirm the crossing point.

We could have used more than three rays to locate the image. All the light striking the mirror from a certain point on the object will be reflected to cross at a corresponding point on the image. The other rays are more difficult to construct, but they all contribute to forming the image. Even if several of the rays are obstructed, even one of the three that are easy to draw, the image will be formed just the same by the light that does manage to get through. We could mask half the mirror and we would still get an image of the whole candle, and the image would be the same size as before we masked the mirror—it would just be dimmer, since not as much light would have gotten through. We can therefore go ahead and construct a ray even if the mirror in our drawing is not large enough to reflect it. A ray that misses the mirror will still act as a tool to locate the image formed by all the rays that do hit the mirror.

FIGURE 2-9 *All the rays striking the mirror pass through the same image as the three rays that are easy to draw.*

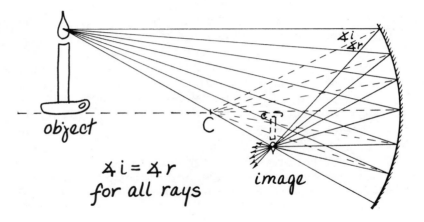

We have only traced rays originating from the tip of the candle. Light actually comes from all points on the candle. For every point on the candle, there is a corresponding point on the image where reflected light rays meet. Light reflected from a drip on the candle forms rays that come in closer to the axis of the mirror and are reflected to cross at a drip on the image. The image looks just like the candle except that it is upside down, or **inverted,** and not necessarily the same size.

The farther the object from the mirror, the closer the image will be to the focal point. In fact, the focal point is the image of a far distant object. Light from a far distant object comes in nearly parallel and the place it crosses is the focal point. As the object approaches the mirror, starting from infinity, the image grows larger and moves away from the focal point toward the center of curvature. The image and object both reach the center of curvature at the same time. At that point, they are both the same distance from the mirror and they are both the same size.

This kind of image, formed by light actually crossing at a point, is called a **real image.** It is lifelike in appearance and can be mistaken for the real thing. A number of magic tricks are based on this illusion. It is entertaining to watch someone reach for a coin that is not there. This magical illusion can be produced with a magnifying mirror, using a coin hidden in a shoe box. If the coin is suspended

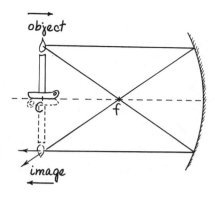

FIGURE 2-10 *The object and image meet at the center of curvature.*

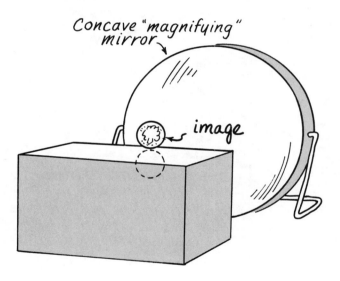

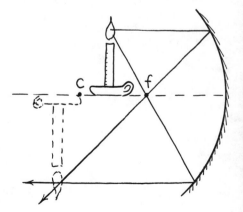

FIGURE 2-11 *A real image is formed of a coin, hidden in a shoe box, by placing it at the center of curvature of a magnifying mirror.*

with tape inside the shoe box and the open side of the shoe box is turned away from view (toward the mirror), all that can be seen is the reflected image of the coin. If the coin is arranged at the center of curvature of the mirror, the image will appear to stand on the top edge of the shoe box. It will be the same distance from the mirror as the actual coin and the same size, as shown in Figure 2-11.

When the object moves closer to the mirror than the center of curvature, but not as far as the focal point, as shown in Figure 2-12, the image moves to a point beyond the center of curvature. The image is still inverted, but is larger and further from the mirror.

As the object approaches the focal point, its image moves still further away, always increasing in size. When the object finally reaches the focal point, the reflected light goes out in a parallel beam,

FIGURE 2-12 *The image is beyond the center of curvature when the object is between the center of curvature and the focal point.*

as in a search light. Since parallel light rays never cross, we say the image has receded to infinity. In becoming infinitely far away, the image also becomes infinitely large.

Strange as it might seem to say, the image can go even *beyond infinity*. The image approaches infinity as the object approaches the focal point, and it clearly must go somewhere when the object crosses over the focal point. If you are wondering where this strange place beyond infinity is physically located, the answer is that just beyond infinity is negative infinity. After the real image has receded to an infinitely large distance on the same side of the mirror as the object, a virtual image comes in from infinity from the opposite direction. Positive distances are measured on the same side of the mirror as the object; negative distances are behind the mirror.

While real images are upside down in the real space in front of the mirror, virtual images are right side up in the negative space behind the mirror. Although light never really passes behind the mirror, it does come away from the mirror at the same angle as though it did.[2] For a concave mirror, the virtual image is always larger and farther away from the mirror than the object. This is how a concave mirror magnifies your nose. You move your nose to a point within the focal length of the mirror and observe the "reflection" or virtual image behind the mirror. It is magnified at the same time it is further from the mirror than the object (see Figure 2-13).

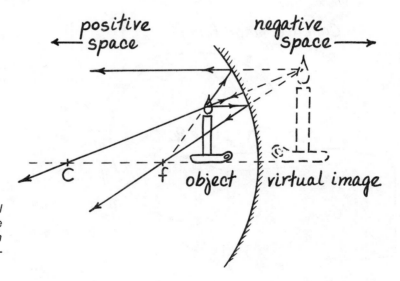

FIGURE 2-13 *A magnified virtual image is formed in negative space, behind the mirror, when the object is within the focal distance of a concave mirror.*

[2] A concave mirror is an optical device that converges light rays. When the object is very far away, the light that comes in parallel converges to cross at the focal point. As the object approaches the mirror, the light rays from the object diverge more and more as they strike the mirror. The mirror is still strong enough to converge them so they cross at a real image, however, as long as the object is beyond the focal point. The more strongly the incident light diverges, the further away is the point at which the reflected rays converge. When the object reaches the focal point, the incident light is so strongly diverging that it is all the poor mirror can do to converge the rays so that they go out parallel. When the object is within the focal length, the mirror still makes the rays more converging (less diverging) but they still diverge. They cannot be converged enough to cross but they appear to come from a point behind the mirror in the same way that light comes away from the virtual image formed by a plane mirror. In this case, however, the image distance is greater than for a plane mirror. The light rays are less diverging and therefore appear to come from a point farther away.

▶ Plane Mirrors

A plane mirror can be thought of as a concave mirror having an infinite radius of curvature. The longer the radius of curvature, the flatter the surface. People used to think that the surface of the earth was flat because its radius of curvature was 4,000 miles. A still longer radius of curvature would mean a still flatter surface. A plane mirror would therefore have an infinite radius of curvature.

Since half of infinity is still infinity, a plane mirror also has an infinite focal length. All real objects are therefore within the focal length of a plane mirror and produce virtual images that are right side up and behind the mirror (in negative space).

▶ Convex Mirrors

A convex mirror may be thought of as a spherical mirror having its center of curvature located on the negative side of the mirror. Just as for a concave mirror, the focal point is half the radius of curvature:

$$f = \frac{c}{2}$$

Parallel light incident on a convex mirror is reflected away from, instead of toward, the focal point.

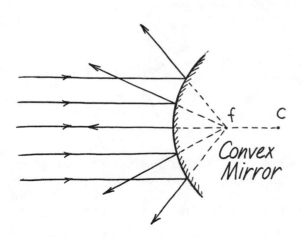

FIGURE 2-14 *Parallel light is reflected by a convex mirror so that it goes away from the focal point.*

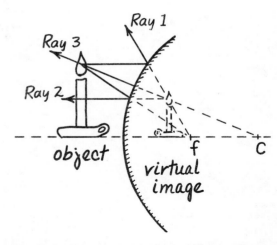

FIGURE 2-15 *The image formed by the convex mirror is virtual, erect, and smaller than the object.*

Just as with concave mirrors, the focal point may be regarded as the image of a far distant object. Objects viewed in a Christmas tree ball appear as virtual images between the surface and the focal point. The closer the object, the closer the virtual image appears to the surface. The further the object, the further the image recedes to a point halfway to the center of the ball. These images can be located with three rays as before, but keep in mind that the roles of approaching and departing rays are reversed. A light ray approaching parallel to the axis (ray 1) is reflected away from the focal point. A light ray which approaches the focal point (ray 2) is reflected out parallel. The virtual image is located at the place where these two departing rays would have crossed had they come from inside the mirror. This position is confirmed by a third ray (ray 3), which approaches the center of curvature and is reflected back on itself. The virtual image is right side up, smaller than the object, and closer to the mirror.

▶ The Object-Image Formula

Ray diagrams are valuable tools for understanding an optical system. It is also helpful to have an algebraic formula relating object and image distances so that the accuracy of results will not depend on the precision of drawing tools. We shall derive such a formula from the ray diagram of a concave mirror. One common convention is to call the height of the object a, the height of the image b, the object distance p, and the image distance q.

FIGURE 2-16 *One set of similar triangles formed by a ray diagram relates the object distance, p, and focal length, f, to the object and image heights, a and b.*

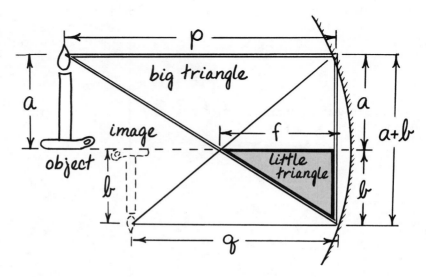

Two sets of similar triangles are formed by such a ray diagram. The first set, shown in Figure 2-16, is made up of a little triangle having the focal length f and the image height b as sides and a big triangle having the object distance p and the sum of object and image

FIGURE 2-17 *The ratio of the sides of the little triangle is equal to the ratio of the sides of the similar big triangle. The equality of these ratios reduces part of the geometry of a ray diagram to an algebraic expression.*

$$\frac{b}{f} = \frac{(a+b)}{P}$$

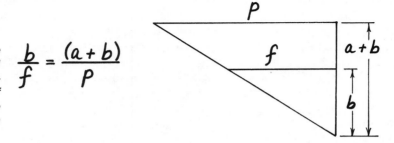

heights $a + b$ as sides. These triangles are geometrically similar because they have parallel sides and therefore the same angles. The similar sides of similar triangles are proportional.

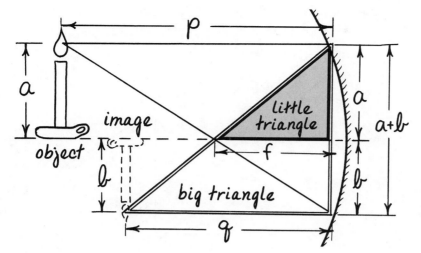

FIGURE 2-18 *Another set of similar triangles is formed by a ray diagram relating the image distance, q, and focal length, f, to the object and image heights, a and b.*

The second set of similar triangles, shown in Figure 2-18, is made up of a little triangle having the focal length f and object height a as two sides and a big triangle having the image distance q and the sum of the object and image heights $a + b$ as two sides. Having parallel sides, these two triangles are also geometrically similar. We can again set up a proportionality between similar sides of similar triangles.

$$\frac{a}{f} = \frac{(a+b)}{q}$$

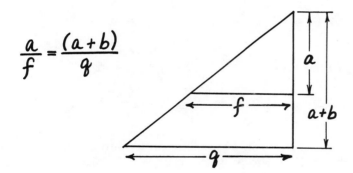

FIGURE 2-19 *The ratio of the sides of the little triangle is also equal to the ratio of the sides of the big triangle for the two similar triangles that relate the focal length to the image distance.*

There is an easy way to eliminate a and b from the two proportionalities in Figures 2-17 and 2-19. We simply add the two equations together:

$$\frac{a}{f} + \frac{b}{f} = \frac{a+b}{p} + \frac{a+b}{q}$$

We can factor $1/f$ from the left side of this expression and $(a + b)$ from the right side.

$$\frac{1}{f}(a+b) = (a+b)\left(\frac{1}{p} + \frac{1}{q}\right)$$

Dividing both sides of the equation by $(a + b)$ gives the **object-image formula:**

$$\boxed{\frac{1}{f} = \frac{1}{p} + \frac{1}{q}}$$

This equation is useful in many problems. It says that the reciprocal of the focal length is equal to the sum of the reciprocals of the image and object distances.

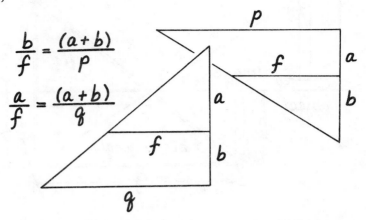

FIGURE 2-20 *The equality of ratios based on both sets of similar triangles can be combined to eliminate either (a + b) or the focal length f.*

The focal length, f, can be eliminated from the same two geometrical proportions to yield a relationship between the sizes and distances of the object and image. These two proportions, restated, can each be cross-multiplied to get everything out of the denominator:

$$bp = f(a + b)$$
$$aq = f(a + b)$$

Since the right sides are equal, so are the left sides.

$$bp = aq$$

Dividing both sides of this equation by a and p,

$$\frac{b}{a} = \frac{q}{p}$$

We have shown that the ratio of the image and object heights b/a is equal to the ratio of the image and object distances q/p. The ratio of the image height b to object height a, defined as the magnification M, tells us how much bigger the image is than the object. The preceding equation says, if you look at it just right, that the image is as much bigger as it is farther away. In other words, the size is proportional to the distance.

$$M \equiv \frac{b}{a} = \frac{q}{p}$$

(The three-lined equal sign $\equiv$ means, as you recall, "defined as.") This is the **magnification formula.** It says that the magnification, which is defined as b over a, is equal to q over p.

▶ Applications of the Object-Image Formula

Just to see how the image-object formula works, let us use it to find the position of the image formed by a concave mirror when the object is at the center of curvature. Of course, we already know from our considerations with ray diagrams that the image will be found also at the center of curvature in this case. That was the situation in the example we discussed where the image of the coin in the shoe box was made to stand on the top of the shoe box. It is instructive, however, to see how this works out in the image-object formula. The image distance p in this case is equal to the distance to the center of curvature c, which is twice the focal length.

$$p = c = 2f$$

Solving the object-image formula for $1/q$,

$$\frac{1}{q} = \frac{1}{f} - \frac{1}{p}$$

Inserting our given value for the object position,

$$\frac{1}{q} = \frac{1}{f} - \frac{1}{2f}$$

Reducing both fractions to the same common denominator, $2f$,

$$\frac{1}{q} = \frac{2}{2f} - \frac{1}{2f} = \frac{1}{2f}$$

Taking the reciprocal of both sides, we see that the image is also at twice the focal length—which is the center of curvature:

$$q = 2f$$

We can also see that the magnification formula tells us that the image and object are the same size:

$$M \equiv \frac{b}{a} = \frac{q}{p}$$
$$= \frac{2f}{2f}$$
$$= 1$$

The magnification M is equal to unity in this case. Thus, the object and image are the same size.

CHECK QUESTIONS

1. An object is located at three times the focal length of a concave mirror, thus $p = 3f$. (a) Where will the image be located in terms of the focal length? (b) What will the magnification be in this case?

   Answer: (a) $\frac{3}{2}f$, (b) $\frac{1}{2}$

2. An object is located 15 cm from a mirror. The mirror forms a real image of this object at a point 30 cm from the mirror. (a) What is the focal length of the mirror? (b) What is the magnification in this case?

   Answer: (a) 10 cm, (b) 2

As another example of the object-image formula, let us consider the special case of the plane mirror. We already know that the image formed in a plane mirror is located the same distance behind the mirror as the object is in front, but we can show that this is the case starting with the image-object formula:

$$\frac{1}{f} = \frac{1}{p} + \frac{1}{q}$$

Remembering that the focal length of a plane mirror, as well as its radius of curvature, is infinite, the reciprocal of the focal length is zero.[3]

$$\frac{1}{\infty} = \frac{1}{p} + \frac{1}{q}$$

$$0 = \frac{1}{p} + \frac{1}{q}$$

Solving for the image distance q in terms of the object distance p,

$$-\frac{1}{q} = \frac{1}{p}$$

We find that the image is as far behind the mirror as the object is in front:

$$q = -p$$

This agrees with our common experience as we stand before the bathroom mirror each morning.

We can also find from the magnification formula that the image and object are the same size:

$$M \equiv \frac{b}{a} = \frac{q}{p}$$

$$= \frac{-p}{p}$$

$$= -1$$

The magnification in this case is negative unity. Since from the definition of magnification,

$$b = Ma$$
$$= (-1)a$$
$$= -a$$

the image height is minus the object height. The minus means that the image is right side up. The derivation of the formula used the implied assumption that the image was both real and upside down. A negative image distance places the image in negative space behind the mirror, making it virtual; a negative image height measures the image above the axis instead of below it.

---

[3] The reciprocal of larger and larger numbers is closer and closer to zero. One one-millionth is smaller than one one-thousandth. One one-billionth is smaller than one one-millionth, and so on.

► Overview of Reflection

We have used the law of reflection to understand the formation of images by flat and curved mirrors. In so doing, we have treated light as though it consisted of rays, in much the same way as the ancient Greeks thought of it.

We first described the positions of images formed by mirrors under various conditions, and then we derived the image positions by means of ray diagrams. Three rays were used to locate the crossing place of all the other rays in the system. Two of the three rays were associated with parallel light and with the focal point. The third ray was associated with the center of curvature and was found to be perpendicular to the mirror.

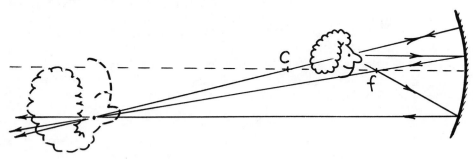

FIGURE 2-21 *Two of the three rays that can be used to locate an image are associated with the focal point. The third is associated with the center of curvature.*

We found that real images are formed by light actually crossing at a point after being reflected from the mirror. Such images can be captured on a screen. Virtual images are formed by light reflected at such an angle that it appears to come from a point behind the mirror. Such images are said to be virtual because they exist in negative space behind the mirror, where light never really travels.

Although we derived the object-image and magnification formulas from the geometry of a concave mirror ray diagram, these formulas can be made to work for convex and plane mirrors as well. All you need to do is let the focal length be negative in the case of a convex mirror and infinite in the case of a plane mirror. These modifications allow the formulas to apply to all three kinds of mirrors.

► Snell's Law

When a light ray interacts with an object, it can bounce off by reflection or it can penetrate the surface and pass into the object by **refraction.** Just as there is a relationship, the law of reflection, that allows us to predict the angle of the reflected light, there is a relationship that allows us to predict the path refracted light will take inside a transparent object. The law of refraction is a little more complicated

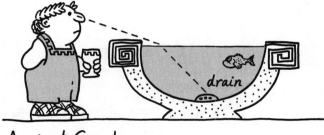

Ancient Greek

FIGURE 2-22 *Light is bent away from the normal when it passes from water into air, allowing the bottom of a fountain to be seen when full, although it could not be seen when empty.*

$$\angle i \geq \angle r$$

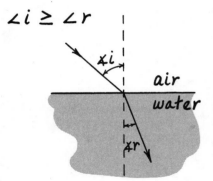

FIGURE 2-23 *The angle of incidence is greater than the angle of refraction for light passing from air into water. Both the angle of incidence and the angle of refraction are measured to the normal.*

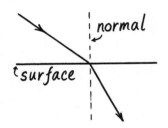

FIGURE 2-24 *The "normal" is defined as a line that is drawn perpendicular to the surface at the point where a ray of light is refracted.*

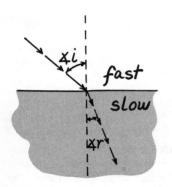

FIGURE 2-25 *Light is deflected toward the normal when it is slowed down.*

than the law of reflection. The ancient Greeks knew about both reflection and refraction, and they could predict angles for reflection; but they did not understand the relationship between the angles for refracted light. They did know, however, that light is bent toward the surface (or away from the normal) when it passes from water into air. They reported that they could see the bottom of a fountain when it was full of water although they could not see it from the same angle when the fountain was empty.

They also knew that light going the other way, incident on the water from air, would be refracted away from the surface by the same amount. In terms of angles measured to a line drawn perpendicular to the surface (the normal), they knew that the angle of incidence for light coming from air and going into water would be greater than the angle of refraction, but that was the extent of their knowledge. They may have tried a proportionality between the two angles, but if they did, they found that it did not work. It was not until about the sixteenth century that a man by the name of Willebrod Snell noticed that while the angles themselves are not proportional, their sines are:

$$\sin \angle i \propto \sin \angle r$$

Snell did not understand why the sines of the angles are proportional; he just observed it as an experimental fact. He noticed that the proportionality constant is different for different materials, and he called it the **index of refraction,** n, of the material. In terms of this proportionality constant, **Snell's law** can be written

$$\sin \angle i = n \sin \angle r$$

No one knows why Snell chose the letter n for his proportionality constant (perhaps it was because n is the second letter of the word *index* and is also the second letter of Snell's name), but this letter can be taken as a reminder that the angles are measured from the *n*ormal rather than from the surface itself. (If Snell had measured to the surface, his law would be expressed as a proportionality between the cosines of the angles.)

We now understand that light bends as it passes into more dense material because it slows down. Imagine how a person on roller skates would change direction as he came across a rug at an angle, one foot slowing down before the other. Light rays are deflected toward the normal in the same way when they come across a surface that slows them down.

The index of refraction turns out to be inversely proportional to the velocity of light in the material. The proof of Snell's law in terms of the velocity of light is easy, but we will leave that for a later course. For our purposes, we can think of Snell's law in the same way it was discovered, as a proportionality that just seems to work.

CHECK QUESTION

Light is incident on water from air. The angle of incidence, 23.6°, has a sine of 0.400. The angle of refraction in water is found to be 17.5°, which has a sine of 0.301. What is the index of refraction of water?

Answer: $n = 1.33$

► Lenses

A light ray bends as it enters glass and bends again as it leaves. If a piece of glass has just the right shape, it can bend parallel rays of light so that they all cross to form a focal point just as with the spherical mirror. It turns out that a spherical convex lens provides the same kind of positive focal point as does a concave mirror.

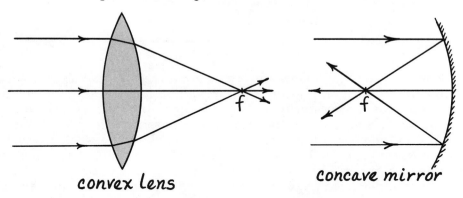

convex lens   concave mirror

FIGURE 2-26 *A convex lens has a positive focal length in the same sense that a concave mirror has a focal length.*

Light that comes in parallel to the axis and strikes the top of the lens is bent toward the first normal. Since the first normal is directed downward, the light is bent downward at the first surface. After the light passes through the lens, it strikes the second surface and is bent away from the normal as it emerges into air. The second normal is directed upward, however, and the light is bent downward again. Light coming in parallel below the axis is also deflected toward the first normal and away from the second, but the directions of the normals are reversed, and the light is bent upward each time. All rays that make reasonably small angles with the surface pass through the same point after being deflected by the lens. This is called the **focal point of the lens.**

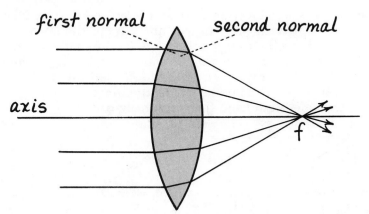

first normal   second normal

axis

f

FIGURE 2-27 *Parallel light is brought to focus by being bent toward the first normal and away from the second in a convex lens.*

Lenses have two focal points. Parallel light coming from the opposite direction is brought to a focus at a point the same distance on the opposite side of the lens.

► Lens Images

A real image is formed by a lens in much the same way as by a mirror. The light from the object is directed by the lens to cross at the image position. Any number of rays can be traced through the lens

FIGURE 2-28 *Three light rays may be constructed to locate the image.*

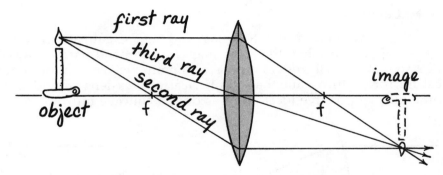

by use of Snell's law. However, there are three rays that are particularly easy to trace. Light coming from the tip of the object parallel to the lens axis is necessarily deflected so as to pass through the focal point on the far side of the lens. Light passing through the focal point on the near side of the lens is deflected and goes out parallel, since the angle of incidence is the same as if the light had *originated* at the focal point. The third ray passes through the center of the lens and is not deflected at all. This ray enters the lens at the same angle to the normal as it leaves; the first surface is curved up and the second surface is curved down. The amount that this ray is deflected toward the first normal is equal to the amount that it is deflected away from the second normal. The net effect is to displace the ray slightly to one side but to leave its direction unchanged. As long as the lens is thin compared to the focal length, we can ignore this tiny sideways displacement within the lens.

The first two rays described are also displaced within the lens. The rays are actually deflected at both surfaces, but only a small error is involved in pretending that the entire deflection takes place at the center of the lens. We use this thin lens approximation when drawing thin lens diagrams, making just one bend in the ray at the center of the lens.

The formula for object and image distances derived for a mirror also applies to lenses. The object and image distances are again called p and q; and the object and image heights are labeled a and b. In a similar manner two sets of similar triangles are formed by the parallel rays and the rays passing through the focal points, together with a line having the same length as the object and image heights. The little triangle is cut from the similar big triangle by the **lens axis.**

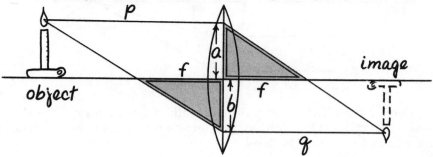

FIGURE 2-29 *Two sets of similar triangles are formed by a ray diagram of a convex lens.*

In this case, however, the two sets of similar triangles are on opposite sides of the lens.

Two proportions are formed by the ratio of sides of these similar triangles. Once again, the object and image heights a and b can be eliminated by the sneaky trick of adding these two proportions. Dividing both sides of this equation by f produces the same relation-

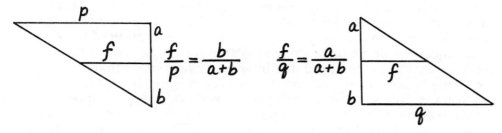

$$\frac{f}{p} = \frac{b}{a+b} \qquad \frac{f}{q} = \frac{a}{a+b}$$

FIGURE 2-30 *The ratios of the sides of the big and little triangles are equal for two sets of similar triangles that relate the image and object distances to the focal length.*

ship as for a mirror. That is, the reciprocal of the focal length is equal to the sum of the reciprocals of the object and image distances:

$$\frac{1}{p} + \frac{1}{q} = \frac{1}{f}$$

This relationship is called the **lens equation,** although we found it convenient to derive it first for the mirror. With this equation we can predict the position of an image produced by either a lens or a mirror.

Notice that the image is larger in one of the lens diagrams and smaller in the other. Just as we did with the mirror, we define the magnification as the ratio of the image height to the object height. It turns out to be equal to the ratio of the image to the object distance.

$$M \equiv \frac{b}{a} = \frac{q}{p}$$

When the object is far away, the image is real, inverted, small, and close to the focal point. As the object approaches the focal point, the image remains real and inverted but grows larger and moves away from the focal point. When they are both two focal lengths distant from the lens, the image is the same size as the object at that point. The object is on one side and the image is on the other. As the object gets increasingly closer to the focal point, the image grows ever larger and goes increasingly further away from the lens. The image becomes infinitely large and is located at infinity when the object is at the focal point. This is how a penlight works. Light from the object is deflected outward in a parallel beam by a lens built into the penlight bulb.

When the object passes over the focal point, the image goes beyond infinity, coming in at a negative infinity on the same side of the lens as the object. The image is now erect, larger, and further away from the lens than the object, but virtual. When the object is within the focal distance of the lens, the light rays diverge so greatly

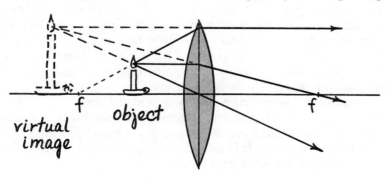

virtual image

object

f   f

FIGURE 2-31 *A magnifying glass is a convex lens used to form a virtual image of an object within its focal distance.*

that the lens cannot converge them enough to make them cross. The rays of light leave the lens still diverging. They diverge as though they had come from a point in space behind the location of the object. Since this point in space, the virtual image, is always further from the lens than the object, the image is always larger than the object. This is how a convex lens is used as a magnifying glass.

You can only see things with your naked eye if they are a certain distance away. When things get too close, they look blurred. Look at a friend's nose and then get closer and closer. The point at which the nose becomes blurred is called your **near point.** A jeweler can examine a diamond very close to his eye with the aid of a little magnifying glass mounted in a cylinder, called a loupe (pronounced loop). The loupe not only makes an enlarged image, but also moves it back to beyond the jeweler's near point so he can see it.

▶ **Optical Instruments**

Your eyes and a camera work on the same principle. The lens in the front of your eye forms a real image of whatever you're looking at on the retina in the back of your eye. In a camera the image is projected on film instead of retinal tissue. The image is upside down in both cases but that fact is compensated for in both cases. You simply turn camera film around to look at it, and your brain has learned to turn around the images it receives from your retina.

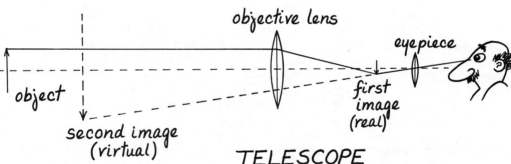

FIGURE 2-32 *A telescope uses one lens to form a real image and another as a magnifying glass to examine the real image.*

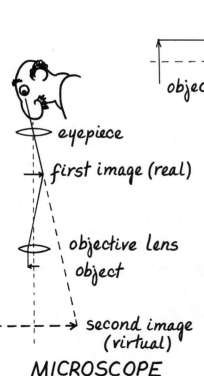

FIGURE 2-33 *A compound microscope uses one lens to form an enlarged real image and another to magnify it again.*

A telescope also uses a lens to form a real image of distant objects. The real image is not caught on a light-sensitive screen, such as a retina or photographic film, but is projected in space to be examined by another lens used as a magnifying glass. The second lens, called the eyepiece, is positioned so that the image produced by the first lens is within its focal point. The eyepiece forms an enlarged virtual image of the real image. When you look through a telescope, you are looking at an image of an image.

A compound microscope forms a real image of a close object. Since the image is further from the lens than the object, it is larger. A second lens, called the eyepiece, forms a virtual image of the first image, further enlarged. It is called a compound microscope because it enlarges an already enlarged image.

The problems in this chapter are on the optics of reflection and refraction. The law of reflection states that the angle of incidence equals the angle of reflection:

$$\angle i = \angle r$$

A concave mirror brings light to a focus at a point halfway to the radius of curvature.

$$f = \tfrac{1}{2}c$$

The focal point and center of curvature of a mirror can be used to trace three rays to locate an image. Two of these rays are associated with parallel light and the focal point. Incident parallel light is reflected so that it passes through the focal point:

$$\| \longrightarrow f$$

Incident light that passes through the focal point is reflected so that it goes out parallel:

$$f \longrightarrow \|$$

The third ray is associated with the center of curvature of the mirror. Incident light that passes through the center of curvature is reflected back on itself:

$$c \longrightarrow c$$

An image may also be found from the object-image formula, which states that the reciprocal of the focal length equals the sum of the reciprocals of the object and image distances.

$$\frac{1}{f} = \frac{1}{p} + \frac{1}{q}$$

Magnification, defined as the ratio of the image height to the object height, is equal to the ratio of the image distance to the object distance:

$$M \equiv \frac{b}{a} = \frac{q}{p}$$

The law of refraction is **Snell's law,** which states that the ratio of the sine of the angle of incidence to the sine of the angle of refraction is a constant, called the index of refraction of the material:

$$\frac{\sin \angle i}{\sin \angle r} = n$$

Three rays may also be used to locate the image formed by a lens. Two are associated with parallel light and the focal points, while the third passes through the center of the lens and is undeflected. The image may also be found from the object-image formula, the same as for a mirror.

These formulas assume that the focal length is positive for a converging optical system in which the image is real and inverted. A negative focal length can be used for a diverging mirror or lens. Negative image distance q and image size b indicate that the image is virtual and right side up.

You will need to know the information summarized above to solve the problems. Try to do the examples yourself before looking at the sample solution. If you need to look, test yourself later to make certain you can do it without looking before moving on to the Essential Problems.

2-1 To produce the illusion of a ghostly candle burning under water, place the candle in front of a window pane and a glass of water an equal distance on the other side. A dark background and a screen to hide the candle from direct view helps to perfect the illusion.

Given: The candle is placed 35 cm in front of the window glass. In understanding the position of the virtual image, you decide to trace rays that make an angle of 30° and 45° with the surface of the glass.

Find: (a) What will be the angle of reflection for these rays?
(b) Draw several reflected rays showing the position of the virtual image.
(c) How far behind the window pane should the glass of water be placed?

Your Solution

Discussion

In a famous old magic book entitled *Stage Magic and Scientific Diversions*, Albert Hopkins describes several illusions based on superimposing a virtual image on a real object. This book is probably responsible for the commonly held notion that many magic tricks are done with mirrors. Actually, very few illusions are done with mirrors, but the ones that are, are most striking.

Sample Solution 2-1

(a) $\angle i = \angle r$

$\angle r_1 = \angle i_1$
$= (90° - 30°)$
$= \boxed{60°}$

$\angle r_2 = \angle i_2$
$= (90° - 45°)$
$= \boxed{45°}$

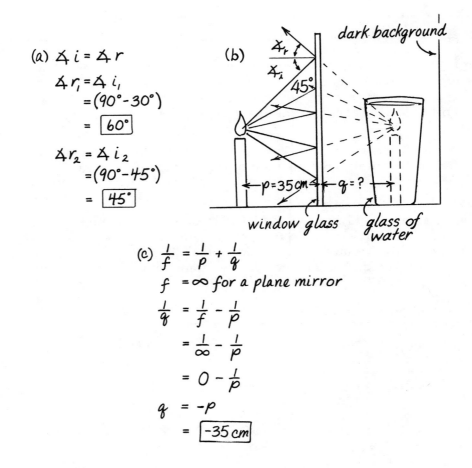

(b)

dark background

window glass   glass of water

(c) $\dfrac{1}{f} = \dfrac{1}{p} + \dfrac{1}{q}$

$f = \infty$ for a plane mirror

$\dfrac{1}{q} = \dfrac{1}{f} - \dfrac{1}{p}$

$= \dfrac{1}{\infty} - \dfrac{1}{p}$

$= 0 - \dfrac{1}{p}$

$q = -p$

$= \boxed{-35\,cm}$

Discussion

The solution to part (a) is based on the law of reflection. The angle of incidence and the angle of reflection are, however, usually measured to the normal rather than to the surface itself. Since the normal is defined as a line that is perpendicular to the surface, the angle of incidence is found by subtracting the given angle from 90°.

The drawing in part (b) shows that the position of the virtual image is independent of the angle of incidence.

Part (c) is based on the object-image formula applied to a plane mirror, where the focal length is infinite. The image and object distances are equal in this case.

2-2   The light source in a commercial movie projector is an electric carbon arc. Electric current passing through the arc produces a white-hot crater in the end of the positive carbon rod. Light from this white-hot crater is concentrated into an image by a concave mirror before being projected onto the screen by a lens.

Given: The white-hot crater of the carbon arc is 20 cm from the mirror, while the image formed by this crater must be near the projection lens some 80 cm from the mirror. The image must be about 5.0 cm in diameter to illuminate each frame on the 35-mm movie film.

Find: (a)   What focal length should the mirror in this projector have?
     (b)   How large should the crater be to produce the proper size image?

Your Solution

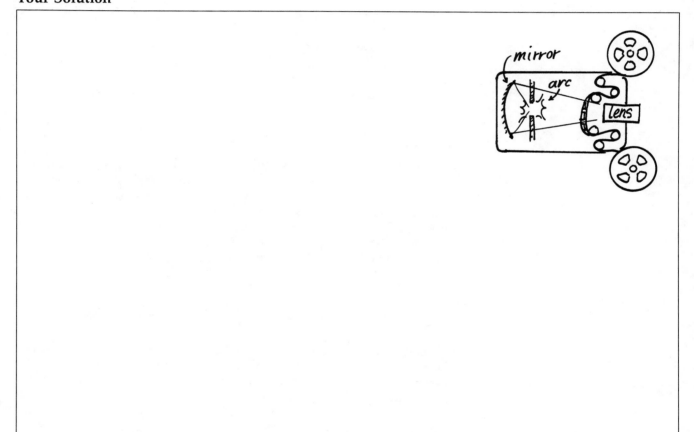

Discussion

Light from the carbon arc is imaged on the projection lens so that it will be totally out of focus on the screen. You want a nice crisp image of the film on the screen, but you do not want to see any hint of the structure of crater in the carbon arc on the screen. The image produced by the mirror is an object for the projection lens, but making this object distance zero causes the object-image formula to blow up, since the reciprocal of zero is infinite. The result is that light from the carbon arc fills the screen evenly (in the absence of film).

Sample Solution 2-2

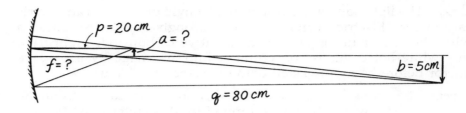

(a) $\dfrac{1}{f} = \dfrac{1}{p} + \dfrac{1}{q}$

$\quad = \dfrac{q}{qp} + \dfrac{p}{qp}$

$\quad = \dfrac{q+p}{qp}$

$f = \dfrac{qp}{q+p}$

$\quad = \dfrac{(80\,cm)(20\,cm)}{(80\,cm + 20\,cm)}$

$\quad = \boxed{16\,cm}$

(b) $M \equiv \dfrac{b}{a} = \dfrac{q}{p}$

$a = b\,\dfrac{p}{q}$

$\quad = (5\,cm)\,\dfrac{(20\,cm)}{(80\,cm)}$

$\quad = 1.25\,cm$

$\quad \cong \boxed{1.2\,cm}$

Discussion

The ray diagram is not called for in the problem and might not be part of your solution. While not required, however, ray diagrams are always helpful for understanding what is going on and for checking the numerical results to see if they look reasonable. Arrows are frequently used in such diagrams to represent the object and the image.

An alternative to solving the object-image formula algebraically for f would be to insert data in it as it stands and find the reciprocal of f:

$$\frac{1}{f} = \frac{1}{p} + \frac{1}{q} = \frac{1}{(0.2\ \text{m})} + \frac{1}{(0.8\ \text{m})} = 5\ \text{m}^{-1} + 1.25\ \text{m}^{-1} = 6.25\ \text{m}^{-1}$$

This is called the power of the mirror, and a reciprocal meter is called a diopter. Taking the reciprocal of $6.25\ \text{m}^{-1}$ produces the same result.

The result of part (b) must be rounded off because the image height is known to only one significant figure. The rule of thumb of significant figures would have us round the answer off to 1. cm, but that would imply an error of ± 0.5 cm, or $\pm 50\%$. That is less accurate than we know the height ($b = 5$ cm ± 0.5 cm, which is an error of $\pm 10\%$), so we stretch the rule a bit and report the answer to two significant figures. Three significant figures would, however, be stretching the rule too far.

2-3 A woman is looking at the virtual image of her nose in a concave shaving mirror. She wants to know where to put her nose in order to produce an image a certain distance behind the mirror.

Given: The shaving mirror has a focal length of 40 cm and the woman wishes to produce an image at -20 cm (behind) the mirror. She has a 3.5-cm nose.

Find: (a) Where should she place her nose to produce the image at the desired location?
(b) What will be the magnification?
(c) How large will the image of her nose be?

Your Solution

Discussion

This problem illustrates the use of a concave mirror as a magnifying mirror. Shaving mirrors are sometimes concave so that you can see defects in your face on an enlarged scale.

In order to use a concave mirror to produce a magnified virtual image, the object must be placed inside the focal length. Light from the object will be converged by the mirror but not enough to actually cross and produce a real image. Light from the object will be so strongly diverging that the converging effect of the mirror can only make the reflected light appear to come from a point further behind the mirror than the object is in front.

Sample Solution 2-3

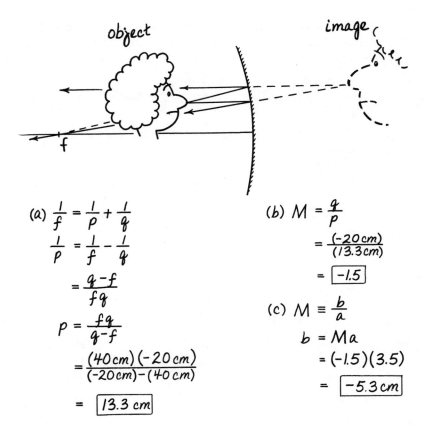

(a) $\dfrac{1}{f} = \dfrac{1}{p} + \dfrac{1}{q}$

$\dfrac{1}{p} = \dfrac{1}{f} - \dfrac{1}{q}$

$\phantom{\dfrac{1}{p}} = \dfrac{q-f}{fq}$

$p = \dfrac{fq}{q-f}$

$ = \dfrac{(40\,cm)(-20\,cm)}{(-20\,cm)-(40\,cm)}$

$ = \boxed{13.3\ cm}$

(b) $M = \dfrac{q}{p}$

$ = \dfrac{(-20\,cm)}{(13.3\,cm)}$

$ = \boxed{-1.5}$

(c) $M \equiv \dfrac{b}{a}$

$b = Ma$

$ = (-1.5)(3.5)$

$ = \boxed{-5.3\ cm}$

Discussion

Here again, while a ray diagram is not required by the problem, it helps to show how the image is formed and allows you to see if the result is reasonable.

Part (a) can be worked by solving the object-image formula for p and then inserting the data. Parts (b) and (c) use the magnification formula. The negative results in parts (b) and (c) come about because the image is virtual and therefore right side up. The formulas were derived from the geometry of a real image, which is upside down. Negative upside down is right side up.

2-4 Convex mirrors are sometimes used at blind intersections to provide a wide-angle view of oncoming traffic.

Given: A convex mirror having a focal length of -1.0 m is placed at a blind intersection. An oncoming car is 3.0 m from the mirror.

Find: (a) What is the position of the virtual image of the oncoming car produced by the convex mirror?
   (b) What is the magnification?
   (c) Draw a ray diagram showing the three rays associated with parallel light, the focal point, and the center of curvature of the mirror.

Your Solution

Discussion

This problem illustrates the use of a convex mirror to produce a wide-angle view. Since the images of real objects are demagnified, more images fit within the field of view than would fit within the field of view of a plane mirror.

Just as is the case for a concave mirror, the focal point is the image of a far distant object. But instead of concentrating parallel light to cross at the focal point, a convex mirror diverges parallel light so that it is reflected away from the focal point. Reflected light "crosses" at the focal point because that is where it appears to come from.

Sample Solution 2-4

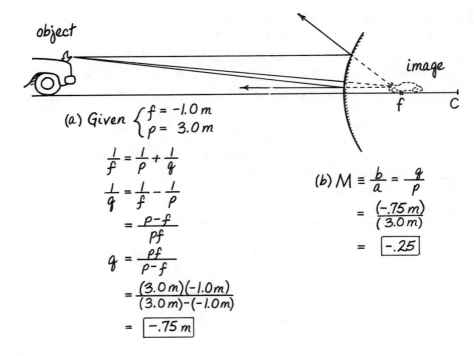

(a) Given $\begin{cases} f = -1.0\,m \\ p = 3.0\,m \end{cases}$

$$\frac{1}{f} = \frac{1}{p} + \frac{1}{q}$$

$$\frac{1}{q} = \frac{1}{f} - \frac{1}{p}$$

$$= \frac{p-f}{pf}$$

$$q = \frac{pf}{p-f}$$

$$= \frac{(3.0\,m)(-1.0\,m)}{(3.0\,m)-(-1.0\,m)}$$

$$= \boxed{-.75\,m}$$

(b) $M \equiv \frac{b}{a} = \frac{q}{p}$

$$= \frac{(-.75\,m)}{(3.0\,m)}$$

$$= \boxed{-.25}$$

Discussion

The negative image distance in part (a) and negative value for the magnification in part (b) mean that the image is in virtual space (behind the mirror) and right side up. The formulas were derived from a ray diagram for a real image, which assumes the image is on the same side of the mirror as the object and that the image is upside down. Negative image distances are just measured in the other direction, on the opposite side of the mirror as the object. Negative magnification means that the image height will be negative if the object height is positive. Negative upside down is right side up. Virtual images are in general right side up.

2-5   A merman is looking at a handsome young woman. He is underwater while she is standing on a rock at the edge of the sea.

Given: Light from her nose on the way to his eye is incident at an angle of 60° to the normal. Light from his chin on the way to her eye makes an angle of 38° to the normal. The index of refraction of water is $n = 1.33$.

Find: (a) What is the angle of refraction for light from her nose on the way to the merman's eye?

   (b) At what angle does the light from his chin leave the water's surface on the way to the handsome young woman's eye?

Your Solution

Discussion

This problem illustrates the use of Snell's law both for light going from air into water and for light going from water into air. Light can travel both ways, although the form of Snell's law that we have been considering is only set up for light going from air into a medium having a higher index of refraction.

Sample Solution 2-5

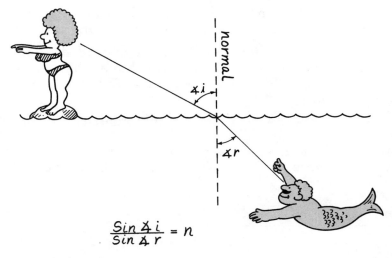

$$\frac{Sin \angle i}{Sin \angle r} = n$$

(a) $Sin \angle r = \frac{Sin \angle i}{n}$

$= \frac{Sin (60°)}{1.33}$

$= 0.651$

$\angle r = \boxed{41°}$

(b) $Sin \angle i = n \, Sin \angle r$

$= (1.33) Sin (38°)$

$= 0.819$

$\angle i = 55°$

Discussion

Part (a) is worked by solving Snell's law for the sine of the angle of refraction. The sine of 60° is found by looking in a trig table. After dividing by the index of refraction to get the sine of the angle we are looking for, the angle itself can be found by again referring to a trig table.

Part (b) is found by solving Snell's law for the angle of incidence, although the light from the merman's chin is actually going the other way in this case. If you can figure out the path that light will take going one way, the same path will work for light going in the reverse direction. This rule even works for the most complicated optical system. It is called the principle of reversibility of optics.

The reason we must reverse the direction of the light ray in part (b) is that we have only concerned ourselves with the simplest form of Snell's law, the form which assumes that light is passing from air or some other medium having an index of refraction close to that of a vacuum, namely $n = 1.00$, into a medium having a greater index of refraction. A more general form of Snell's law, good for light passing from any medium having an index of refraction n_1 into any other medium having an index of refraction n_2, says that the first index of refraction times the sine of the angle in that medium equals the second index of refraction times the sine of the angle in the second medium.

$$n_1 \sin \angle_1 = n_2 \sin \angle_2$$

For light going from water, $n_1 = 1.33$ and $\angle_1 = \angle i$, into air, $n_2 = 1.00$ and $\angle_2 = \angle r$, this general form reduces to our simplified form but the index of refraction of water is replaced by its reciprocal.

$$n \implies \frac{1}{n}$$

2-6 A college student is amazing her little brother with a demonstration of real images. She uses a strong magnifying glass (about seven power) to project an image of a candle on a piece of paper several centimeters away.

Given: The lens has a focal length of 3 cm. The candle flame is about 1 cm tall and is held 5 cm from the lens.

Find: (a) Predict the position of the image and approximate size by drawing a ray diagram showing three rays that a little brother could understand.

(b) What is the image distance from the object-image formula?

(c) How large should the image be from the magnification formula?

Your Solution

Discussion

A good test of your understanding of a subject is to try to explain it to a child. If you don't have any around the house, go out and try your neighborhood playground. It takes some planning, however, to get even a very smart child through an explanation involving a ray diagram. Start with the image of a bright nearby object and show how the image distance changes with the object distance. Then shift to a discussion of the focal point by using the magnifying glass to burn your initials in a piece of wood. Shift back to a discussion of real images by showing how things look upside down when you hold the magnifying glass far away from your eye. Socratic teaching is difficult, but it is fun.

Sample Solution 2-6

(a):

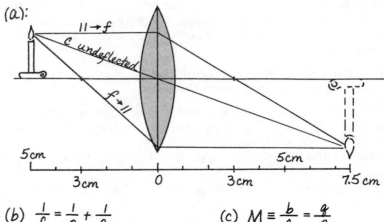

(b) $\dfrac{1}{f} = \dfrac{1}{p} + \dfrac{1}{q}$

$\dfrac{1}{q} = \dfrac{1}{f} - \dfrac{1}{p}$

$\quad = \dfrac{p-f}{pf}$

$q = \dfrac{pf}{p-f}$

$\quad = \dfrac{(5\,cm)(3\,cm)}{(5\,cm)-(3\,cm)}$

$\quad = \boxed{7.5\,cm}$

(c) $M \equiv \dfrac{b}{a} = \dfrac{q}{p}$

$b = a\,\dfrac{q}{p}$

$\quad = (1\,cm)\,\dfrac{(7.5\,cm)}{(5\,cm)}$

$\quad = \boxed{1.5\,cm}$

Discussion

Part (b) is worked by solving the object-image formula for the image distance. This algebra was done in an earlier example, but it is best to repeat it for each problem. You could memorize the general solution of every physics problem, but it is far better for you to know how to obtain it.

Part (c) is solved by the magnification formula and the definition of magnification. Magnification is defined as b over a, but it is equal to q over p. Solving for the image height and substituting the numerical data gives the answer.

2-7 When a physics instructor focuses on an exam paper, the lens in his or her eye forms a real image of the paper on the retina in the back of the eye. The light-sensitive retina then translates the patterns of light and dark areas to the brain, which in turn translates them into a grade.

Given: An exam paper is 20 cm long and is located 28 cm from a physics instructor's eye. The eyeball is 3.0 cm in diameter.

Find: (a) What is the focal length of the lens in the instructor's eye?
(b) How large will the image of the exam paper be when projected by the lens on the back surface of the eyeball?

Your Solution

Discussion

The eye works very much like a camera in that a real image is formed on a light-sensitive film. This image, like all real images, is upside down. The brain is used to that, however, and only becomes confused if the image is right side up.

One of the classical experiments in psychology is based upon wearing special glasses that invert the image so that it is in fact right side up on the retina. It takes the brain between one and two days to learn to flip this image. After that, a person wearing the glasses can function normally. Everything is fine until they take off the glasses.

Sample Solution 2-7

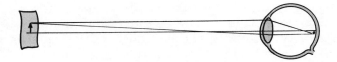

(a) $\dfrac{1}{f} = \dfrac{1}{p} + \dfrac{1}{q}$

$\quad = \dfrac{q+p}{qp}$

$\quad f = \dfrac{qp}{q+p}$

$\quad = \dfrac{(28\,cm)(3.0\,cm)}{(28\,cm)+(3.0\,cm)}$

$\quad = \boxed{2.7\,cm}$

(b) $M \equiv \dfrac{b}{a} = \dfrac{q}{p}$

$\quad b = a\,\dfrac{q}{p}$

$\quad = (20\,cm)\,\dfrac{(3.0\,cm)}{(28\,cm)}$

$\quad = \boxed{2.1\,cm}$

Discussion

The focal length of the lens is found from the object-image formula. The trick is to recognize that the diameter of the eyeball is the image distance.

This problem is useful on examinations to discriminate between people who do the example problems and those who do not. People who see this problem for the first time tend to confuse it with a mirror situation and take the focal length as one-half the radius of curvature. Those who have thought about it some realize that the back side of the eyeball is not a reflecting surface but is the location of the image.

2-8 A detective is examining evidence with a magnifying glass. He holds the glass, a double convex lens, close enough to the evidence so that it produces an enlarged virtual image. He adjusts the position of the magnifying glass so that the image occurs at his distance of most distinct vision.

Given: The magnifying glass has a focal length of $+15$ cm. The detective adjusts its position to produce an image at -22 cm from the lens (on the same side of the lens as the object).

Find: (a) How far from the evidence should the detective hold the lens?

(b) What is the magnification?

Your Solution

Discussion

This problem illustrates the use of a converging lens to produce a virtual image. The object must be located within the focal length if the image is to be virtual.

Sample Solution 2-8

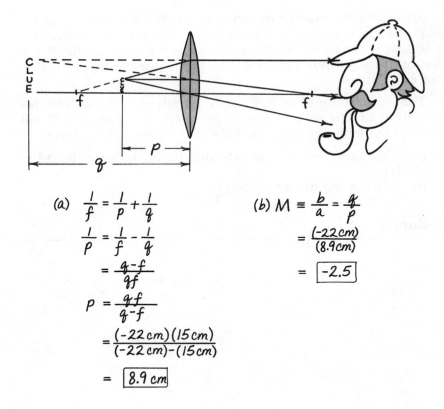

(a) $\dfrac{1}{f} = \dfrac{1}{p} + \dfrac{1}{q}$

$\dfrac{1}{p} = \dfrac{1}{f} - \dfrac{1}{q}$

$\phantom{\dfrac{1}{p}} = \dfrac{q-f}{qf}$

$p = \dfrac{qf}{q-f}$

$ = \dfrac{(-22\,cm)(15\,cm)}{(-22\,cm)-(15\,cm)}$

$ = \boxed{8.9\ cm}$

(b) $M \equiv \dfrac{b}{a} = \dfrac{q}{p}$

$ = \dfrac{(-22\,cm)}{(8.9\,cm)}$

$ = \boxed{-2.5}$

Discussion

Parts (a) and (b) use the object-image formula and the magnification formula. Just as was the case with mirrors, a negative magnification means the image is right side up, since the formulas are derived from the geometry of a real-image ray diagram. The only difference between the sign conventions for a lens and for a mirror is the positive direction for image location. Since a real image is produced by a mirror on the same side of the mirror as the object, that is considered to be the direction of positive image distances. A lens produces a real image on the opposite side from the object, so positive image distances are measured in that direction for a lens.

2-9 In attempting to understand the nature of virtual images, a student sets up a little experiment on the kitchen table. A lump of clay is used to stand a small pocket mirror up vertically in the middle of the table. Two pencils are stuck into two smaller lumps of clay, one to act as the object and the other to locate the image. The object pencil is stood up a short distance in front of the mirror, and the image pencil is moved around behind the mirror until it lines up with the image even when viewed from different angles.

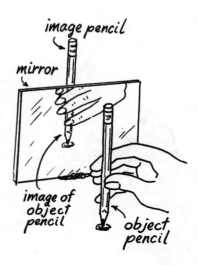

Given: The object pencil is placed 20 cm from the mirror. Viewed from one angle, the light reflected from the mirror makes an angle of 15° with the normal. Viewed from a larger angle, the reflected light makes an angle of 25° with the normal.

Find: (a) What is the angle of incidence for these rays?
(b) Draw a top-view diagram showing the incident and reflected rays. Indicate the position of the image by showing where the reflected rays seem to come from after they leave the mirror.
(c) How far behind the mirror should the image pencil be placed to coincide with the image formed by the mirror?

2-10 Just for fun, a physics student takes a magnifying mirror off the dressing table and uses it to form a real image of a light bulb on the wall.

Given: The focal length of the mirror is 68 cm. The student finds it necessary to hold the mirror 102 cm from the light bulb to produce a nice sharp image on the wall some distance away. The light bulb is 10 cm tall.

Find: (a) How far is the wall from the mirror?
(b) How large will the image be on the wall?

2-11 In examining the inside surface of his eyeglasses one day, a young man notices a magnified reflection of his eye. He starts wondering what the radius of curvature might be for this partially reflecting concave surface.

Given: When he holds his eye 2.0 cm from the inside surface of the lens, the reflected image is 3.5 cm on the other side of the surface (in negative space).

Find: (a) What is the magnification?
(b) What is the focal length of the reflecting surface?
(c) How far is the center of curvature from the surface of the eyeglass?

2-12 A shoplifter is shopping for convex mirrors one day. He spots a wide-angle mirror in the corner of a store and wishes to judge its focal length.

Given: He notices that his image is half as large as he is when he stands 2.0 m from the mirror. He also notices that the image is virtual, since it is right side up.

Find: (a) What is the magnification?
(b) How far is the virtual image on the other side of the mirror?
(c) What is the focal length of the mirror?

2-13 A wicked magician has trapped an Arabian princess in a crystal cube. Light from the magician's evil eye is deflected toward the normal as it enters the crystal due to the increased index of refraction.

Given: Light from the evil eye is incident at an angle of 45° from the normal and is refracted at an angle of 25°. Light from the evil chin is incident at an angle of 75° from the normal.

Find: (a) What is the index of refraction of the crystal?
(b) What is the angle of refraction for light from the magician's chin?

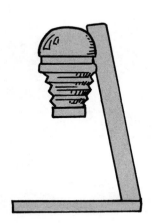

2-14 A photographic enlarger is basically a high-quality projector used to expose photosensitive print paper with a real image of the negative. The size of this image can be adjusted by changing the relative distances between the negative, or object, and the lens and the print paper, or image.

Given: A normal lens for enlarging 35-mm film has a focal length of 5.0 cm. A photographer who wishes to enlarge a feature 2.5 cm tall on the negative places the negative 8.3 cm from the lens.

Find: (a) Predict the position of the image and approximate size by drawing a ray diagram showing three rays.
(b) What should be the distance from the lens to the print paper from the object-image formula?
(c) How large will the image of the desired feature be from the magnification formula?

2-15 The type of view camera most frequently used by professional photographers to take portraits in the studio uses a large film format, usually 4 in. by 5 in. The image can therefore be much larger compared to the grain of the film than is possible for a small camera.

Given: An object that is 3.2 ft tall is positioned 4.5 ft from the lens of a large-format view camera. The photographer wishes the image to be 3.6 in. tall on the film and needs to select a lens of the proper focal length to produce this result.

Find: (a) What image distance is needed to produce the proper magnification?

(b) What focal length does the lens need to produce the proper image distance?

2-16   A jeweler's loupe is a small magnifying glass of short focal length that is held up close to the eye so that the jeweler can take a very close look at a gem or a watch part, holding the object much too close to be seen with the naked eye. The lens forms a virtual image of the object far enough away for the jeweler to see it.

Given: The magnifying glass has a focal length of 2.2 cm. The jeweler is looking at a gem that he positions just far enough away from the lens so that the image distance is -22 cm from the lens (on the same side of the lens as the gem).

Find: (a) How far from the lens should the jeweler hold the gem?

(b) What will be the magnification?

MORE INTERESTING PROBLEMS

2-17   A physics instructor is playing with a magnifying mirror while watching television. He notices that he can use the mirror to project a real image of the TV screen on the surface of a beer can that happens to be close at hand.

Given: The beer can is 25 cm from the mirror, and the TV set is 3 m across the room. The TV screen is 20 cm tall.

Find: (a) What is the focal length of the mirror?

(b) How tall is the image projected on the beer can?

2-18   A woman is admiring herself in the convex surface of a fun-house mirror. She notices that her image seems far away when she looks at it with one eye closed because the image is so small.

When she looks at the image with both eyes, however, her binocular vision tells her that the image is really closer behind the surface of the mirror than she is in front.

Given: The woman is 2.5 m from the mirror. The mirror has a focal length of -0.75 m. The woman is 1.78 m tall.

Find: (a) Draw a ray diagram showing the position and approximate size of the image using three reflected rays.
(b) Find the position of the woman's image using the object-image formula.
(c) How tall is the woman's image?

2-19 A movie theater projectionist must know how to select the proper projection lens to fill the screen with an image of the movie film.

Given: The picture on the movie film is 3.4 cm wide. In a particular theater the screen is 10 m wide and is located 30 m from the projector.

Find: (a) What magnification is needed to fill the projection screen?
(b) What object distance is needed to produce this magnification?
(c) What should be the focal length of the lens?

2-20 One way to find the focal length of a lens is to form a real image of a far-distant object. Unfortunately, you find yourself wishing to know the focal length of a double convex lens that you are holding in your hand while standing in a storeroom where there are no far-distant objects. The furthest thing away is an overhead light fixture. You form an image of this light fixture on the floor and decide to approximate the distance from the lens to the floor as the focal length even though you know that the light from the fixture is not really parallel.

Given: The distance from the lens to the light fixture is about 2 m. The distance from the lens to the floor is 35 cm.

Find: (a) What is the focal length of the lens from the object-image formula?
(b) How much error is involved in forgetting about the lens equation and just assuming the object distance to be infinite?
(c) What is the percentage error? (How great is the error compared to the best value?)

2-21 A common type of Christmas tree ornament is a glass ball having a mirrored surface. You can use such an ornament as a convex mirror to reinforce your understanding of virtual images. The radius of curvature is, naturally, half the diameter of the ball.

Given: You are looking at a Christmas tree ball having a diameter of 10 cm. You compare the image position of your nose to the image position of the fireplace and decide that the fireplace image is probably near the focal point. Your nose is 30 cm from the surface of the ball, and the fireplace is 5.00 m across the room.

Find: (a) What is the focal length of the mirrored surface?

(b) What is the image position of your nose?

(c) How much error is involved in assuming the fireplace image to be the focal point of the mirror?

(d) Draw a diagram showing parallel light from a far-distant object being reflected in accordance with the law of reflection and show that it comes away from the focal point.

2-22 A man is admiring himself in the shiny concave surface of a half-spherical stainless steel mixing bowl. He notices that his image is upside down and smaller than his object when he holds the bowl at arm's length.

Given: The bowl has a diameter of 30 cm. The distance from the bowl to the object is 75 cm, and the object size is 6 cm.

Find: (a) What is the focal length of the mirrored surface of the bowl?

(b) What is the position of the image?

(c) What is the size of the image?

2-23 A magician is practicing before a mirror. He needs to be able to see both his left hand, in which he has palmed a coin, and his right hand, which is holding an imaginary coin. His left arm is vertical while his right arm is horizontal.

Given: The distance from his eye to the tip of the fingers of his open left hand is 1.03 m. The distance from his right fist to his left arm is 1.15 m. He is standing 2.0 m from the mirror.

Find: (a) What is the angle of incidence for light reflected from the magician's open left hand to his eye?

(b) What is the minimum size of the mirror needed to see both hands at once?

(c) How far is it from the magician to his image?

ANSWERS *(Rounded off to the proper number of significant figures)*

2-9	(a) 15° and 25°; (c) 20 cm.
2-10	(a) 2.04 m; (b) 20 cm.
2-11	(a) $-1.75 \cong -1.8$; (b) 4.7 cm; (c) 9.3 or 9.4 cm.
2-12	(a) -0.5; (b) 1.0 m; (c) -2.0 m.
2-13	(a) 1.67; (b) 35°.
2-14	(b) 12.6 cm $\cong$ 13 cm; (c) 3.79 cm $\cong$ 3.8 cm.
2-15	(a) 5.1 in.; (b) 4.6 or 4.7 in.
2-16	(a) 2.0 cm; (b) -11.
2-17	(a) 23 cm; (b) 1.7 cm.
2-18	(b) -0.58 m; (c) 41 cm.
2-19	(a) $294 \cong 2.9 \times 10^2$; (b) 10.2 cm; (c) 10.2 cm or 10.1 cm.
2-20	(a) 30 cm; (b) 5 cm; (c) 17%.
2-21	(a) -2.5 cm; (b) -2.3 cm; (c) 0.4% or 0.5%.
2-22	(a) 7.5 cm; (b) 8.3 cm; (c) 7 mm.
2-23	(a) 14°; (b) 51.5 cm $\times$ 57.5 cm; (c) 4.0 m.

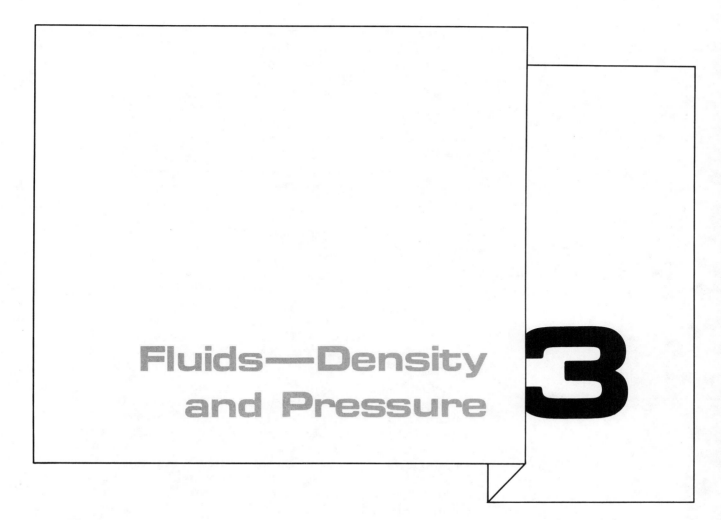

Fluids—Density and Pressure 3

▶ Density

Suppose you are interested in buying a water bed but are worried that it might be too heavy for your floor. One water bed manufacturer suggests that an easy way to test the strength of your floor is to have a party and arrange things so that most of your friends wind up standing in the part of the room where you plan to put the bed. If the floor holds, you would be safe to go out and buy a water bed. It is a simple matter to figure out how many friends you would need to invite to your party.

One cubic foot of water weighs 62.4 pounds. Some people find it surprising that water is that heavy, but we can make it seem reasonable by thinking of the size and weight of a half gallon of milk. A carton of milk has a base of about $3\frac{1}{2}$ inches square and holds milk up to a level of about 7 inches. Each carton weighs about 4 pounds. A cubic foot of milk would fill about 15 such cartons, since it occupies a volume a little more than three times as great in both width and depth and a little less than twice as high. That much milk would thus be about 15 times as heavy as one carton, or about 60 pounds.

Since milk and water weigh about the same for the same volume, we can make this comparison. (You can test this for yourself by putting a few drops of milk into a glass of water. If they sink, milk is heavier than water. If they float, it is lighter. Try it and see.)

FIGURE 3-1 *Finding the weight of a water bed is one application of the idea of density.*

79

Sea water is a bit heavier than fresh water. That is because the salt dissolved is heavier than the water it displaces. One cubic foot of sea water weighs about 64 pounds. We say that sea water is more **dense** than fresh water.

Weight density, D_{wt}, is the ratio of weight, wt, to volume, V:

$$D_{wt} = \frac{wt}{V}$$

The weight of a water bed can be derived from this definition by solving the equation for weight:

$$wt = D_{wt} \cdot V$$

Since the water bed is rectangular, its volume is length times width times height:

$$wt = D_{wt}(l \cdot w \cdot h)$$

Let us assume that the water bed is about 7 feet long ($l = 7$ ft), 5 feet wide ($w = 5$ ft), and 1 foot deep ($h = 1$ ft):

$$wt = (62.4 \text{ lb/ft}^3)(7 \text{ ft})(5 \text{ ft})(1 \text{ ft})$$
$$= 2,200 \text{ lb}$$

We calculate that a water bed weighs a little over a ton. To simulate this weight you would need to invite 22 friends who weigh 100 pounds each, or 11 friends who weigh 200 pounds each. It is true that fat friends tend to eat more, but you would need fewer of them.

If this experiment shows that the weight is a bit much for your floor, you might consider using something that is lighter than water. Ice, for example, is less dense than water, which is why ice floats. One cubic foot of ice only weighs 57 pounds. Ice, however, tends to be lumpy. Alcohol would be better, since it is even less dense than ice. One cubic foot of alcohol weighs only 50 pounds. Alcohol also has more social status than water, and it won't freeze in the winter if heating fuel becomes scarce.

We have been thinking of the density of substances in terms of their weight per unit volume. In the metric system, however, substances are usually measured in terms of their mass rather than their weight. We should therefore define a **mass density** for use in the metric system.

$$D_{mass} \equiv \frac{Mass}{Volume}$$

It turns out that the mass density is in more common use than weight density. We will assume that when someone says density they mean mass density and that when they mean weight density they will specifically say so.

These two kinds of density are related by the same proportionality constant that relates mass and weight, namely, the acceleration due to gravity, g. Weight, as you recall, is a force related to mass by

$$wt = mg$$

If we divide both sides of this expression by volume, V,

$$\frac{wt}{V} = \frac{m}{V} g$$

we have the proportionality between weight density and mass density:

$$D_{wt} = D_m g$$

The mass density of water turns out to be 1 kilogram per liter in the metric system because of the way in which a kilogram was originally defined.[1] A gram of water was originally the amount of water that would fit in a volume 1 centimeter on a side. A kilogram is therefore the amount of water that will fit into 1,000 cubic centimeters of volume.

Density is often used to help identify substances. A geologist uses density to differentiate between certain rocks that look alike. One of the reasons gold was used as money is that it is nearly the most dense of all substances and could therefore be easily identified. If a merchant suspected that the gold had been thinned out with a less valuable substance, he had only to measure both its weight and volume. Add almost anything to gold, and you will make it less dense. Platinum and platinum-type metals are the only exception, but they were not discovered until 1735. They were, of course, promptly used to adulterate gold—until they became more valuable. By the 1920s platinum was about eight times as valuable as gold.

Gold would therefore be one of the worst possible choices for use in your water bed. One of the best possible choices would be air. Not only is it cheap (at least now), but its density is only 1.3×10^{-3} kilograms per liter. Air lacks mass, however, and it is the inertia of the water that gives the undulating wave motion some people seem to enjoy. Besides, air lacks status. Nobody would invite somebody over to see their new air mattress. Air mattresses tend to go down about half way through the night.

TABLE 3–1

Substance	Weight Density $D_{wt} = \frac{wt}{V}$ (English units)	Mass Density $D = \frac{m}{V}$ (metric units)
Water, pure	62.4 lb/ft³	1 kg/liter
Water, salt	64	1.03
Ice	57.2	0.917
Alcohol	50	0.80
Wood, pine	22–31	0.35–0.50
Rock salt	136	2.18
Aluminum	168	2.70
Iron	491	7.87
Silver	655	10.5
Lead	708	11.3
Mercury, 39°C	842	13.5
Gold	1,200	19.3
Platinum	1,340	21.5
Air, 0 °C sea level, moist	0.080	1.29×10^{-3}

[1] In the English system nearly the same thing is done by defining the **specific gravity** of a substance, which is the density of the substance divided by the density of water. The specific gravity turns out to be the same number as the mass density of the substance, but without units.

CHECK QUESTIONS

1. You add a bit of alcohol to some fruit punch to make it lighter. To test the mixture, you weigh one quart of it on a scale. You find that one quart weighs 2.0 lb, and you happen to remember that a quart is one-thirtieth of a cubic foot ($\frac{1}{30}$ ft³). What is the weight density of the alcohol-punch mixture?

    Answer: 60 lb/ft³

2. A king wishes to know the density of a metal crown. The crown has a mass of 7.5 kg and a volume of 0.50 liter. What is the density of the metal?

    Answer: 15 kg/liter

▶ **Pressure**

One of the advantages of a water bed is that it distributes the force of your weight evenly over your body, that is, it is soft. The larger the area over which you distribute the force, the less it is concentrated at any particular point. This idea of concentration of force is called **pressure.** Pressure is defined as the ratio of force, F, to area, A:

$$\text{Pressure} \equiv \frac{\text{Force}}{\text{Area}}$$

Just as you would expect, pressure is something that is directly proportional to force but inversely proportional to area. The greater the force, the greater the pressure; but the *smaller* the area, the greater the pressure. The discomfort you experience when your air mattress goes down is due to the same force being distributed over a smaller area.

It is said that a law was passed in Houston, Texas, to prohibit women from walking on the grass in shoes with those highly pointed heels that were fashionable back in the 1950s. Story has it that the law was passed because of the huge pressure generated by the tiny area of those heels when worn by an ordinary Texas-sized woman.

We can calculate this pressure if we assume all her weight, say about 125 pounds, is on one heel and that the heel comes down to a point about $\frac{1}{8}$ inch across. Even without knowing the exact shape of the area presented to the ground, we can calculate the pressure approximately:

$$P = \frac{F}{A}$$

$$= \frac{125 \text{ lb}}{(\frac{1}{8} \text{ in.})^2}$$

$$= (125) \left(\frac{8}{1}\right)^2 \text{ lb/in.}^2$$

$$= 8{,}000 \text{ lb/in.}^2$$

FIGURE 3-2 *Keep off the grass.*

The story goes that this law was passed because the fire department was spending too much of its time capping the oil wells produced by this huge pressure.

In this modern time, such a law might seem unfairly discriminatory. Men have worn high-heeled boots in Texas for a long time. Let us resolve this question in retrospect by calculating the pressure exerted by an ordinary 200-pound Texas-sized man wearing walking boots with 2 by 3 inch heels:

FIGURE 3-3 *A Texan walking boot has a reasonably large heel.*

$$P = \frac{F}{A}$$

$$= \frac{200 \text{ lb}}{(2 \text{ in.})(3 \text{ in.})}$$

$$= \frac{200}{2 \times 3} \text{ lb/in.}^2$$

$$= 33 \text{ lb/in.}^2$$

Certainly the 33-pound pressure of the boot heel is vastly different than the 8,000-pound pressure of the woman's heel—not because of the force, which is actually greater for the man, but because of the difference in area over which the force is concentrated.

The relationship between pressure and area applies to many situations in our daily lives. Desk edges are rounded to increase the area of contact and decrease the pressure in case you bump against them. Automobile tires give a cushioned ride by supporting the entire weight of the automobile with an air pressure inside of only 30 pounds per square inch. Ice skates concentrate weight on an area so thin and small that the pressure is great enough to actually melt the ice under the skate so that you glide along on a thin film of water, which instantly refreezes as soon as the pressure of the skate is removed.

One pressure we constantly experience is air pressure, which results from the weight of the atmosphere above us. Air pressure is the most common example of fluid pressure, but it is a bit difficult to visualize because air is invisible. The easiest example of fluid pressure to consider is the special case where the fluid happens to be a liquid such as water.

CHECK QUESTION

Air pressure of 15 lb/in² is pushing in on the side of an empty can having an area of 100 in². What is the force produced on the side of the can?

Answer: 1,500 lb

▶ Liquid Pressure

Let us apply the idea of pressure to the example where we were worried about the floor beneath a water bed. The pressure on the floor is the weight of the water (ignoring the rest of the bed) divided by the area on which it is resting:

$$P \equiv \frac{F}{A}$$

$$= \frac{2,200 \text{ lb}}{(5 \text{ ft})(7 \text{ ft})}$$

$$= 60 \text{ lb/ft}^2$$

It is not an accident that the pressure in pounds per square foot is close to the weight density of water in pounds per cubic foot. It would be right on 62.4 lb/ft³, as a matter of fact, if it were not for the round-off error introduced into our rough calculations. The pressure acting on a square foot of floor is due to the weight of the water directly over that area, and we had water to a depth of 1 foot in our water bed. Thus, we had a single cubic foot of water over each square foot of area.

If we had stacked two water beds, one on top of the other, we would have twice as much pressure on each square foot of floor, since there would be the weight of twice as much water above it. The same argument can be extended to any number of water beds in the stack and to any amount of area of the floor.

There is an old European folk tale in which a queen tested a princess by putting a pea under her mattress just to see if she was sensitive enough to be disturbed by such a small discomfort. In some versions, the queen placed the pea under a whole stack of mattresses. Great concern is expressed for the pressure exerted on the princess by the pea. No one ever thinks about the pressure exerted on the poor pea by the weight of all those mattresses.

Let us construct a modern version of this story, giving more consideration to the pea. Assume that the queen places the pea under a stack of water beds, and contemplate the pressure acting on the pea. Since the pressure is due to the weight of the beds, it would seem reasonable to assume that the pressure would be directly proportional to the number of beds, or what would be nearly the same thing, the combined height, h, of whatever fluid is in the water beds.

$$P \propto h$$

We shall show that the proportionality constant in this relationship turns out to be the weight density D_{wt} of the fluid:

$$\boxed{P = D_{wt}h}$$

We can see that this is the case by considering the definition of weight density:

$$D_{wt} \equiv \frac{wt}{V}$$

The weight of the water is therefore the product of weight density times the volume:

$$wt = D_{wt}V$$

Since this weight is the force causing the pressure on the floor, we can substitute $D_{wt}V$ for the force in the definition of pressure. Thus, the pressure due to this weight is the weight density times the volume, divided by the area.

$$P \equiv \frac{F}{A}$$

$$= \frac{D_{wt}V}{A}$$

$$= \frac{D_{wt}(A \cdot h)}{A}$$

The volume of the water divided by the area of the bed is the height of the water bed:

$$\boxed{P = D_{wt}h}$$

This statement says just what we started out to prove. The pressure exerted by the water bed is equal to the weight density of the water times the height. If we had two water beds stacked on top of each other, we would have twice the height of water and thus twice the pressure—which is just what we expected.

CHECK QUESTIONS

1. A stack of water beds 5.00 ft high is placed on top of a pea. The weight density of water is 62.4 lb/ft³. What is the pressure on the pea in lb/ft²?

   Answer: 312 lb/ft²

2. The water in a toilet flush tank is 10 in. deep. The weight density of water is 62.4 lb/ft³ = 0.0361 lb/in.³. What is the pressure of the water at the bottom of the flush tank?

   Answer: 52 lb/ft² = 0.36 lb/in.²

The same argument could have been made for any fluid, whether liquid or gas. It may seem strange, but the pressure in a fluid does not depend on the total volume of fluid, only on its *height*. The force acting on a dam, for example, is independent of the amount of water backed up behind it; it depends only on the height of the water right next to the dam. The force acting on two dams facing each other would be the same if there were 50 kilometers of water in a reservoir between them or if there were only 1 centimeter of water between them. In either case, the height of the water is the same and the area over which the pressure of the water is exerted is the same. We may think of the role of a dam as being to hold in place only the layer of water right next to the dam. That layer holds the next layer in place, and so on.

FIGURE 3-4 *The force of the water pushing against two dams facing each other is independent of the distance between them.*

It becomes a little easier to believe this strange fact when you think about the pressure that acts on your ears when you are swimming. You can predict from the relationship between pressure and height that the pressure on your ears under water in a swimming pool is independent of the size of the pool. Your ears will feel the same pressure when you swim 1 meter beneath the surface of a tidal pool at the beach as when you swim 1 meter under the waves of the ocean itself. The fact that there are thousands of miles of ocean out there has no effect on your ears.

A man named Blaise Pascal designed some funny looking vases to illustrate the independence of pressure and volume. The vases were different sizes and shapes but were all connected at the bottom. In spite of the differences, water would rise to the same level in each vase—demonstrating the fact that the pressure at the bottom of each vase was the same independent of the volume of water above. A pressure difference would cause water to flow from one vase to another and make the levels different. Thus, the old saying, "Water seeks its own level," is just an application of a much broader rule that **pressure equals weight density times height** independent of volume.

This relationship between pressure, weight density, and the height of a fluid is an example of a deceptively simple rule with powerful ramifications. Archimedes' principle, which itself has many unexpected results, can be entirely derived from the simple fluid pressure rule.

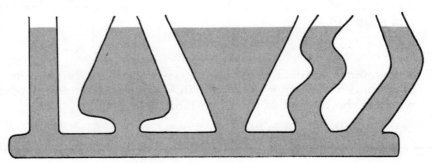

FIGURE 3-5 *The Pascal vases demonstrate that pressure in a fluid is independent of volume.*

▶ Archimedes' Principle

It is said that Hieron, king of the ancient Greek city of Syracuse, suspected a goldsmith of adding silver to the pure gold that had been refined for the purpose of making a crown. Hieron knew that the gold content of the crown could be determined from its density. It was no problem to weigh the crown, but he could not think of a way to measure the crown's volume without melting it down. So he assigned Archimedes the task of finding the crown's volume using geometry. The story goes that while Archimedes was swimming in public baths, he thought of a way of measuring the crown's volume from the buoyant force which would act on it if he were to place it under water. The picture of Archimedes running through the streets naked, shouting, "I found it! I found it!" has a special place in this legend.

Archimedes' principle can be stated, **The buoyant force is equal to the weight of the fluid displaced.** Buoyant force is the force with which the fluid pushes up on the object. If a rock is put under water, it seems to weigh less than it does in air. The water helps to lift up the rock and therefore decreases the rock's apparent weight. "If the rock were withdrawn and the resulting cavity allowed to fill with water, the latter would be in equilibrium under the action of its own weight and the external forces formerly exerted by the surrounding water on the rock."[2] The water in the cavity may be thought of as the displaced water, since it has the same volume as the rock. The surrounding water pushes up with the same force on the rock as it would on the water in the cavity, because it is not in the nature of water to care what it is pushing on. The buoyant force acting on the rock is thus equal to the weight of the displaced water. The same consideration holds for any fluid, whether liquid or gas.

It is interesting, however, to derive Archimedes' principle from the fluid pressure rule. The buoyant force is derived from the difference in the pressures acting on the bottom and the top of the object. It is convenient to select a rectangular shaped object to start with so that the pressure on the sides produces obviously symmetrical opposing horizontal forces that cancel each other out and do not enter into the buoyant force calculation. The pressure on the bottom of the object is greater than the pressure on the top because the bottom is at a greater depth in the fluid than the top.

The buoyant force is found by solving the definition of pressure

$$P \equiv \frac{F}{A}$$

for force to get

$$F_{\text{top}} = P_{\text{top}}A$$

and

$$F_{\text{bottom}} = P_{\text{bottom}}A$$

The buoyant force is the difference in the two forces acting on the bottom and on the top.

$$
\begin{aligned}
F_{\text{buoyant}} &= F_{\text{bottom}} - F_{\text{top}} \\
&= P_{\text{bottom}}A - P_{\text{top}}A \\
&= (D_{\text{wt}}h_{\text{bottom}})A - (D_{\text{wt}}h_{\text{top}})A
\end{aligned}
$$

Pressure is equal to weight density times height (according to the fluid pressure rule). We factor out weight density and area:

$$F_{\text{buoyant}} = D_{\text{wt}}A(h_{\text{bottom}} - h_{\text{top}})$$

This difference in the heights, $h_{\text{bottom}} - h_{\text{top}}$, is just the height of the body. The area, A, of the bottom times this height is the volume of the object:

$$\boxed{F_{\text{buoyant}} = D_{\text{wt}}V}$$

But the weight density of the fluid times the volume of the object is just the weight of the fluid that would be there if the object wasn't.

$$F_{\text{buoyant}} = \text{wt of fluid displaced}$$

[2] *Van Nostrand's Scientific Encyclopedia*, 4th ed. (1968) p. 253.

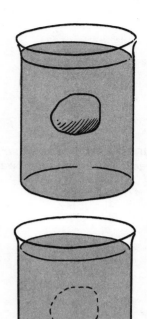

FIGURE 3-6 *Water pushes up on a rock the same as it would on water of the same volume.*

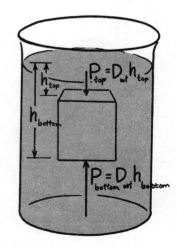

FIGURE 3-7 *Archimedes' principle can be derived by thinking of the buoyant force in terms of the difference in pressures acting on the bottom and top of an object.*

Thus, we have proved Archimedes' principle for a rectangular box. This proof can be generalized by approximating the shape of any object with a bunch of elongated boxes having rectangular horizontal and vertical surfaces. Archimedes' principle holds for each box, no matter how many there are. Choose an infinite number of boxes, and you can generalize Archimedes' principle for any body of any shape.

FIGURE 3-8 *A boat sinks until it displaces its own weight of water.*

If the buoyant force acting on an object is less than the object's weight, the object will seem to weigh less than it would in air but it will still sink. If the buoyant force equals the weight of the object, the object will float. A boat floats because it can displace more than its own weight of water. It sinks into the water until it displaces whatever water it needs to balance its weight. The more heavily a boat is loaded, the more water it must displace in order to float. One way to think of a submarine is as a boat that can adjust the amount of buoyant force acting on it so that it can just barely float or sink. The amount of water displaced by the submarine can be adjusted by allowing water to come into, or forcing water out of, ballast tanks on the sides of the submarine.

Dirigibles and balloons float in air by displacing their own weight of air. We tend to disregard the changes in air pressure over short distances, yet the air pressure on the bottom of a helium balloon is sufficiently greater than the pressure on the top so that the buoyant force more than compensates for the weight of the balloon. The helium in the balloon does not do anything other than hold out the sides of the balloon. Adding helium to the balloon without allowing it to expand would only make the balloon heavier. If it were some inherent tendency of helium to go up that made helium balloons float, balloon merchants would have a tough time holding their tanks of compressed helium down. (Think how nice it would be, though, to solve your parking problems by filling your tires with helium. Just put it in under enough pressure to make your car float, and you could just tie it to a flag pole when you went to class.)

FIGURE 3-9 *Archimedes' principle also works when the fluid is air.*

CHECK QUESTION

A block of aluminum has a volume of 1.0 ft³ and a weight in air of 168 lb. (a) What is the buoyant force acting on it under water if the weight density of water is 62.4 lb/ft³? (b) How much does the block seem to weigh under water?

Answer: (a) 62 lb, (b) 106 lb

► **Pressure in a Gas**

Our discussion so far has been mostly concerned with pressure in a liquid. We have seen how liquid pressure is proportional to the depth in the fluid, and how this relationship leads to the buoyant force relationship known as Archimedes' principle. We then saw how this relationship applies as well to balloons floating in air, where the fluid is a gas instead of a liquid.

Before leaving the topic of pressure, we should develop the general rule for pressure in a gas related to other variables such as temperature, volume, and the number of molecules of the gas that are present. This relationship is called the "ideal gas law" because it only works for an "ideal" gas. Fortunately, the behavior of most gases approximates that of an ideal gas for ordinary temperatures and pressures. We can build up this general pressure rule for a gas by constructing a mental model based on proportion.

Pressure and Amount of Gas

The relationship between pressure and the amount of gas, for instance, would seem to be one of direct proportion. We know that gas is made up of molecules and that pressure in a gas is due to the collision of these molecules with a surface. When you blow up a balloon, you increase the pressure inside by putting in more air. If you think of the air in the balloon as a bunch of tiny little molecules flying around in constant motion, you could picture the pressure in the balloon as due to tiny forces of impact continuously produced by the air molecules bouncing off the inside of the balloon. This mental picture of pressure makes a proportional model seem reasonable. The pressure, P, on the inside surface of the balloon should be proportional to the number of molecules, N, in the balloon, everything else being the same, since doubling the number of molecules in the balloon would double the number of forces of impact on the inside surface of the balloon in any given length of time:

$$P \propto N$$

In saying that the pressure is proportional to the number of molecules, we must be careful to specify that everything else stays constant. Our balloon would have to have a very tough surface, sort of like a beach ball, if its volume were not to change with increased pressure. We would also have to keep the temperature constant so that each molecule would strike the surface with the same force of impact. We can consider the effect of allowing volume and temperature to change in a moment, but first we must look more carefully at the proportionality between pressure and the number of molecules in the balloon.

The pressure in our proportionality must be the kind of pressure that is equal to zero if the number of molecules is equal to zero. This is not the kind of pressure you would read on a tire gauge. When a tire gauge reads zero, when the tire is flat, there is still air in the tire. It is just that the air pressure inside the tire is no greater than that on the outside. When inflated, the pressure read on a tire gauge is understood to be the pressure over and above atmospheric pressure. The only difference between this gauge pressure and the absolute pressure used in our proportionality is the zero point. Whereas the gauge pressure is measured relative to the atmosphere, the absolute pressure is measured relative to a vacuum. To get from gauge pressure, P_{gauge}, to absolute pressure, P (with no subscript), just add the atmospheric pressure. At sea level, atmospheric pressure is 14.7 lb/in.²

$$\text{Absolute pressure} = P = P_{gauge} + 14.7 \text{ lb/in.}^2$$

Pressure and Volume

Let us now consider the relationship between pressure in a gas and the volume of the container. Instead of putting more air into our balloon or beach ball, we hold the number of molecules and their temperature constant and squeeze the volume down to one-half its original size. Here again, we expect a proportional relationship—but not a direct proportion. The relationship between pressure and volume should be an inverse proportion, since the pressure goes up when the volume goes down. Although the relationship is inverse, we still expect it to be basically proportional, because decreasing the volume increases the density of molecules in the same way as does adding gas. That is, if you try to crowd the same number of molecules into half the volume, the same number of collisions occur at the surface as if you had doubled the number of molecules in the original volume. This inverse proportionality can be expressed in terms of a reciprocal:

$$P \propto \frac{1}{V}$$

Pressure and Temperature

Pressure of a gas also increases with the temperature of the gas. Temperature is a measure of the internal energy of the gas molecules. The higher the temperature, the more rapidly the molecules strike the surface and the harder they hit. The relationship between pressure and temperature is linear, that is, equal increases in one bring about equal increases in the other. A graph of pressure against temperature produces a straight line. This relationship is like a proportion except that the line does not pass through the origin. The pressure is not equal to zero at 0 degree Celsius (formerly called *centigrade*). But if you decrease the temperature one degree, from 0 degree Celsius to −1 degree Celsius (°C), the pressure will decrease by one part in 273 from the original pressure. If you decrease the temperature two degrees say, from 0 °C to −2 °C, the pressure in the gas decreases two parts in 273. This leads us to wonder what would happen if we were to decrease the temperature 273 degrees, from 0 °C to −273 °C.

Any real gas becomes a liquid before −273 °C, but we could imagine an "ideal gas" that would follow the linear relationship all the way. For such a gas the pressure would equal zero at this temperature. This temperature is called **absolute zero,** because it forms a natural zero point just as a vacuum forms a natural zero point for pressure measurement. Pressure in a gas is directly proportional to temperature as long as temperature is measured on a scale that starts at absolute zero.

$$P \propto T$$

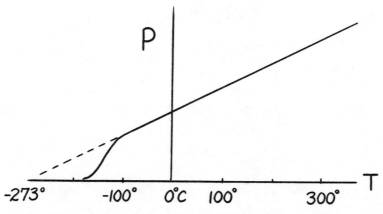

FIGURE 3-10 *A graph of absolute pressure against temperature can be extrapolated to zero pressure at −273 °C.*

An absolute temperature scale that starts at absolute zero and has the same-size degrees as the Celsius scale is called the **Kelvin scale.** The freezing point of water is 273 K and the boiling point of water is 373 K on this scale. Temperatures are converted from degrees Celsius to degrees Kelvin by adding 273 to the Celsius reading.

We have seen that pressure of a gas is directly proportional to the number of molecules of the gas, the temperature of the gas, and inversely proportional to the volume of the container holding the gas. These three relationships can be combined into a single statement:

$$P \propto \frac{NT}{V}$$

A proportionality can always be expressed in terms of an equality sign and a proportionality constant. With a proportionality constant k, called **Boltzmann's constant,** the pressure in an ideal gas can be expressed as

$$P = k \frac{NT}{V}$$

This proportionality constant turns out to have the same value for all gases:

$$k = 1.83 \times 10^{-23} \text{ newton} \cdot \text{meter/molecules} \cdot \text{K}$$

The proportional relationship between pressure, volume, temperature, and the amount of gas is called the **ideal gas law** and is usually written

$$\boxed{PV = kNT}$$

All gases obey this relationship well at reasonably high temperatures and low pressures, where the gas is far from the liquid state. An ideal gas, a hypothetical gas that obeys the relationship exactly at all pressures and temperatures, is an approximation of most real gases under most conditions.

The problems in this chapter are concerned with pressure in liquids and gases. The **weight density** of a fluid is defined as the ratio of the weight to volume:

$$D_{wt} = \frac{wt}{V}$$

Mass density is defined as the ratio of mass to volume:

$$D_{mass} = \frac{mass}{V}$$

The density of water is 1 gram per cubic centimeter because the original definition of a gram was that it be the mass of a cubic centimeter of pure water. A liter of water, which is 1,000 times as big as a cubic centimeter, therefore has a mass of 1 kilogram.

Pressure is defined as the ratio of the force to area:

$$P \equiv \frac{F}{A}$$

Pressure of a liquid is proportional to the height of the liquid, the proportionality constant being the weight density of the liquid:

$$P = D_{wt}h$$

Archimedes' principle states that the buoyant force acting on an object is equal to the weight of the fluid displaced.

$$F_{buoyant} = \text{wt of fluid displaced} = \underset{(fluid)}{D_{wt}}\underset{(object)}{V}$$

Pressure in a gas is directly proportional to the absolute temperature and the number of molecules of the gas, and inversely proportional to the volume.

$$PV = kNT$$

The value of the proportionality constant is the same for all gases:

$$k = 1.83 \times 10^{-23} \text{ newton} \cdot \text{meter/molecules} \cdot \text{K}$$

The pressure in this relationship is the absolute pressure relative to a vacuum. Atmospheric pressure must be added to gauge pressure to get absolute pressure:

$$P_{absolute} = P_{gauge} + P_{atmosphere}$$

Atmospheric pressure is 14.7 pounds per square inch in British units:

$$P_{atmosphere} = 14.7 \text{ lb/in.}^2$$

Absolute temperature is found by adding 273 to the Celsius temperature.

$$K = °C + 273$$

The information from the chapter has been summarized here for easy reference. You should understand it well enough so that you don't have to look it up to do the following problems. It is important to understand it, not as "a bunch of formulas" but as meaningful representations of real conditions.

3-1 A suitcase filled with books is only about half as heavy as the same suitcase filled with water.

Given: A suitcase 50 cm long by 30 cm wide by 15 cm thick is filled with books having an average density of 0.50 kg/liter.

Find: What is the mass of the suitcase and contents?

Your Solution

Discussion

This problem illustrates the relationship between mass and volume. It also involves a conversion between cubic centimeters and liters, units of measure discussed in Chapter 1.

Sample Solution 3-1

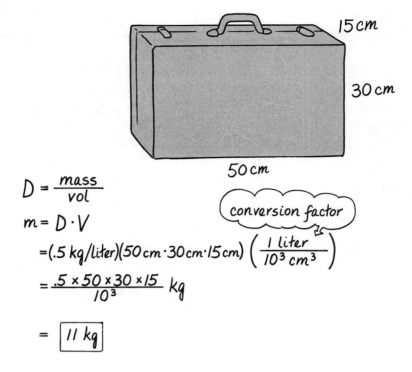

$$D = \frac{mass}{vol}$$

$$m = D \cdot V$$

$$= (.5 \; kg/liter)(50 \, cm \cdot 30 \, cm \cdot 15 \, cm) \left(\frac{1 \; liter}{10^3 \; cm^3} \right)$$

conversion factor

$$= \frac{.5 \times 50 \times 30 \times 15}{10^3} \; kg$$

$$= \boxed{11 \; kg}$$

Discussion

Eleven kilograms weigh about 24 lb. Do you have a suitcase that small? You might find it interesting to get some books and try it out.

3-2 An automobile's weight is supported by the air pressure inside the tires spread out over the area that all four tires present to the ground.

Given: A 1946 Lincoln complete with driver weighs 5,000 lb. The inflation pressure recommended by one tire manufacturer is 28 lb/in.²

Find: The area that each of the four tires presents to the ground.

Your Solution

Discussion

This problem illustrates the relationship between pressure and force. It shows how a large force can be supported by a reasonably small pressure acting over a large area.

Sample Solution 3-2

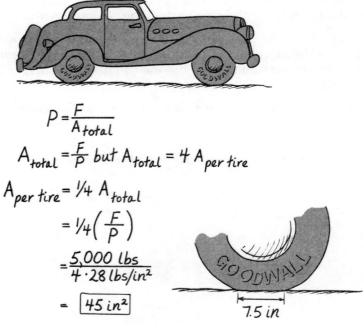

$$P = \frac{F}{A_{total}}$$

$$A_{total} = \frac{F}{P} \text{ but } A_{total} = 4\, A_{per\ tire}$$

$$A_{per\ tire} = \frac{1}{4}\, A_{total}$$

$$= \frac{1}{4}\left(\frac{F}{P}\right)$$

$$= \frac{5{,}000\ lbs}{4 \cdot 28\ lbs/in^2}$$

$$= \boxed{45\ in^2}$$

7.5 in

If each tire is about 6 inches wide, this corresponds to a length of about 7.5 inches touching the ground.

Discussion

The sample solution interprets the answer in terms of length of each tire on the ground, assuming a reasonable width. This is more work than is called for by the problem, and your solution might well not contain this information. Sometimes people get interested in a problem, however, and carry it a bit further than they really need to.

You might, for example, go out and measure the tires on your own automobile. With a ruler and a pressure gauge, you should be able to figure out the weight of your automobile.

3-3 The height of water necessary to produce a given pressure is sometimes called the water head.

Given: A common water pressure available at a kitchen faucet is 60 lb/in.^{2}. Water has a weight density of 62.4 lb/ft^3.

Find: How much water head is necessary to produce this pressure?

Your Solution

Discussion

This problem illustrates one of the most common applications of the fluid pressure rule to our daily lives. Running water has become so much a part of our lives that we tend to take it for granted. Think of all that water up there someplace next time you turn on a faucet.

Sample Solution 3-3

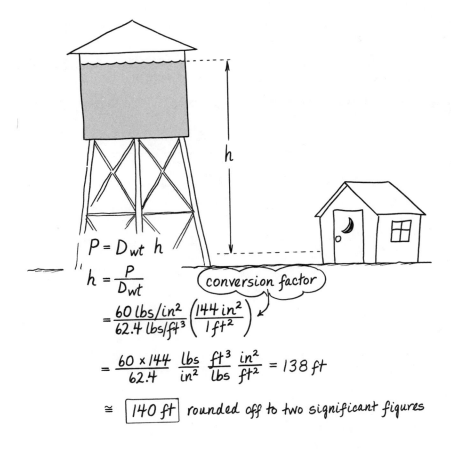

$$P = D_{wt}\, h$$

$$h = \frac{P}{D_{wt}}$$

conversion factor

$$= \frac{60\ lbs/in^2}{62.4\ lbs/ft^3} \left(\frac{144\ in^2}{1\ ft^2} \right)$$

$$= \frac{60 \times 144}{62.4}\ \frac{lbs}{in^2}\ \frac{ft^3}{lbs}\ \frac{in^2}{ft^2} = 138\ ft$$

$$\cong \boxed{140\ ft}\ \text{rounded off to two significant figures}$$

Discussion

The answer is rounded off to two significant figures, because we don't know the pressure to any greater accuracy than that. We certainly don't know the pressure to the closest tenth of a pound per square inch, since the pressure in the water main changes in that range depending on the demand for water at different times of the day. You might even interpret 60 lb/in.^{2} to be anything between 55 and 65 lb/in.^{2}, in which case the answer would only be good to one significant figure. This would be physically reasonable. A house up the hill 14 ft higher would have 10% less pressure. If you want to write 60 lb/in.^{2} in such a way that people know that the zero after the six really has zero value and is not there just to hold the decimal place (which is the custom widely used outside scientific circles), you need to use scientific notation. You would say the pressure is 6.0×10^1 lb/in.^{2}.

3-4
A wooden plank sinks into water up to the point where the buoyant force is equal to its weight.

Given: A plank floating in water is 2.0 m long and 15 cm wide. The density of the water is 1.00 kg/liter. The mass of the plank is 10 kg.

Find: How far does the water come up on the edge of the plank?

Your Solution

Discussion

Floating objects are applications of Archimedes' principle. They are often difficult to figure out. If you have difficulty with this problem, stop and try to figure out the volume of the object under water from the buoyant force. You can then get the thickness from the area.

Sample Solution 3-4

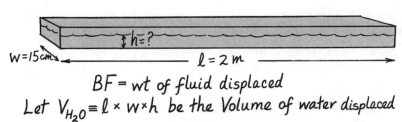

$$BF = \text{wt of fluid displaced}$$

Let $V_{H_2O} \equiv \ell \times w \times h$ be the Volume of water displaced

Then from $D_{wt} \equiv \frac{wt}{vol}$, the weight of the water displaced by the plank is

$$wt = D_{wt} \times V$$
$$= D_{H_2O} \times V_{H_2O}$$
$$= D_{H_2O} \times \ell \times w \times h$$

This weight is also the weight of the plank since the buoyant force BF is holding the plank up.

Thus $wt_{plank} = D_{H_2O} \times \ell \times w \times h$

The only thing we don't know is h. Solving,

$$h = \frac{wt_{plank}}{D_{H_2O} \times \ell \times w}$$

$$= \frac{(10 \text{ kgwt})}{(1 \text{ kgwt/liter})(2 \text{ m})(.15 \text{ m})} \left(\frac{10^3 \text{ cm}^3}{1 \text{ liter}}\right)\left(\frac{1 \text{ m}}{10^2 \text{ cm}}\right)^2$$

$$= \boxed{3.3 \text{ cm}}$$

Discussion

The buoyant force acting on a floating object is the object's weight. The weight of the fluid (water in this case) is just the density of the fluid times the volume of that part of the object under water. In this case, the thickness h of that volume is the thing we don't know. We therefore put it in algebraically and solve for it as an unknown.

3-5 A balloon merchant is filling balloons to atmospheric pressure from a tank of compressed helium.

Given: Each balloon contains 0.50 ft³ at atmospheric pressure and temperature. The tank at the same temperature contains 5.0 ft³ of helium at 2,000 lb/in.². Atmospheric pressure is 14.7 lb/in.².

Find: How many balloons can the merchant fill?

Your Solution

Discussion

This is an application of the ideal gas law. Temperature is assumed to remain constant, since the gas in the balloons quickly adjusts to room temperature after the balloons have been filled.

Sample Solution 3-5

$$P_1 V_1 = k N_1 T_1 \text{ for tank}$$

$$P_2 V_2 = k N_2 T_2 \text{ for balloons}$$

so $P_1 V_1 = P_2 V_2$ since $N_1 = N_2$ and $T_1 = T_2$

Solve for V_2 and then find the number of balloons that make up a total volume $= V_2$

$$V_2 = \frac{P_1 V_1}{P_2}$$

$$= \frac{(2000 \, lb/in^2)(5.0 \, ft^3)}{(14.7 \, lb/in^2)}$$

$$= 680 \, ft^3$$

So number of balloons is

$$680 \, ft^3 \times \frac{1 \text{ balloon}}{0.50 \, ft^3} = \boxed{1.36 \times 10^3 \text{ balloons}}$$

Discussion

The ideal gas law applies to the gas that starts out in the tank, for which we use subscript 1, and then is put into the balloons, for which we use subscript 2. The number of molecules is the same before and after, hence $N_1 = N_2$. Also, the temperature is the same. Thus $T_1 = T_2$. Since the ideal gas law constant k is obviously the same for the gas when it is in the tank as when it is in the balloons, the right-hand sides of the two ideal gas law expressions are equal. The two left-hand sides are therefore equal, and this equality is solved for the volume V_2 of the gas after it is put into the balloons, which can then be expressed in terms of balloon volumes.

3-6   A rectangular fish aquarium is filled with water having a weight density of 62.4 lb/ft³.

Given: The inside dimensions of the aquarium are 18 in. by 10 in. by 12 in.

 Find: What is the weight of the water?

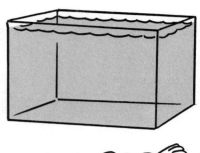

3-7   Your outstretched hand supports the weight of the air above it. (Of course, your hand is partly supported by air pressure from below.)

Given: Air pressure is 14.7 lb/in.². Someone's hand is about 3 in. by 7 in. when held palm up.

 Find: What is the weight of the air supported by that person's hand?

3-8   The pressure on your ears increases as you swim deeper under water.

Given: You are at the bottom of a 9-ft swimming pool examining the drain. The weight density of water is 62.4 lb/ft³.

 Find: What is the water pressure on your ears in pounds per square inch?

3-9   An empty coal barge is floating in the water. The water comes up to a place on the side of the barge called the water line.

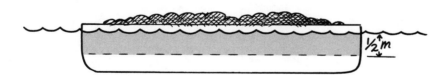

½ m

Given: The water line on a barge moves up 0.5 m when it is loaded with coal. The barge has an area of 10 m long by 3 m wide. The density of water is 1.00 kg/liter.

 Find: How much mass of coal was loaded into the barge?

3-10   The air pressures recommended for automobile tires are listed by their manufacturers for the correct pressure when the tire is cold. When you have been driving for a while, the tires warm up and the air pressure inside increases. (You should *not* adjust the pressure downward when the tires are hot, as the recommended pressure is given in anticipation of the fact that it will be higher when the tire is hot.)

Gauge pressure

Given: The recommended pressure for a particular tire is 25 lb/in.² (above atmospheric). You fill the tire when the temperature is 20 °C.

Find: (a) What are the initial pressure and temperature on absolute scales in this case?

(b) What should the gauge pressure be after the tire has been driven for an hour on the freeway and has reached a temperature of 50 °C?

MORE INTERESTING PROBLEMS

3-11   A rectangular fish aquarium is filled with fresh water and one large fish. The fish does not swim; he hangs motionless in the water, neither sinking to the bottom nor floating to the surface.

Given: The aquarium is 22 cm long, is filled to a height of 15 cm, and is 10 cm thick. The fish is 7 cm long, 2 cm high, and 2 cm thick.

Find: (a) What is the average density of the fish? How do you know?

(b) Find the combined mass of the water and the fish.

3-12   A metal crown seems to weigh less under water than in air because of the buoyant force of the water.

Given: A crown weighed in air is found to have the same weight as a 5.0-kg mass, while the same crown under water seems to weigh only as much as a 4.4-kg mass.

Find: (a) What is the buoyant force acting on the crown in kilograms?

(b) What is the volume of water displaced?

(c) Calculate the density of the metal with which the crown is made.

3-13   The gravitational field of a neutron star is so great that ordinary matter is squeezed down to the density of nuclear matter. An ordinary student falls asleep in class and dreams of awaking on a neutron star.

Given: The volume of a hydrogen nucleus is about 1×10^{-41} liter and has a mass of 1.67252×10^{-27} kg. The student has a mass of 65 kg.

Find: (a) What is the volume of the student back home on earth assuming the density is about 1 kg/liter (the same as that of water here on earth?
(b) What is the density of nuclear matter?
(c) What would be the volume of the student if she really were on a neutron star?

3-14 Air pressure can be demonstrated by Magdeburg hemispheres, matching steel hemispheres that can be fitted together with air tight edges and evacuated.

Given: Air pressure is 14.7 lb/in.², and the hemispheres have a diameter of 3.5 in. (The area of a circle is $A = \pi r^2$).

Find: (a) What is the cross-sectional area of the vacuum chamber?
(b) What is the force necessary to pull the hemispheres apart when they are fully evacuated?

3-15 A balloon will float in air if it weighs less than the air it displaces. It is common to use a gas lighter than air, such as helium or hydrogen, to hold the sides of the balloon out so that it will displace a great deal of air.

Given: The density of air is 1.29 kg/m³, while the density of helium is 0.18 kg/m³ and that of hydrogen is 0.09 kg/m³. A given balloon has a mass of 50 kg and is inflated to a volume 200 m³.

Find: (a) What is the mass of the helium necessary to fill the balloon?
(b) What is the buoyant force acting on the balloon?
(c) How much force is necessary to hold the balloon down if it is filled with helium?
(d) How much force would be necessary to hold the balloon down if it were filled with hydrogen?

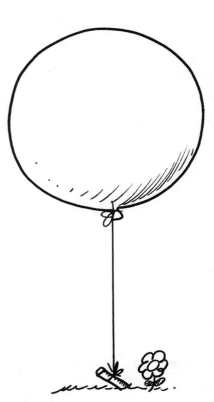

ANSWERS

3-6   78 lb.
3-7   3 × 10² lb.
3-8   3.9 lb/in.² ≅ 4 lb/in.²
3-9   1.5 × 10⁴ kg.
3-10 (a) 40 lb/in.², 293 K; (b) 29 lb/in.² (gauge).
3-11 (a) 1 kg/liter; (b) 3.3 kg.
3-12 (a) 0.6 kg; (b) 0.6 liter; (c) 8 kg/liter.
3-13 (a) 65 liter; (b) 2 × 10¹⁴ kg/liter; (c) 3 or 4 × 10⁻¹³ liter.
3-14 (a) 9.6 in.²; (b) 141 lb ≅ 1.4 × 10² lb.
3-15 (a) 36 kg; (b) 258 kg(wt); (c) 172 kg(wt) ≅ 1.7 × 10³ N;
          (d) 190 kg(wt) ≅ 1.9 × 10³ N

Force **4**

So Christopher Robin took hold of Pooh's front paws and Rabbit took hold of Christopher Robin, and all Rabbit's friends and relations took hold of Rabbit, and they all pulled together. . . .

Winnie-the-Pooh*

FIGURE 4-1 *The concept of force is as sophisticated as it is simple.*

The concept of force plays a central role in our everyday understanding of cause and effect. If you were trying to move someone who was stuck in a tight place, you would certainly apply a force. If you found that the force was not great enough, you would get help.

The force is greater, according to our common sense, when everyone pulls together than when one person pulls alone.

If some inanimate object beside you suddenly started to move, you might well look to find the cause. You would probably become alarmed if nothing seemed to be moving the object. Objects that move by themselves are generally ascribed to the presence of demons or witchcraft. We don't believe much in supernatural or magical powers these days, because we have acquired so much confidence in certain kinds of force that we feel are natural. These kinds of force may not be the whole story. Science has shown an interest of late in paranormal or "unnatural" phenomena, such as psychokinesis and extrasensory perception. The success of science to date, however, is based on explaining the vast diversity of almost all physical phenomena in terms of a very few well-behaved forces of nature. This success is underscored by the fact that many people refuse to believe in paranormal and other phenomena that have not been explained.

There are, in fact, only two forces that explain all of our normal everyday experiences. They are the electrical and gravitational forces. (Magnetic forces may be regarded as another aspect of electrical forces.) The only other forces of nature that are now known exist only inside the atomic nucleus. They are so short-range that they could never affect the motion of objects around us.

Gravitational forces between objects are so small that they demonstrate their effects only when one of the objects is astronomically large. Since the only astronomically large object around us is the earth, the effect of gravitational forces becomes quite simple. Every object around us has weight, and that weight is just proportional to the mass of the object as long as we stay near the surface of the earth. Thus, gravitational forces seem quite natural and uniform.

The only other force that has much of an effect in our daily experience is electrical. Perhaps you find it hard to believe that almost all of the diverse physical phenomena around you are governed by only two kinds of force. There are, however, certain underlying similarities in the way things behave.

▶ Hooke's Law, A Way of Conceptualizing Force

As an example of the way things behave similarly, consider the relationship between the stretch or strain in a solid object and the force applied to it. It turns out that the stretch or strain is always nearly proportional to the force, as long as you stay within certain limits. This fact was first noticed by Sir Robert Hooke, an acquaintance of Isaac Newton, and is therefore known as **Hooke's law.**

Picture yourself holding one end of a fishing line and a fish holding the other end. The harder you two pull against each other, the more the fishing line stretches. If you pull twice as hard, the fishing line stretches twice as much—up to a point. If you get too near the breaking point of the fishing line, it will stretch to its limit and will not spring back after the force is released. Snap! But as long as you remain well within what is called the **elastic limit** of the fishing line, you will find that the stretch is very nearly proportional to the force. This will be true whether the fishing line is made out of cotton,

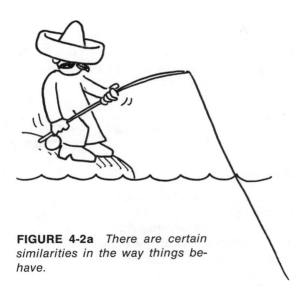

FIGURE 4-2a *There are certain similarities in the way things behave.*

nylon, or even steel. The reason that the same proportional rule works for all materials is that all solids are held together by the same kind of electrical forces acting between atoms (see the following Highlight); the main point of our discussion is that the proportion exists as an experimental fact.

You could do some experiments to confirm Hooke's law for yourself. You might use rubber bands hanging from a drawer pull or a long string hung from the ceiling. A bunch of coat hangers all made out of the same kind of wire make good weights. You could use a ruler to measure the stretch of the rubber band or string and see if it isn't proportional to the number of coat hangers used to cause the stretch. On the other hand, you might prefer to go fishing.

FIGURE 4-2b *The harder a fish pulls, the more the fishing line stretches.*

We can use the stretch of the fishing line as a way of visualizing the concept of force. You can think of the force acting on your hand in terms of the stretch in the line. The amount of force is something proportional to the amount of stretch, but the direction of the force is important too. The direction of the force must be along the fishing line, since any sideward force would make the line bend. Of course, the fishing line is actually pulling in two directions. It is pulling one way on your hand; it is pulling the other way on the fish. When you use the concept of force, you need to direct your attention to one object at a time. If you are thinking of your hand, the force is along the fishing line in the direction of the fish. If you are thinking of the fish, the force is along the fishing line in the other direction.

Suppose you start thinking of the force acting in the fishing line. The fishing line is pulling in two opposite directions, holding you and the fish together. The word **tension** is frequently used to refer to this pair of balanced forces. If you found yourself grasping a tree branch to keep from being pulled into the water, as a matter of fact, it would be proper to think of the balanced forces acting on your body in terms of tension. This concept of tension turns out to be useful in many force problems.

HIGHLIGHT 4-1   Forces in Solids

The forces holding a solid object together are electrical in nature. Electrical attraction holds negatively charged electrons in orbit around positively charged nuclei. Atoms are held together into molecules by electrical forces caused by the sharing or displacing of their electrons. Atoms are also held *apart* by the electrical forces of repulsion between positively charged nuclei. A solid object holds its shape because the electrical forces of attraction are just balanced by the electrical forces of repulsion.

When a fish gives a little tug on your fishing line, it pulls the atoms a little further apart. The positively charged nuclei are not quite as close together as before, and the forces of repulsion are not quite as strong. The attractive force is partially unbalanced so that you feel the tension in the line. The harder the fish pulls, the greater the stretch in the line and the greater is the imbalance between the attractive and repulsive forces between atoms—up to a point. If the displacement between atoms is too great, attractive forces begin to decrease as rapidly as repulsive forces. The line breaks, and you consider yourself a true sportsperson. You used a line light enough to cast a maximal distance, and the fish had a fighting chance. Had you used a heavier line, more atoms would have been exerting a force through any cross-section of the line, and there would not have been as much stretch per given amount of force. Your chances of recovering the fish would not have depended on your skill at playing the line to keep its stretch below the elastic limit. You could have just reeled the fish in with no sense of fair play.

Forces of attraction and repulsion are complicated functions of distance; but as long as the stretch remains well within what is called the **elastic limit** of the fishing line, the force and the stretch are nearly a simple proportion. Twice as much force pulling on the line, called the **tension** in the line, produces twice as much stretch. Three times as much force produces three times as much, and so on. The individual atoms in the fishing line are near the position where the forces of attraction and repulsion balance each other out; and even though the graph of force versus distance is a complicated curve, it is a smooth curve. The elastic limit is just a small enough force or distance for this smooth curve to look like a straight line. When you get outside of this region, the fishing line begins to stretch permanently and break because the forces of attraction and repulsion between individual atoms get on to the nonlinear part of the curve, where equal changes in one do not produce equal changes in the other to compensate. As long as you remain inside the elastic limit, however, the proportion works very well.

Hooke's law works as well for bending, twisting, and compression of solids as it does for stretching. A coil spring, for example, can be made to stretch over great distances by small forces applied to the ends, but most of the actual give in the spring comes from the twisting of the wire in the coils of the spring. Very little of the stretch in the spring comes from stretching the wire.

Coil springs are often used in physics labs to study Hooke's law because they can be stretched over such long distances and still spring back to their original shape. Such coil springs are frequently provided with hooks at either end for ease of applying forces. Many people have come away from a physics course believing that Hooke's law got its name from this convenient structure on the ends of a coil spring.

The proportionality between the force F applied to the bottom hook(e) of the spring and the distance x by which the spring stretches may be written with a proportionality sign:

$$F \propto x$$

It looks more official, however, with an equality sign and a proportionality constant:

$$F = kx$$

This proportionality constant k is called the **spring constant** of the spring.

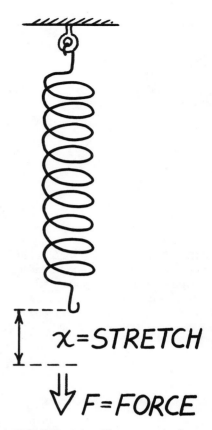

FIGURE 4-3 *The stretch in a spring is proportional to the force applied.*

CHECK QUESTIONS

1. A force of 3.0 N stretches a spring by 6.0 cm. What is the spring constant?

   Answer: $k = 0.50$ N/cm

2. How far would the same spring stretch if a force of 2.3 N were applied to it?

   Answer: $x = 4.6$ cm

Hooke's law for small displacements is universally applicable to all solid objects. It works for everything from building beams to eyelashes. An ant crawling across the floor makes the floor sag by an amount proportional to the ant's body weight. An ordinary-sized ant might be so light that the sag would be too small to measure, but the sag is still there, at least in principle. The small force you can apply to a 10-in. steel I beam like those used in the construction of large buildings will make the beam compress, bend, and twist by an amount that can easily be measured with an optical instrument called an interferometer.

Of course, Hooke's law is only an approximation of any real situation. Such things as frictional losses in a spring, "creep" in a rubber band, and thermal expansion enter into any real situation to limit the applicability of our simple model. In many cases, however, Hooke's law works so well that it is used as a measure of force. Fishermen frequently keep a spring scale in their tackle box to weigh their catch. A simple spring scale is nothing more than a spring within a housing and a pointer to show how much the spring has stretched.[1]

A spring scale is also a useful device for conceptualizing force. If you think of something as being pulled by a rope with a spring scale in between, the scale gives the strength of the force while the rope gives the direction of the force.

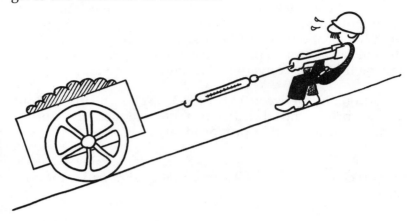

FIGURE 4-4 *Forces have both magnitude and direction. The magnitude can be visualized as the reading on a spring scale, while the direction is along a rope that would exert the force.*

▶ Vectors

Force is a quantity that has both magnitude and direction. Both the strength and direction of a force must be considered when combining it with another force. This kind of quantity is called a **vector.** Other quantities, which are fully described by magnitude alone, are called **scalars.** A scalar quantity is measured by "how much," while a vector must be described by "how much" and "which way."

The distinction becomes important when adding things together. If you add 1 liter of beer to a half-full 1-liter stein, obviously it runs over. We know from experience that an amount of beer is a scalar quantity. One liter added to half a liter always results in one and a half liters, even if part of it is in someone's lap.

Force is a different story. Suppose you and a friend of yours are both applying a force to a cart. You are applying a force of 100 lb. Your friend, being somewhat stronger, is applying a force of 150 lb. The resultant, or vector sum, of these two forces is 100 lb + 150 lb = 250 lb *only* if you agree with your friend on the direction you are pushing. If you and your friend disagree and decide to push in opposite directions, the resultant is 100 lb − 150 lb = − 50 lb in the direction you are pushing (or + 50 lb in the direction your friend is pushing).

FIGURE 4-5 *Forces in the same direction add. Forces in opposite directions subtract.*

---

[1] A spring scale is not the only way to measure force. A more accurate way to measure force is to balance it with a known force in the opposite direction. Meat market scales marked "no springs" are of the balance type.

HIGHLIGHT 4-2

Vectors add when they are in the same direction and subtract when they are in opposite directions. It is fortunate that forces are vectors and not scalars. Otherwise, nothing could ever be stationary. You are able to remain in your seat only because the upward and downward forces acting upon your body are balanced. The forces exerted on you toward the right are balanced by forces acting upon you to the left. If the forces in one direction added to, rather than subtracted from, the forces in the other direction, you would find yourself in a state of accelerated motion.

Stationary objects can only remain so when all of the forces acting on them are balanced. A spring scale supporting a 10-lb weight can only remain stationary if the force acting on the bottom hook is balanced by an equal force pulling upward on its top ring. Most people would correctly guess that the reading on the scale in this situation should be 10 lb. A few people might guess that the reading should be 20 lb, reasoning that that would be the total force acting on the scale. The 10 lb acting upward, however, does not add to the 10 lb downward. The forces subtract because they are in opposite directions. A 124-lb woman standing on a bathroom scale would be startled if the scale were to read 248 lb. Fortunately, bathroom scales are not constructed to read the *numerical sum* of the forces acting on them. Bathroom scales are also not constructed to read the *vector sum* of the forces acting on them. The same person would be nearly as startled if the scale were to read 0 lb. Scales are constructed to measure only the force applied to one side; they assume that an equal but opposite force is applied to the other side to hold the whole scale stationary.

Many people who have no difficulty understanding a simple vertically supported spring scale have considerably more difficulty with the same scale supported horizontally with a 10-lb weight at each end by means of strings and pulleys, as shown in Figure 4-6. They tend to think that the reading should now be 20 lb, because that is the total weight being held up, but that is not the case. The 10 lb on one side cannot possibly add to the 10 lb on the other, as their weights are pulling in *opposite* directions on the scale. The vector sum of these two forces is zero, just as the force pulling up on the ring of the vertical scale exactly balances the force pulling down on the hook. The reading on the horizontal scale is 10 lb, just as it was on the vertical scale.

Let us now consider a similar situation, in which the pulley and weight on one side of the horizontal scale is replaced with a hook in the wall. A force applied to the ring by a string attached to a hook has the same effect as attaching a weight to it over a pulley. In both cases, the force is just that necessary to hold the ring stationary. The scale's ring does not care what is at the other end of the string. All that the ring can feel is the tension in the string, which is the same in both cases. The scale between the weight and the hook in the wall looks a great deal like the first force situation with the scale between the weight and a hook in the ceiling. The only difference is that we are using a pulley to pull horizontally. We know that the tension in the string, however, is the same on both sides of the pulley. Otherwise the pulley would turn.

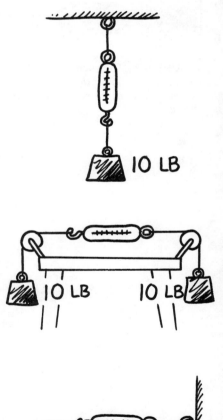

FIGURE 4-6 *Three similar force situations, all having the same tension in the cord.*

CHECK QUESTIONS

Two forces, one 500 N (newtons) and the other 700 N, are acting on the same object. What is the vector sum of these two forces (a) if they are acting in the same direction, and (b) if they are acting in opposite directions?

Answers: (a) 1,200 N, (b) 200 N

FIGURE 4-7 *The resultant of two forces has a direction that is a combination of the two original directions.*

We have only considered examples so far in which the forces have all been along the same line, either in the same or opposite directions. It is easy to find cases, however, in which one force pulls off at some angle relative to the other. You might be pulling, for example, on the cart in a direction perpendicular to the direction your friend is pulling. Maybe the cart is sloshing around in the mud, and you want it to go off to the side as well as forward. The resulting force, called the **resultant,** is a force equivalent to the combined efforts of both you and your friend. It would be somewhat in your direction and somewhat in his. We shall see that the resultant of 100 lb in your direction and 150 lb in your friend's direction would, for example, be a force of 180 lb at an angle of about 34° off to your side from your friend's direction. Since your friend is pulling harder than you, you would certainly expect the resultant to be more in your friend's direction. Since you are both pulling, you would expect the resultant to be larger in this case than either of your forces pulling separately. It is clear that we need a general rule for figuring out the magnitude as well as the direction of the resultant of any two forces.

▶ Vector Spaces

One useful convention for conceptualizing the addition of vectors is to draw a picture in which all the vectors are represented by arrows. We let the length of the arrows be proportional to the magnitude of the vectors and their direction be the same as that of the vectors.

One way to look at the representation of vectors by arrows is to think of yourself as operating in a special mathematical space where direction is the same as in real space but where the length of the arrows corresponds to the magnitude of whatever quantity you wish to represent. Such mathematical spaces are called **vector spaces.** The force arrows, for example, allow length to stand for strength.

The general rule for finding the resultant is to move the tail of one vector up to the head of the other. This is demonstrated in Figure 4-8. The resultant then starts at the tail of the vector you did not move and ends up at the head of the one you did move. The maximum length of a resultant is achieved when the two original vectors are in the same direction, and the minimum length occurs when the original vectors are in opposite directions. Any other angle produces a resultant whose magnitude is between these two extremes and whose direction is somewhat in the direction of one vector and somewhat in the direction of the other. It doesn't matter which vector is moved and which is left stationary. If we draw a parallelo-

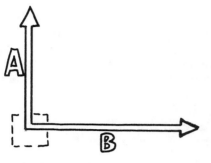

Original Vectors

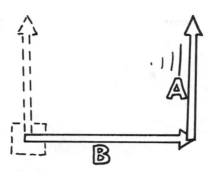

One moved to the head of the other

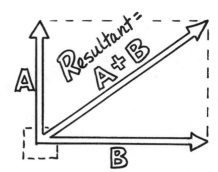

FIGURE 4-8 *A resultant is formed by moving one vector to the head of the other.*

gram, the sides of which are the original vectors, the resultant will have the same length and be along the direction of the diagonal of the parallelogram.

This parallelogram rule works for adding forces as well as any other vectors. One vector quantity for which the rule obviously works is displacement. If you walk 150 paces east and 100 paces north, you get to the same place as if you had walked 100 paces north first and then 150 paces east. Displacement is thus a vector quantity in that it has both magnitude and direction.

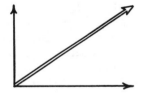  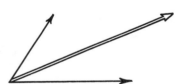

Sometimes it is convenient to draw arrows representing force on a picture depicting the real thing. Force space diagrams are shown superimposed on real space. Such diagrams are useful. They demonstrate what force is acting on what object. Sometimes people lose the distinction, however, between the real objects and the force space diagram and make the mistake of thinking that the strength of the force has something to do with the length of the rope, or whatever is applying the force. This is a failure to distinguish the model from reality itself.

FIGURE 4-9 *A resultant is somewhat in the direction of each of the original vectors and is smaller when they tend to be in the opposite directions but larger when they tend to be in the same direction.*

FIGURE 4-10 *A force space diagram superimposed on a real space diagram.*

▶ Addition of Perpendicular Vectors

The **Pythagorean theorem** gives the magnitude of the resultant when two vectors at right angles to each other are added together. Suppose you and a friend are building a snowman. In the process of making the body, you find yourselves both rolling the same snowball across the snowy field. The snowball is still too small for you both to get on the same side and push in the same direction, so you push north while your friend pushes east. If the snowball rolls more east than north, it is evident that your friend is pushing harder than you are.

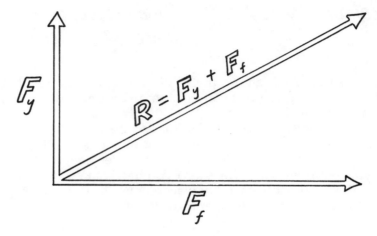

FIGURE 4-11 *The resultant of the forces is force that would be equivalent to your force $\mathbf{F}_y$ and your friend's force $\mathbf{F}_f$ combined.*

Suppose you push with a force of 30 lb while your friend pushes with 60 lb of force. The snowball will roll twice as far in your friend's direction as it does in yours. The path of the snowball will be the direction of the resultant, which is defined as a force equivalent to you and your friend's forces combined. The resultant in this case is the hypotenuse of a right triangle whose sides are equal to the individual forces exerted by you and your friend. The Pythagorean theorem tells us that the square of the hypotenuse of a right triangle is equal to the sum of the squares of the sides.

$$R^2 = F_y{}^2 + F_f{}^2$$

where R stands for the magnitude of the vector $\mathbf{R}$. F_y and F_f are equal to 30 lb and 60 lb, the magnitudes of $\mathbf{F}_y$ and $\mathbf{F}_f$. (It is always a good idea to write a vector, such as $\mathbf{R}$, different from a scalar, such as R, so that you can keep track of what rules are appropriate for addition.) Taking the square root of both sides, we find that the magnitude of it is

$$
\begin{aligned}
R &= \sqrt{F_y{}^2 + F_f{}^2} \\
&= \sqrt{(30 \text{ lb})^2 + (60 \text{ lb})^2} \\
&= \sqrt{900 \text{ lb}^2 + 3{,}600 \text{ lb}^2} \\
&= \sqrt{4{,}500 \text{ lb}^2} \\
&= 67 \text{ lb}
\end{aligned}
$$

The direction part of $\mathbf{R}$ can be found from the definition of the tangent of an angle (the ratio of the side that is opposite the angle to the side adjacent). If we measure the angle θ (theta) from your friend's direction to $\mathbf{R}$, the definition gives us

$$\tan \theta = \frac{F_y}{F_f}$$

In our case,

$$\tan \theta = \frac{30 \text{ ft}}{60 \text{ ft}} = 0.50$$

The angle whose tangent has that value can be found from a trigonometry table or a handy slide rule:

$$\theta = 27°$$

Thus, the resultant **R**, which is the vector sum $\mathbf{F}_y + \mathbf{F}_f$, is a force of 67 lb east by 27° north (27° north of east).

CHECK QUESTIONS

A force of 100 N toward the north is added to a force of 100 N toward the east. What is the (a) magnitude and what is the (b) direction of the resultant?

Answer: (a) 141 N, (b) 45° east of north

► Resolving a Force into Components

The equivalence between the resultant **R** and the combination of $\mathbf{F}_y$ and $\mathbf{F}_f$ works both ways. When we already know the resultant, we can find the two perpendicular forces that will add together to produce it. This is called **resolving** the initial force into **components.** Suppose that we were trying to find out how hard you and your friend were pushing from the observation that the snowball rolls 27° north of east, the same as though it had an applied force of 67 lb in that direction. We want to know how much of this 67 lb is in the northward direction and how much is in the eastward direction.

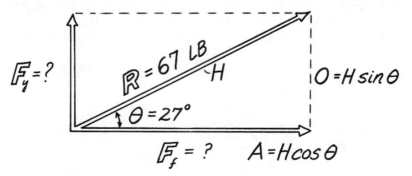

FIGURE 4-12 *The resultant can be resolved back into components using trigonometry.*

The northward force $\mathbf{F}_y$ is found from the sine of the angle θ (since the definition of the sine is the ratio of the side opposite, O, the angle to the hypotenuse, H), and the magnitude of force F_y is the same as the length of the side of the triangle opposite θ:

$$F_y = R \sin \theta$$

The eastward force $\mathbf{F}_f$ is found from the cosine of the angle θ (since the definition of the cosine is the ratio of the adjacent side, A, to the hypotenuse, H), and your friend's force is the side of the triangle adjacent to θ:

$$F_f = R \cos \theta$$

Given the magnitude and direction of the force acting on the snowball as 67 lb at 27°, we can calculate your force as

$$
\begin{aligned}
F_y &= R \sin \theta \\
&= (67 \text{ lb})\sin 27° \\
&= (67 \text{ lb})(0.47) \\
&= 30 \text{ lb(north)}
\end{aligned}
$$

while your friend's force is

$$
\begin{aligned}
F_f &= R \cos \theta \\
&= (67 \text{ lb})\cos 27° \\
&= (67 \text{ lb})(0.89) \\
&= 60 \text{ lb(east)}
\end{aligned}
$$

We therefore have a way of getting back to our starting place.

▶ Sailboats

One ship drives east and another drives west
With the self-same winds that blow,
'Tis the set of the sails and not the gales
Which tells us the way to go.
Ella Wheeler Wilcox, *Winds of Fate*

To illustrate the great power of the reasoning by which a force is broken into components, let us investigate the ability of a sailboat to sail upwind. Christopher Columbus would have given a great deal for this secret. He had a square rigger and had to wait for the wind to blow from east to west before he could make any progress from Spain toward America. Since the prevailing winds are from the other direction, he had to spend most of his time with his sails furled and his sea anchor out. It was well known that the wind filling the sails exerts a force that is directly perpendicular to the sail, and thus it seemed that only when the wind was behind him could he make any progress.

FIGURE 4-13 *Square-rigged vessels could only go in the direction of the wind.*

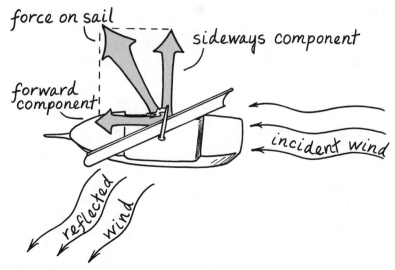

FIGURE 4-14 *One component of the force on the sail is in the forward direction.*

It seemed to make little sense for him to change the direction of his sail. Not only would the sail catch less wind but the force of the wind, acting perpendicular to the sail, would not be in the direction he wanted the boat to go. Only the component in the forward direction would make the boat go. The other component would act against the keel and tend to tip the boat over. Thus, such an arrangement seemed to be full of disadvantages.

However, had Columbus seen this diagram (with one slight modification) it would have blown his mind. Just reverse the roles of the incident and reflected winds, and the forces stay the same. The force on the sail resulting from the wind bouncing off is still perpendicular to the sail, and there is still a forward component.

A sailboat cannot go directly into the wind, but it can angle back and forth to get upwind.[2] It can actually go faster tacking upwind

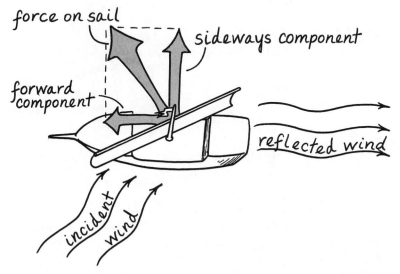

FIGURE 4-15 *One component of the force is still in the forward direction even though the direction of the wind is reversed.*

[2] Only an experienced helmsperson would be ordered to sail a close-hauled vessel "full and by," as close to the wind as possible. An inexperienced person would be given the order "by and large," to sail slightly off the wind since this tack would leave the sailor in less danger of being "taken aback." Hence today a person speaking "by and large" about a subject is not speaking directly to the point. William Morris, *Dictionary of Word and Phrase Origins* (New York: Harper and Row, 1962).

than when going downwind. The maximum velocity when going downwind, neglecting friction, would be the velocity of the wind. But the faster the boat goes upwind, the harder the wind seems to blow. It is common for the speed of an ice boat, a type of sailboat on ice skates, to be four or five times that of the wind.

▶ Equilibrium Problems

There are many instances in modern society where forces must be balanced to equal zero. Some engineer would be very embarrassed if the forces acting on a building or bridge did not balance. An imbalanced force always causes acceleration, and acceleration is usually not so good for buildings and bridges. If you want something to stay put, you have to arrange things so that the sum of all the forces acting on it is zero. When the forces acting on an object are all balanced, the object is said to be in equilibrium. We will be considering situations in this chapter involving static equilibrium. We will assume that an object is stationary to start with, and we will examine the conditions necessary to keep it that way.

The simplest equilibrium problems are those in which all the forces act along the same line in the same or opposite directions. Consider, for example, a friend of yours who steps on a bathroom scale one morning to discover that the pull of gravity on his body is too great. Not satisfied, he purchases another scale. This scale also produces too high a reading, 200 lb. His solution is to stand on both scales at the same time. If he stands evenly on both feet, what will be the reading on each scale?

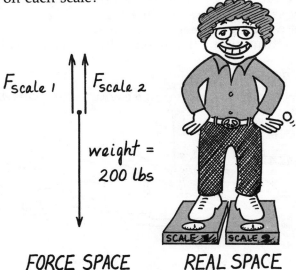

FIGURE 4-16 *The force of each leg is only half the weight.*

FORCE SPACE   REAL SPACE

Your friend's weight will be evenly divided between the two scales. When he is just standing there, he has to be in equilibrium. The sum of all the forces acting on him has to be equal to zero, since he is not moving.

Sum of the forces = 0

If there were a greater force in the upward direction than in the downward direction, for example, he would go up. Since he is going neither up nor down, the forces in these two opposite directions must be equal. We can express this condition of equilibrium in vector notation, using the Greek letter sigma Σ to stand for "sum of":

$$\Sigma \ \mathbf{F} = 0$$

In equilibrium problems it is important to focus attention on one object and consider all the forces acting on that object alone. The object in this case is your friend. The sum of the forces $\Sigma \ \mathbf{F}$ is therefore the two forces $\mathbf{F}_1$ and $\mathbf{F}_2$ by which the scales are pushing up, and the weight of your friend $\mathbf{wt}$, which is pulling down:

$$\mathbf{F}_1 + \mathbf{F}_2 + \mathbf{wt} = 0$$

Since, as we initially stated, your friend is standing evenly on both scales, these two forces are equal and in the same direction:

$$\mathbf{F}_1 + \mathbf{F}_2 = 2\mathbf{F}$$

Including this fact in our condition of equilibrium,

$$2\mathbf{F} + \mathbf{wt} = 0$$

and solving for the force

$$\mathbf{F} = -\frac{\mathbf{wt}}{2}$$

we see that the reading in each scale is equal to half the weight. (The minus sign comes from the fact that we are dealing with vector quantities and the force provided by each scale is in the opposite direction from the weight. If we call the weight positive, we are assigning downward as the positive direction. The upward force on your friend would therefore be negative.)

Having 200 lb divided between two legs is one thing, but this result would be more important to your friend if he were being hung by his thumbs. Would it be kinder to hang him by one thumb or by both thumbs? The force upward must equal the force of his entire weight downward. The force acting on one thumb would be twice as

FIGURE 4-17 *The force on each thumb is only half the weight.*

FIGURE 4-18 *The force on each thumb is greater than half the weight, since it also has a horizontal component.*

great as if the force were divided between the two thumbs. Now consider the relative merits of hanging with one's arms straight up as against hanging with one's arms separated by some angle. The former turns out to be easier on the thumbs. When the arms are stretched out at some angle, it is the upward component of their force that is equal to half the weight. But now there is also a horizontal component because of the angle involved. The condition of equilibrium applies to both the horizontal and vertical components of force independently. This means the horizontal forces by themselves must equal zero and vertical forces by themselves must equal zero.

The condition of equilibrium may be stated in vector form as before:

$$\Sigma \ \mathbf{F} = 0$$

The only way this condition can be satisfied is for the sum of the force components in the upward direction to equal the sum of the forces in the downward direction:

$$\Sigma \ F_\uparrow = \Sigma \ F_\downarrow$$

(Note that the vector notation is dropped because we are now talking about components along only one direction.) If the upward and downward forces weren't equal, no amount of sideward force could bring the forces into balance. In the same way, the sum of the forces to the left, $\Sigma \ F_\leftarrow$, must equal the sum of the forces to the right, $\Sigma \ F_\rightarrow$:

$$\Sigma \ F_\leftarrow = \Sigma \ F_\rightarrow$$

We can now apply these equilibrium conditions to the situation at hand. The sum of the downward forces, $\Sigma \ F_\downarrow$, is nothing more than the weight of your friend wt, since that is the only force acting in the downward direction. All the other forces are components of the forces in the arms.

Drawing a force diagram, and letting θ be the angle between the arms and vertical, the vertical components are the sides adjacent to the angle. The vertical equilibrium condition becomes

$$F_{\Sigma y} = 0$$
$$F_1 \cos \theta + F_2 \cos \theta - \text{wt} = 0$$
$$F_1 \cos \theta + F_2 \cos \theta = \text{wt}$$

The horizontal part of the equilibrium condition says that the pull to the right equals the pull to the left.

$$F_{\Sigma x} = 0$$
$$F_1 \sin \theta - F_2 \sin \theta = 0$$
$$F_1 \sin \theta = F_2 \sin \theta$$

Assuming the angles are the same on both sides, the horizontal part shows that the forces in the two arms are the same.

$$F_1 = F_2$$

We let F stand for both F_1 and F_2:

$$2F \cos \theta = \text{wt}$$

$$F = \frac{\text{wt}}{2 \cos \theta}$$

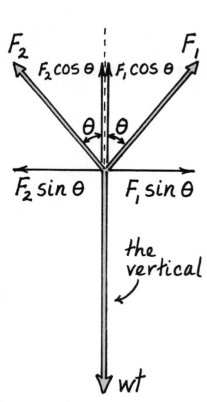

FIGURE 4-19 *The horizontal components of the two upward forces balance each other and the two upward components balance the weight.*

This answer applies as well to the situation in which your friend's arms were straight up. In that case, the angle θ is zero and the cosine of zero degrees is equal to 1. The result is the same as when we were considering the force applied to your friend by the two scales placed side by side:

$$F = \frac{wt}{2(1)}$$

Any larger angle produces a larger force in the two arms since any angle larger than 0° has a cosine smaller than unity (1). The cosine of the angle is in the denominator of our general solution. Making it smaller thus has the effect of making the result larger:

$$F = \frac{wt}{2(\cos\,\theta)}$$

We might see what happens, for example, when the angle θ is equal to 60°. The arithmetic is very easy in this case because the cosine of 60° equals 0.5. The result is reasonable, because the total angle between the two upward forces adds up to the same 120° as remains between either of the two upward forces and the weight. Everything is therefore symmetrical for this particular angle, and you expect all three forces to be the same.

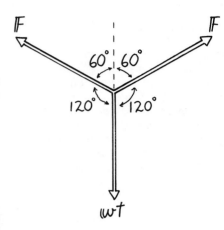

FIGURE 4-20 *When the angle to the vertical is 60°, all three forces are equal, as you might expect from the fact that all three angles are equal.*

CHECK QUESTION

An object weighing 180 N is supported by two equal forces, each making an angle of 25.8° with the upward vertical. The cosine of 25.8° is 0.900. How great are the two upward forces?

Answer: 100 N

▶ Picture Frame Problems

We have been considering a class of problems in which something is held up by two equal forces that pull upward and outward. The force situation is nearly the same in another class of problems, in which two equal forces pull upward and *inward* to support the weight of an object. A common object supported in this way is an ordinary picture frame. In looking at this other class of problems, we will see how the same force space diagram can apply to two different real space situations.

Before we proceed with this discussion, let us go back to make certain that you understand why each of the upward forces must be greater than half the weight as long as the angle to the vertical is greater than 0°. We simplify the situation by looking at only half the picture frame problem. Let's begin by having only one upward and outward force. Assume we have a monkey in the zoo holding on to a rope with one hand and the side of the cage with the other. If the second hand exerts only a horizontal force to pull him over to the side of the cage, none of this force helps to support his weight. The first

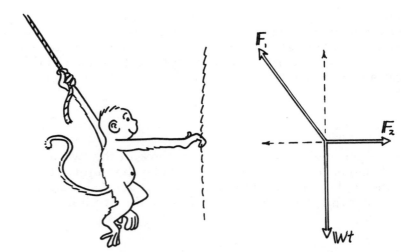

FIGURE 4-21 *The force with which the monkey shown in part (a) holds on to the rope must be greater than its weight since the force in that arm must balance the horizontal force with which he holds on to the zoo cage as well as to lift its weight. The force diagram shown in part (b) depicts the upward and sideward components of this force.*

hand, which is holding on to the rope, is therefore supporting all of the monkey's weight as well as balancing the sideward force exerted by the second hand. The vector sum of these two perpendicular forces is by necessity greater than either of the two forces taken alone, just as the hypotenuse of a right triangle is always greater than either of the two legs.

The monkey hanging from the rope represents half of the force problem discussed in the previous section where a weight is supported by two equal forces pulling upward and outward. The horizontal force exerted by the monkey's second hand is replaced by the horizontal component of the other upward and outward force in the situation where we had your friend hanging by his thumbs. The monkey also represents half of the force situation involved in supporting a picture frame. The only difference is that it represents the other half.

A picture frame is generally supported by a single wire attached to the two sides of the frame and stretched over a nail in the wall. The frame is usually balanced so that the tension in the wire will be the same on both sides of the nail, since too great an imbalance in this tension would cause the wire to slip over the nail.

FIGURE 4-22 *(a) One way to look at a picture frame problem is to think of it as two problems like the monkey hanging from the rope. (b) Another equally valid way is to use a single vector diagram that turns out to be nearly the same as we used for the friend hanging by the thumbs.*

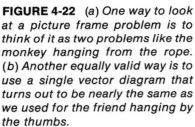

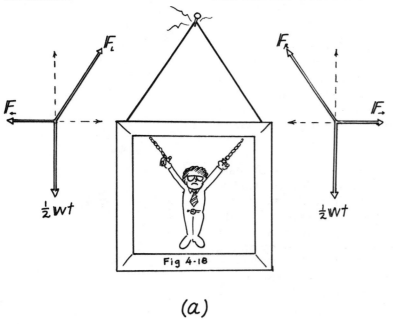

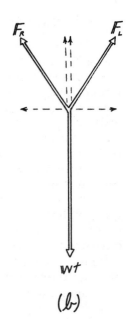

(a)

(b)

One way to look at a picture frame is to say that each end of the wire is holding up one side of the picture frame. That would be to look upon it as two monkey problems. The wire on the left of the nail is lifting up half the weight of the frame and pulling it toward the right. The wire on the right is lifting the other half of the frame and pulling toward the left. The sideward force of one balances the sideward force of the other, and together they lift the whole picture frame.

Another equally valid way of treating the picture frame problem uses a single force space diagram. Looking upon the frame as a single object, we see one weight pulling down and two forces pulling upward and sideward. The upward components work together to balance the weight; the sideward components work in opposite directions to balance each other. The force space diagram looks very much like the diagram we used for your friend being hung by his thumbs. The only difference is that the wire that is physically on the left of the nail applies a force to the frame that is depicted on the right in the force space diagram. The wire on the right of the nail, on the other hand, pulls up and to the left. It is therefore properly shown as a vector in force space which is upward and toward the left.

This reversal of position between force space and real space results from the picture frame (extended over real space) being represented as a point in force space. While the two ends of the wire are attached to the frame at two points physically located some distance apart, they both are acting on the same object. When making force diagrams, it is essential to focus your attention on one object at a time and consider the forces acting on that object alone.

▶ Real Space Versus Force Space

Let us illustrate this trick of representing something that is extended over real space with a point in force space by considering the forces acting on the various parts of a rope, which is tied in a leverage knot. Teamsters use this kind of knot to tie things down on flat-bed trucks. You are likely to see it on trucks driving down the highway. The operation of the knot is not hard to understand if you are willing to ignore the fact that it is extended over real space.

A leverage knot is formed by making a loop in the rope and passing the free end of the rope, called the "bitter end" of the line, around a hook on the truck bed and back up through the loop. The rope in the loop then acts like a pulley; the truck driver can get nearly three times as much force in the part of the rope going up and over the load, called the "standing part" of the line, as he must apply to the bitter end.

If we look only at the parts of the rope going into the knot, we see one part going up and three parts going down. The one part going up, the standing part, must therefore exert a force on the knot which is equal in magnitude but opposite in direction to the three forces that are pulling down. To the extent that friction can be ignored, the three parts of the rope pulling down on the knot all exert equal forces. As the force on the bitter end becomes much greater than the part of the rope passing from the loop and going around the hook on the truck bed, the latter part of the rope will slip through the loop. As this part of the rope becomes tighter than the other part of the rope going to the hook, the rope will slip around the hook. In practice

FIGURE 4-23 *A leverage knot is used by teamsters to hold down loads on the back of flat-bed trucks. A greater force will be exerted on the standing part of the line than is exerted by the truck driver on the bitter end.*

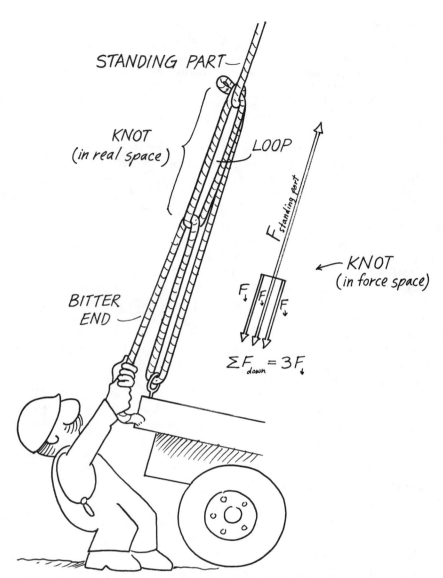

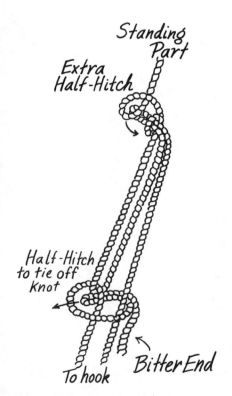

FIGURE 4-24 *The leverage knot shown in Figure 4-23 can be tied by using the bitter end to make a half-hitch around the two parts of the line going to the hook just below where it passes through the loop. The knot can be made more secure by using the standing part to put an extra half-hitch on the bend which forms the loop.*

there is considerable friction as the rope slides through the loop and around the hook in the truck bed, but the truck driver can largely compensate for this friction by pulling on the bitter end with one hand and helping the rope slide around the hook with the other.

This is a useful knot for tying down camping gear to the luggage rack on your station wagon, or whatever. It can be tied off by making a simple half-hitch on the other two downward parts of the line with the bitter end just below the loop. To be more secure, you can put an extra half-hitch on the bend forming the loop with the standing part. Made with this extra half-hitch (see Figure 4-24), this knot will hold the load on a truck from coast to coast without slipping.

After tying the knot you only have two parts of the rope pulling down and one part pulling up. The force situation gets simpler, even though the twists and turns of the rope making up the knot get more complicated. No matter how gnarled the knot becomes, however, you can figure out what is happening outside the knot by looking only at the ropes going into it and representing the whole knot as a point in force space, disregarding how much it might be extended over real space.

The problems in this chapter deal with forces and how they combine.

Hooke's law provides a way of conceptualizing force. Hooke's law states that the force in a spring or almost any other object is proportional to the amount of stretch, twist, bend, or other displacement. The proportionality constant is called the **spring constant** of the object.

$$F = Kx$$

Force has both magnitude and direction, which must be taken into account when two or more forces are added together. Quantities that have both magnitude and direction are called **vectors.** The rules for adding vectors are as follows: (1) Vectors in the same direction add, while vectors in the opposite direction subtract, the direction of the resultant being that of the larger. (2) Perpendicular vectors are

added by Pythagorean theorem. The resultant has a magnitude which is the square root of the sum of the squares of the two original vectors, and a direction given by the angle whose tangent equals the ratio of the two sides.

$$R = \sqrt{A^2 + B^2}$$
$$\tan\theta = \frac{A}{B}$$

A vector is resolved into rectangular components by dropping perpendiculars to the coordinate axes and finding two perpendicular vectors whose resultant would be the original vector. The component adjacent to the angle is the vector times the cosine of that angle, and the other component is the vector times the sine of the angle.

A body is said to be in **equilibrium** if the sum of all the forces acting on it is equal to zero. This condition implies the sum of all the vertical components is equal to zero, and the sum of all the horizontal components is separately equal to zero.

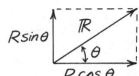

$$\Sigma F_x = 0 \qquad \Sigma F_y = 0$$

4-1 A nylon fishing line is stretched when a fish pulls on it and returns to its original length when the force is removed.

Given: The line stretches from 100 m to 103 m when the fish pulls with a force of 120 N.

Find: (a) What is the spring constant of the fishing line?
(b) How much will this line stretch if the fish pulls with only 50 N?

Your Solution

Discussion

This problem illustrates Hooke's law, a useful tool for conceptualizing the idea of force. The direction of the force is, of course, along the fishing line and the magnitude of the force with which the fish pulls is assumed to be directly proportional to the stretch of the fishing line. The spring constant is the proportionality constant between the force and the amount of stretch.

Sample Solution 4-1

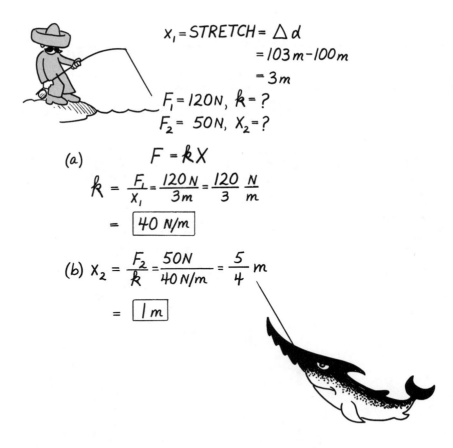

$$x_1 = STRETCH = \Delta d$$
$$= 103\,m - 100\,m$$
$$= 3\,m$$
$$F_1 = 120N, \quad k = ?$$
$$F_2 = 50N, \quad X_2 = ?$$

(a)
$$F = kX$$
$$k = \frac{F_1}{X_1} = \frac{120\,N}{3\,m} = \frac{120}{3}\,\frac{N}{m}$$
$$= \boxed{40\ N/m}$$

(b)
$$X_2 = \frac{F_2}{k} = \frac{50N}{40\,N/m} = \frac{5}{4}\,m$$
$$= \boxed{1\,m}$$

Discussion

Hooke's law $F = kx$ is solved for the proportionality constant k in part (a) and for the stretch x in part (b). The subscripts on F and x are used to show that Hooke's law is being applied in two different situations. There is no subscript on the spring constant k, because it is the same in both cases.

The answer in part (a) is only good to one significant figure, since that is the accuracy with which we know the stretch x_1. This result also limits the accuracy of the answer in part (b) to one significant figure.

4-2   Teamsters use a leverage knot to secure a load to a flat-bed truck. The knot is formed by making a loop in the rope and passing the free end of the rope, called the "bitter end" of the line, around a hook on the truck bed and back up through the loop. The rope in the loop acts like a pulley.

Given: A truck driver pulls on the bitter end with a force of 200 lb.

Find: (a) How much force is applied to the load?
(b) How hard does the rope pull on the truck-bed hook?

Your Solution

Discussion

This is a pulley problem, used to illustrate the addition of forces that all act in the same or opposite directions. We can assume that the tension in all parts of the rope is the same, because any difference in force would cause the sheaves to turn in the blocks. In doing equilibrium problems, remember to focus your attention on all the forces acting on a single object at one time.

Hook

free end

Sample Solution 4-2

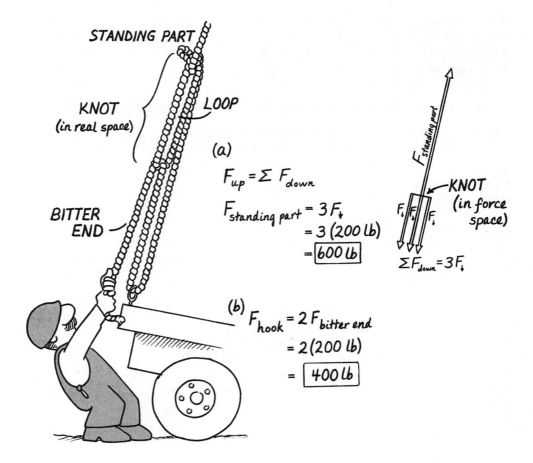

STANDING PART

KNOT
(in real space)

LOOP

BITTER
END

(a)

$$F_{up} = \Sigma\, F_{down}$$

$$F_{standing\ part} = 3F_{\downarrow}$$
$$= 3(200\ lb)$$
$$= \boxed{600\ lb}$$

$F_{standing\ part}$

KNOT
(in force
space)

$F_{\downarrow}\ F_{\downarrow}\ F_{\downarrow}$

$\Sigma F_{down} = 3F_{\downarrow}$

(b)

$$F_{hook} = 2F_{bitter\ end}$$
$$= 2(200\ lb)$$
$$= \boxed{400\ lb}$$

Discussion

This is a useful knot to know if you are ever going to want to tie camping gear or the like to the top of your car. You may wish to find a piece of rope and tie this knot over the back of a chair to better understand how it works. (The figure shows the loop held to the standing part by a simple half hitch. For long distance runs, teamsters double this hitch to hold the loop more securely, but you can tie a loop in the standing part any old way and the knot will function as a leverage knot. The important features are that the bitter end, the end you pull on, passes through the loop, back around a hook of some kind, and then back to the knot.)

The trick to understanding this and many other force situations is to ignore the fact that the knot (or whatever has forces applied to it) is spread out over real space. A force space diagram for part (a), in which the object of interest is the knot, shows three forces pulling in one direction and only one pulling in the other. The three forces in the one direction are assumed to be equal, neglecting the frictional forces with which the rope slides through the loop and around the hook, so that the one force in the other direction is three times as great. Part (b), in which the object of interest is the hook, shows a similar situation, but there are only two equal forces pulling in one direction against the one force in the other direction.

4-3 An elephant is pulling a log through the jungle, one end of a rope being tied to the elephant and the other end being tied to the log.

Given: The rope makes an angle of 30° with the horizontal and is under a tension of 4,000 N.

Find: (a) What is the horizontal component of the force?
(b) What is the vertical component of the force?

Your Solution

Discussion

This is a force problem in which the force is doing two things. Part of the force is pulling up and part is pulling sidewards. You are given the magnitude and direction of the force and must find the components. That is like changing from polar coordinates to rectangular coordinates in force space.

Sample Solution 4-3

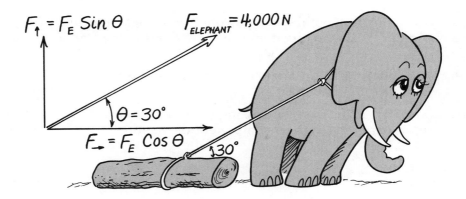

(a) $F_{\rightarrow} = F_E \cos \theta$
$= (4,000 N)(\cos 30°) = 3,464 N$
$\cong \boxed{3,500 N}$

(b) $F_{\uparrow} = F_E \sin 30°$
$= (4,000 N)(.500)$
$= \boxed{2,000 N}$

Discussion

The force in the force space diagram has the same direction as the rope in real space, but the length of the force space arrow has nothing to do with the length of the rope.

The side of the force space triangle adjacent to the angle is clearly the hypotenuse times the cosine of the angle. The other side, which is not adjacent to the angle, is the hypotenuse times the sine, since it is the same length as a side of the right triangle that would be opposite the angle.

4-4 Paul Hew is rolling the lawn. The force he applies to the lawn roller handle performs two functions. It helps to hold the roller on the ground so it doesn't slip, and it pushes the lawn roller forward at a constant velocity.

Given: Paul is pushing with a force of 250 N at a direction of 45° into the ground. The lawn roller weighs 100 N.

Find: (a) What is the horizontal force making the lawn roller go forward?
(b) What is the total vertical force holding the roller against the ground?

Your Solution

Discussion

One of the forces in this problem must be resolved into components, as in the previous problem. The force in the handle is pushing both sidewards and down. The downward component, however, is being aided by a second force, the weight of the lawn roller.

Sample Solution 4-4

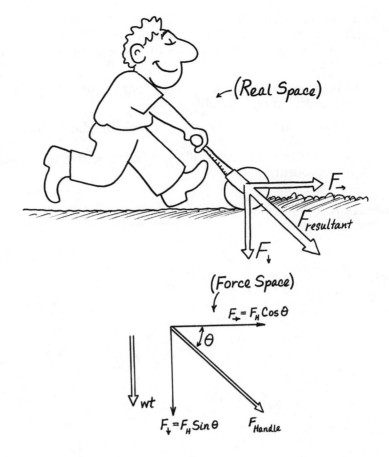

(a) $F_{\rightarrow} = F_H \cos \theta$
    $= (250\,N)(\cos 45°)$
    $= \boxed{177\,N}$

(b) $\Sigma F_{\downarrow} = F_H \sin \theta + wt$
    $= (250\,N)(\cos 45°) + (100\,N)$
    $= \boxed{277\,N}$

Discussion

The two actual forces acting on the lawn roller, the force in the handle and the weight, are shown as double-line arrows in the force space diagram. The components of the force in the handle are shown as single-line arrows. The forward component found in part (a) is the side adjacent the angle and is therefore given by the hypotenuse times the cosine of the angle.

The total downward force acting on the roller found in part (b) is the downward component of the handle force added to the weight of the roller, since both act in the same direction.

4-5 Two men are carrying a trunk. They each support half the weight, but they also pull sidewards against each other to hold the trunk away from their legs.

Given: Each man pulls with a force of 200 N at a direction 12° from the vertical.

Find: (a) How heavy is the trunk?

(b) How hard do the two men pull against each other?

Your Solution

Discussion

There are three forces in this problem: two upward forces and the weight. The problem is simpler than it could be because the two upward forces are equal and make the same angle with the vertical when pulling away from each other. The force space diagram for this problem is the same as for a picture frame problem.

Sample Solution 4-5

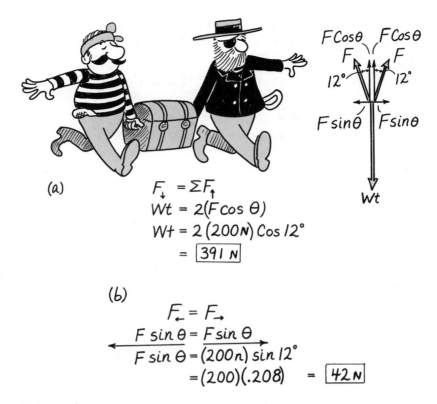

(a)

$$F_\downarrow = \Sigma F_\uparrow$$
$$Wt = 2(F\cos\theta)$$
$$Wt = 2(200N)\cos 12°$$
$$= \boxed{391\,N}$$

(b)

$$F_\leftarrow = F_\rightarrow$$
$$F\sin\theta = F\sin\theta$$
$$F\sin\theta = (200n)\sin 12°$$
$$= (200)(.208) = \boxed{42\,N}$$

Discussion

All the forces are assumed to be balanced in both parts of this problem such that the upward forces are balanced by downward forces and the sideward forces in one direction are balanced by sideward forces in the other direction. Part (a) deals with forces in the upward and downward direction and is solved from the assumption that the one downward force, the weight of the trunk, is equal to the sum of the two upward forces, the upward components of the forces in the two men's arms. Since these are the components which are adjacent the known angle, they are found by multiplying the known forces in the arms by the cosine of the angle.

The sideward components are found in part (b) by multiplying the known forces in the arms by the sine of the angle, since these components are opposite the known angle. (Remember that the side adjacent is associated with the cosine and the side opposite is associated with the sine of the angle.) These sideward components are equal in magnitude but opposite in direction so that they balance each other.

4-6 A concrete bucket on low friction wheels is being held stationary on a ramp. The cable holding the bucket is parallel to the ramp. The force exerted by the ramp on the wheels is perpendicular to the ramp.

Given: The weight of the bucket and contents is 1,000 kg (one metric tonne), and the ramp makes an angle of 25° to the horizontal.

Find: (a) What is the force with which the ramp pushes on the bucket?
(b) What is the tension in the cable?

Your Solution

Discussion

Many balanced force problems are best worked out by resolving the forces into horizontal and vertical components. Incline plane problems, however, are best solved by resolving forces into components that are parallel and perpendicular to the plane. Using a tilted frame of reference frequently reduces the number of forces that need to be resolved into components.

Sample Solution 4-6

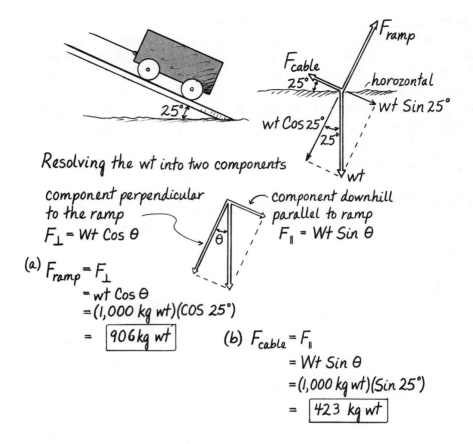

Resolving the wt into two components

component perpendicular
to the ramp
$F_\perp = Wt \cos \theta$

component downhill
parallel to ramp
$F_\parallel = Wt \sin \theta$

(a) $F_{ramp} = F_\perp$
$= wt \cos \theta$
$= (1,000 \text{ kg wt})(\cos 25°)$
$= \boxed{906 \text{ kg wt}}$

(b) $F_{cable} = F_\parallel$
$= Wt \sin \theta$
$= (1,000 \text{ kg wt})(\sin 25°)$
$= \boxed{423 \text{ kg wt}}$

Discussion

The force in the cable is in the direction of the ramp, namely 25° to the horizontal. The force exerted by the ramp is perpendicular to the cable force, assuming that there is no friction in the wheels. We can therefore choose a tilted frame of reference in which these forces are already resolved into components, and then we have only to work out the weight if we want to find the conditions where everything balances. (We will see in Chapter 6 that the forces are balanced even if the object is moving, so long as the velocity is constant.)

The angle between the ramp and the horizontal is the same as the angle between the weight and the perpendicular to the ramp. One way to see this is to consider that both angles are complements of the same angle.

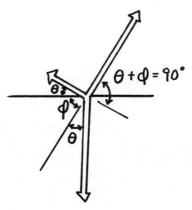

4-7 A mountain climber finds herself dangling from the end of her nylon safety line one morning after she has slipped over the edge of a cliff.

Given: The length of rope between the climber and the edge of the cliff has a spring constant of 8.3 N/cm. The climber's weight is 780 N. The unstretched length of the rope is 3.25 m.

Find: (a) What is the amount of stretch in the rope due to the climber's weight?
   (b) What is the stretched length of the rope?

4-8 Harry, the sign painter, has rigged up a boatswain's chair with which to pull himself up the side of a tall building with a block and tackle. He usually ties the free end of the rope back onto the chair, but one day he decides to tie it onto a flagpole that happens to be sticking out the side of the building.

Given: Harry and his chair weigh 850 N. The rope is secured at the roof of the building, passes through a pulley attached to the chair, back up through a pulley attached to the chair, back up through a pulley attached to the roof, and then down to Harry again.

Find: What is the tension in the rope if
   (a) Harry ties the rope onto the chair?
   (b) Harry ties the rope onto the flagpole?

4-9 A cabin cruiser has a rowboat in tow. The force on the rowboat due to the tow line is somewhat upward, tending to lift the rowboat as well as pull it horizontally.

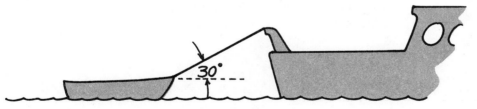
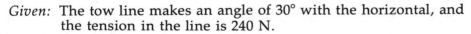

Given: The tow line makes an angle of 30° with the horizontal, and the tension in the line is 240 N.

Find: (a) What is the horizontal force?
   (b) What is the vertical force acting on the rowboat?

4-10 A floor-polishing mop has a weighted head to help increase the downward pressure against the floor. The rest of the downward pressure comes from the force applied to the mop handle.

Given: The head of the floor-polishing mop weighs 11 N, and the components of the force in the handle are 15 N sidewards and 25 N downwards.

Find: (a) What is the net downward force acting on the mop head?
(b) What is the magnitude of the resultant force acting on the mop head due both to its weight and the force of the handle?
(c) What is the direction of the resultant force (sum of weight and handle force)?

4-11   Two monkeys are holding hands in the zoo. They are both hanging from equal lengths of rope, which are attached to the top of the monkey cage some distance apart.

Given: Each monkey exerts a force of 44 lb in holding on to the equal lengths of rope, and the lengths of rope each make an angle of 25° with the vertical.

Find: (a) What is the combined weight of the two monkeys?
(b) With what horizontal force do the monkeys pull on each other in holding hands?

4-12   In one of the nicest known forms of public transportation, people ride in cars drawn up steep hills by cables running through a slot between the tracks. The gripperson who operates such a car activates a cable-gripping mechanism that fits down inside the cable slot.

Given: The weight of one such car fully loaded with passengers is 85,000 N. On one hill the cable and track make an angle of 20° with the horizontal.

Find: (a) What is the force with which the cable pulls on the gripping mechanism?
(b) What is the force between the wheels and the ground, neglecting friction (assuming that the force is perpendicular to the hill)?

MORE INTERESTING PROBLEMS

4-13   A picture frame is supported by a wire hooked over a nail in the wall. The tension in the wire is the same on both sides of the nail.

Given: The tension in the wire is 500 N, and the wire makes an angle of 35° with the horizontal top part of the picture frame.

Find: (a) What is the upward component of the force exerted on the frame by one end of the wire?
(b) What is the sideward (horizontal) component of the force exerted by one end of the wire?
(c) What is the weight of the picture frame?

4-14 Canal boats are pulled along the canal by a team of horses who walk along a tow path on one bank of the canal. The canal boat resists being pulled into the bank, setting its angle away from the bank by using its rudder.

Given: A canal boat is being towed by a rope 10 m long. When the boat is 3.0 m from the tow path, a tension in the rope of 240 N keeps the boat moving at a uniform constant velocity.

Find: (a) What is the angle the rope makes with the canal bank?
   (b) How much of the tow rope force is in the forward direction?
   (c) How much of the tow rope force is sidewards?

4-15 A kind daddy is pushing his little girl on a swing in the park. One fun thing to do is to hold her in equilibrium for a moment (with a horizontal force) before releasing her so she can swing back. The vector sum of daddy's force, the little girl's weight, and the force of the chain is then zero.

Given: The little girl has a weight of 200 N, and the chain makes an angle of 25° to the vertical.

Find: (a) What is the tension in the chain?
   (b) With what force must daddy push horizontally to hold his little girl in equilibrium?

4-16 A strong man is trying to pull a chain out horizontally when the chain has a physics book attached at its midpoint. He finds that the chain is never straight, no matter how hard he pulls on its ends.

Given: The physics book weighs 24 N. The strong man first pulls with a force of 100 N and then, summoning all of his strength, he pulls with 400 N.

Find: What is the angle the chain makes with the horizontal with
   (a) a force of 100 N on each end of the chain?
   (b) a force of 400 N on each end?

4-17 A picture frame is supported by a wire having breaking tension of somewhat more than half the weight of the picture.

Given: The test strength of the wire is 300 N, and the picture, together with the frame, weighs 450 N. The support points on the frame to which the wire is attached are 80 cm apart.

Find: (a) What is the minimum angle the wire can make with the horizontal before it breaks?
   (b) What is the minimum length of wire necessary to support the picture frame?

4-18 Two men are putting a 1,275-cc engine into a 1960 "bug-eyed" Sprite sports car. One of the men is required to guide the engine into its mounts while the other man is lifting the engine with a simple two-strand block and tackle.

Given: The engine weighs 750 N. The man operating the block and tackle weighs 650 N.

Find: (a) Draw a vector diagram showing all the forces acting on the engine.
   (b) How much tension is in the block and tackle line?
   (c) Draw a vector diagram showing all the forces acting on the man who is lifting the engine.

ANSWERS

4-7   (a) 94 cm; (b) 4.19 m.
4-8   (a) 283 N; (b) 425 N.
4-9   (a) 208 N ≅ 210 N; (b) 120 N.
4-10   (a) 36 N; (b) 39 N; (c) 23° from vertical.
4-11   (a) 80 lb; (b) 18.6 lb ≅ 19 lb.
4-12   (a) 29,000 N; (b) 80,000 N.
4-13   (a) 287 N ≅ 290 N; (b) 410 N; (c) 574 N ≅ 570 N.
4-14   (a) 17.5°; (b) 230 N; (c) 72 N.
4-15   (a) 221 N ≅ 220 N; (b) 93 N.
4-16   (a) 6.9°; (b) 1.7°.
And so no force, however great,
Can stretch a cord, however fine,
Into a horizontal line
That shall be absolutely straight.[1]
4-17   (a) 49°; (b) 1.2 m.
4-18   (a)   (b) 375 N;   (c)

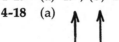

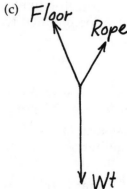

[1] Quoted in the *Oxford Dictionary of Quotations* as an example of accidental meter and rhyme. (2d ed; New York: Oxford University Press, 1955. p. 566) Printed in prose in Whewell's *Elementary Treatise on Mechanics*, 1819.

Velocity and Acceleration

King John: Be Mercury, set feathers to thy heels,
And fly like thought from them to me again.

Bastard: The spirit of the time shall teach me speed.
—Shakespeare, *King John IV*, ii, 174

As sophisticated as they were, the ancient Greeks did not fully develop the concept of velocity. They thought in terms of distance and in terms of time sequence, but they did not combine these two quantities.

It was in Galileo's time that motion through space was described in terms of **displacement,** with both magnitude and direction, and time as something that moved onward forever independent of space.

Today, when we want to describe motion over a distance, it is customary to use displacement, **d**, a vector having both magnitude and direction. Average velocity is defined as a ratio of displacement over time.

$$\mathbf{v} = \frac{\mathbf{d}}{t}$$

This definition agrees with most people's understanding of the word except that there are a few wrinkles that most people never worry about. One is that velocity is a vector quantity. We usually don't have to be concerned with velocity's vector nature because we either understand the direction part very well, or it doesn't make much difference to the problem at hand. We can then deal with the magnitude part alone. The word *speed* is frequently used to refer to the magnitude, or scalar part, of velocity. Thus, the speedometer in your car is properly named. It tells you how fast you are going, but not which way. To know your velocity, however, you need both a speedometer and a compass. A vector quantity, as you recall, is specified by both "how much" and "which way."

The vector nature of velocity becomes important when you add two velocities together. Like forces, they add numerically when they are in the same direction, they subtract when they are in the opposite direction, and they add by the same rule as for forces, sometimes called the parallelogram rule, when they are off at some angle relative to each other.

The main class of physics problems where we will encounter the addition of velocities will be those in which a frame of reference is moving. Suppose you are walking across the deck of a boat for example. If the boat is moving, the combined motion of your walking and the motion of the boat would be the vector sum of those two velocities. If both you and the boat are going in the same direction, you would add the magnitudes of the two velocities (or speeds) to get the combined velocity. If you and the boat are going in opposite directions, you would subtract. If you are walking perpendicular to the boat's direction, you would use the Pythagorean theorem to add the perpendicular velocities just as you would to add two perpendicular forces.

Galileo recognized these rules for changing from one frame of reference to another. He understood the full vector nature of velocity. These rules stood for centuries, only to be revised by Albert Einstein. It was Einstein's genius to recognize that Galileo's rules had to be revised at velocities near the speed of light.

Most people in our present society are intuitively comfortable with Galileo's concept of velocity but find Einstein's interpretation difficult to apprehend. To understand velocity on Galileo's level is, however, no mean chore. The ancient Greeks would have had as much difficulty with it as most people now have with Einstein. It is interesting to speculate how people 200 years from now will understand the concept. The spirit of the time does indeed "teach" us speed, if not in exactly the same sense that Shakespeare's bastard intended.

▶ Velocity as a Vector

Velocity is a vector because it is the ratio of displacement, a vector, over time. Although time is a scalar quantity, the ratio of a vector to a scalar is always a vector.

When a vessel, such as a boat or airplane, is traveling in a moving medium, the velocity relative to the ground is the vector sum of the velocity of the vessel plus the velocity of the medium. When traveling with or against the wind, the ground speed is the sum or difference of the air speed and the wind velocity. Ground speed is important because it indicates when to start looking for a

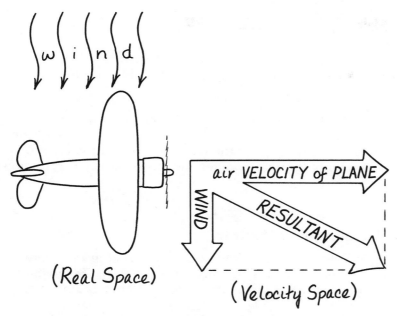

FIGURE 5-1 *An airplane headed in a direction perpendicular to the wind actually goes in a direction that is a combination of the plane's velocity and that of the wind.*

destination. Velocity relative to the ground is even more important if the vessel is heading perpendicular to the wind or water current. If a pilot fails to concern himself with the vector sum of velocities, he will *never* see his destination.

Actual velocity resulting from both air and wind velocities will carry the plane in a direction that is a combination of the two. Another pilot might, for example, need to arrange her velocity relative to the air so that the resultant velocity will be in the direction of her destination. Suppose that she wants to go east while the wind is toward the south. The plane needs to be pointed to the northeast to compensate for the wind. The exact angle θ toward the north depends on her air velocity v_a and the velocity of the wind v_{wind}.

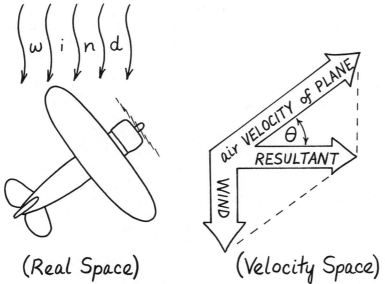

FIGURE 5-2 *To actually go in a direction perpendicular to the wind, the plane must head somewhat upwind.*

One way to think of the air velocity of the plane is to resolve it into two components, one of which compensates for the velocity of the wind. This component is found from the sine of the angle θ, since it has a length equal to the side of the triangle opposite this

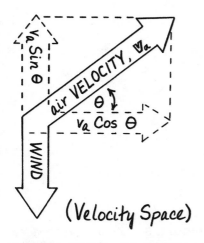

(Velocity Space)

FIGURE 5-3 *The air velocity of the plane may be thought of as being composed of two components.*

angle. The angle θ can be found by setting this component equal to the wind velocity and solving it for θ:

$$v_a \sin \theta = v_{\text{wind}}$$

The resultant is the other component found from the same angle:

$$v_a \cos \theta = \text{resultant}$$

Although the parallel and perpendicular cases will be adequate to illustrate addition of velocities for our purposes, the argument can be extended for any angle of the wind by resolving the wind velocity into components that are parallel and perpendicular to the destination of the craft.

▶ Instantaneous Velocity

Unless velocity is constant, we must distinguish between the concept of instantaneous and average velocity. By instantaneous velocity we mean, as before, the ratio of displacement over time,

$$\mathbf{v} \equiv \frac{\mathbf{d}}{t}$$

except we specify that t be only an instant of time. An instant of time refers to an infinitesimally small bit of time, smaller than anything that we can imagine. Average velocity, on the other hand, refers to the same definition of displacement over time, except that t is some finite interval of time. The distinction only becomes important, as we shall see, when the velocity is changing.

An automobile speeding down the highway at 60 mph will travel a distance of 60 miles if it can maintain this average velocity for a whole hour. It is, however, impossible to maintain this constant velocity exactly for an hour, particularly if the rest of the traffic is traveling at 55 mph. The driver would probably find that his instantaneous velocity is sometimes a bit over and sometimes a bit under the average. Converting from miles to feet and from hours to seconds, we find that 60 mph equals 88 ft/s.

$$60 \, \frac{\text{miles}}{\text{hour}} \left(\frac{5,280 \text{ feet}}{1 \text{ mile}} \right) \left(\frac{1 \text{ hour}}{3,600 \text{ seconds}} \right) = 88 \, \frac{\text{feet}}{\text{second}}$$

The displacement equals 88 ft for each time interval of 1 s. It is much easier to keep the velocity of a car constant for a second than for an hour. We might think that 88 ft/s is our instantaneous velocity. And it might be for many practical purposes, but that is not what is really meant by instantaneous. At an average velocity of 88 ft/s an automobile will travel a distance of 0.088 feet in a millisecond. This, however, is still an average velocity. No matter how small we make the time interval, the displacement will be correspondingly small. The ratio of displacement to time will approach a constant value no matter how small the time interval becomes, as long as it does not actually become zero. In the language of calculus, this is called taking the first derivative of distance with respect to time.[1] As long as the

---

[1] For those who take calculus, this will be an excellent example of what is meant by a derivative.

time interval remains finite, no matter how small it is, the ratio of displacement to time is technically an average velocity. Only when the time is an instant, by definition something smaller than anything we can imagine, is the ratio of displacement to time truly **instantaneous velocity.**

From a practical point of view, the distinction between instantaneous and average velocities can be disregarded as long as the velocity remains constant over the time interval in question. The size of the time interval that can be considered an instant depends on how fast the velocity is changing and how particular you might be in your measurements. If a car is traveling along at 88 ft/s, a second might well be regarded as an instant for all practical purposes. If the car accelerates from 0 to 88 ft/s in 11 s, a second would be too long for judging instantaneous velocity. The velocity at the end of any one second would be 8 ft/s greater than at the start of that second. At that acceleration a millisecond might be a better time interval to use as an instant.

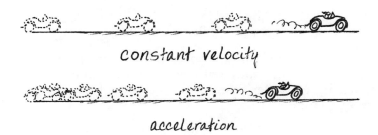

constant velocity

acceleration

FIGURE 5-4 *The difference between constant velocity and constant acceleration is sometimes elusive.*

▶ Acceleration

A concept closely related to velocity is the change of velocity. The two concepts are so closely related, in fact, that people sometimes fail to distinguish between them. Velocity is already a change in something, a change in distance. A change in velocity is something else again. We call it acceleration.

$$\mathbf{a} \equiv \frac{\text{change of } \mathbf{v}}{\text{change of } t}$$

Acceleration is defined as the **rate of change of velocity.** Just like velocity, acceleration is a vector quantity. It has magnitude and direction. The accelerator in an automobile is a pedal that produces an acceleration in the same direction as the initial velocity. The brake is also an accelerator in the sense that it produces an acceleration. The direction of the acceleration is, however, opposite to the direction in which you were originally going. An acceleration in this direction is frequently called a deceleration, but it is an acceleration nonetheless. Even the steering wheel is an accelerator of sorts. The direction of the acceleration it produces is perpendicular to the original direction of motion.

We will worry about the true vector nature of acceleration in Chapter 7, when we deal with motion along a curved path. For now, we can treat acceleration as a scalar quantity by concerning ourselves only with accelerations that are in the same direction as the original velocity, or in the opposite direction by using a negative sign to deal with deceleration.

$$a \equiv \frac{\text{change of } v}{\text{change of } t}$$

Just as with velocity, we may distinguish between average and instantaneous acceleration. Instantaneous acceleration is the preceding definition with the added qualification that the time interval be smaller than anything you can imagine. This distinction, however, may be disregarded as long as the acceleration is constant over the time interval considered.

Concepts of velocity and acceleration are similar in several ways. The displacement we spoke of in defining velocity is really a change in position, just as the time interval is really a change in time. Thus, the definition of velocity can be made to look a great deal like the definition of acceleration.

$$v \equiv \frac{\text{change of position}}{\text{change of time}}$$

Velocity is rate of change of position with respect to time, while acceleration is rate of change of velocity with respect to time. Acceleration is a higher-order change than velocity. It is the rate of change *of* a rate of change.

$$a \equiv \frac{\text{change of } \left(\dfrac{\text{change of position}}{\text{change of time}} \right)}{\text{change of time}}$$

This business of the rate of change of one thing with respect to something else is very common in mathematics at the level of calculus. While we don't need such sophisticated mathematics for this course, we might borrow a symbol to make our notation look a little simpler. The Greek letter delta, Δ, is frequently used to stand for "the change of." That is a shorthand way of saying the final value of something minus the initial value. Using this delta notation, the definition of velocity can be written

$$v \equiv \frac{\Delta d}{\Delta t}$$

People who are familiar only with algebra sometimes want to try to cancel the Δ in the numerator with the Δ in the denominator. That is not possible, because Δd, as it is used here, means $d_f - d_i$. It does not mean Δ times d. In the same way, Δt means $t_f - t_i$.

This delta notation is particularly useful in showing the similarity between the definitions of velocity and acceleration. Using it, the definition of acceleration can be written

$$a \equiv \frac{\Delta v}{\Delta t}$$

Here again, Δv means $v_f - v_i$, and Δt stands for $t_f - t_i$.

The distinction between velocity and acceleration must be clearly understood; these two concepts are frequently confused. The units of velocity and acceleration look a great deal alike. They both have units that involve the ratio of distance and time. A sports car accelerating from 0 to 60 mph in 11 s, for example, has an acceleration that is expressed in units having distance in the numerator and time in the denominator:

$$a \equiv \frac{\text{change of velocity}}{\text{change in time}}$$

$$= \frac{60 \text{ miles/hour}}{11 \text{ seconds}}$$

$$= 5.5 \frac{\text{miles}}{\text{hour} \cdot \text{second}}$$

The difference is that acceleration has time in the denominator twice. The automobile is increasing its velocity at the rate of 5.5 mph/s. At the end of each second, the car is going 5.5 mph faster than it was before. This distinction is clear, because hours and seconds are different units for time. The confusion comes when the same units of time are used for both velocity and acceleration. Since 60 mph equals 88 ft/s, acceleration may be expressed in feet/second/second, or ft/s².

$$a = \frac{88 \text{ ft/s}}{11 \text{ s}}$$

$$= 8.0 \text{ ft/s}^2$$

The acceleration of the car is such that it gains 8.0 ft/s of velocity for each second that it accelerates. At the end of 1 s, starting from rest, it is going 8.0 ft/s. At the end of 2 s, it is going 16 ft/s. At the end of 3 s, it is going at a velocity of 24 ft/s, and so on.

The units "feet per second" and "feet per second squared" sound alike, and people sometimes forget that they are different. But acceleration is as different from velocity as velocity is different from distance.

You might ask yourself how far the car in the preceding example will go in 1 s, starting from rest. By your answer, you can tell if you are one of those people who are failing to make a clear distinction between velocity and acceleration.

While you are thinking about that, we might point out that the instantaneous velocity of the automobile in our example is constantly changing. It starts out at zero, since the car is initially at rest. It then increases as time progresses, reaching a velocity of 8.0 ft/s at the instant that falls at the very end of the first second.

If you said to yourself that the car went a distance of 8.0 ft in the first second, you are absolutely . . . wrong. The car would go that far if it had kept up a constant velocity of 8.0 ft/s for the whole second. The car in our example, however, has an instantaneous value that is always smaller than 8.0 ft/s until the very last instant of that first second. As we shall see, the actual distance traveled in the first second is only 4.0 ft.

► Velocity-Time Space

We can draw a graph of the velocity of some object, plotting the velocity on one axis and time on the other. Such a graph may be thought of as a representation in a special mathematical space, which we may call **velocity-time space.** Velocity is one dimension of this space, and time is the other.

It turns out that distance is represented by area in velocity-time space. Distance is, after all, equal to the product of velocity and time—just by algebraically rearranging the definition of velocity.

$$v \equiv \frac{d}{t}$$
$$d = v \cdot t$$

Since velocity is one dimension in our special mathematical space, and time is the other, the product of velocity and time would be area. The total distance traveled by an automobile at a rate of 20 m/s for 10 s is 200 m, since

$$d = vt$$
$$= (20 \text{ m/s})(10 \text{ s})$$
$$= 200 \text{ m} \left(\frac{s}{s}\right)$$

The distance traveled by an automobile accelerating is still represented by area in velocity-time space. On a graph in velocity-time space the velocity at any instant of time increases along a more or less straight line from an initial to a final value. The initial velocity v_i is zero in this case, and the final velocity v_f is the instantaneous velocity at the time t_f. We can see from Figure 5-6 that the area under the curve (which is a straight line as long as the velocity increases at a constant rate) is less than it was for the constant velocity case. Instead of being the area of a rectangle, the distance traveled under constant acceleration is the area of a triangle in our mathematical space.

Since the distance is less, the *average* velocity is less. By average

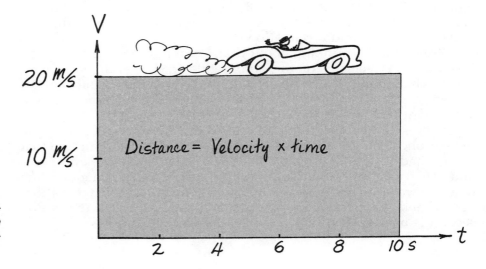

FIGURE 5-5 *The distance traveled at constant velocity is the area of a rectangle on a velocity-time graph.*

FIGURE 5-6 *The distance traveled from rest at constant acceleration is the area of a triangle on a velocity-time graph.*

velocity, we mean the ratio of distance to time where the time interval is larger than an instant.

$$v_{\text{ave}} \equiv \frac{d}{t}$$

As you recall, instantaneous velocities, such as v_i and v_f, are the values of this ratio of distance to time when the time interval is very small. The average velocity is this same ratio when the time interval is large, such as all the way from zero to t_f.

We can show that the average velocity, defined in this way, is the same thing as the average *of the* velocities, as long as the acceleration is constant. This would mean that the average velocity v_{ave} as shown on a graph would have a value halfway between v_i and v_f. Let us go ahead, however, and include a more general case in which the initial velocity is not necessarily zero. Let us work from a numerical example involving two constant velocities. Suppose a car travels 10 m/s for 5.0 s and then somehow instantaneously speeds up to 20 m/s for the next 5.0 s. (The time intervals must be equal for the argument to work.) The distance traveled in the first 5.0 s would be

$$\begin{aligned} d &= vt \\ &= (10 \text{ m/s})(5 \text{ s}) \\ &= 50 \text{ m} \end{aligned}$$

The distance traveled in the next 5.0 s would be twice as great, 100 m, since the velocity would be twice as great. The total distance trav-

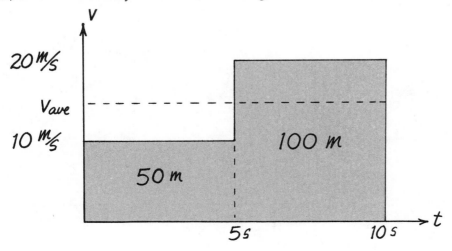

FIGURE 5-7 *The average velocity is the constant velocity that would produce the same displacement as the two different velocities.*

eled would be the sum of the two rectangular areas, as shown in Figure 5-7:

$$d_{total} = 50 \text{ m} + 100 \text{ m}$$
$$= 150 \text{ m}$$

By definition, average velocity would be the total distance traveled over the total time:

$$v \equiv \frac{d}{t}$$
$$= \frac{150 \text{ m}}{10 \text{ s}}$$
$$= 15 \text{ m/s}$$

Our result is halfway between the initial velocity of 10 m/s and the final velocity of 20 m/s. So the average velocity turns out to be the *average of* the velocities. Although the car did not go as far in the first 5.0 s as it would have gone at the average velocity (the distance it did *not* go being represented by the blank rectangle below the v_{ave} line in Figure 5-7), this distance is exactly compensated for by the extra distance that the car did go in the next 5.0 s. This extra distance is represented by the shaded rectangle above the v_{ave} line. The shaded and blank rectangles and the distances they represent, are exactly equal, since we chose equal time intervals. The v_{ave} line is therefore halfway between v_i and v_f.

The preceding example involving two constant velocities is unrealistic, because it is impossible for a car to jump from one velocity to another with a short but infinite acceleration. The same reasoning, however, can be applied, with minor modification, to a constant acceleration situation. Constant acceleration would be represented by a straight line from initial velocity v_i at initial time t_i to final velocity v_f at final time t_f. The distance traveled, represented by the whole shaded area, would be the same as if the car had traveled for the whole time at average velocity v_{av}. Since they have parallel sides, the shaded triangle above the v_{ave} line is similar to the blank triangle below the line. Moreover, these triangles are congruent, and so have equal areas, providing that v_{ave} is halfway between v_i and v_f, since that would give the triangles one equal side. The distance not traveled in the first half of the time is exactly made up for by the extra distance traveled in the second half.

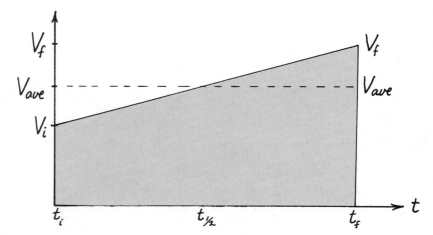

FIGURE 5-8 *If the acceleration is constant, the average velocity is halfway between the initial and final velocities.*

This discussion assumes constant acceleration and will not work otherwise. Average velocity is not the *average of* initial and final velocities unless the area above the average velocity line is equal to the area below. We can, however, use constant acceleration to illustrate physical principles in this course, and the arguments can be extended to nonconstant acceleration by simply breaking the time interval into enough equal chunks so that the acceleration looks constant for each little interval of time. The average velocity is then the average of instantaneous velocities at the end of each little time chunk.[2]

FIGURE 5-9 *Some people think that acceleration is fun.*

CHECK QUESTION

An automobile enters a freeway on-ramp at 50 ft/s and accelerates uniformly up to 80 ft/s in a time of 10 s. (a) What is the automobile's average velocity? (b) How far does the automobile travel in this amount of time? (c) What is the acceleration?

Answers: (a) 65 ft/s, (b) 650 ft, (c) 3.0 ft/s²

▶ Motion Under Constant Acceleration

We now have enough theory to solve any problem involving constant acceleration. We can find time, distance, or average velocity, assuming we know the other two.

$$v_{\text{ave}} \equiv \frac{d}{t} \tag{1}$$

If we don't know the average velocity, we can find it from the initial and final velocities:

$$v_{\text{ave}} = \frac{v_f + v_i}{2} \tag{2}$$

As long as the acceleration is constant, we have seen that the average velocity is the same as the average of the velocities. We can therefore combine definition (1) and equation (2) to obtain distance and time directly in terms of the initial and final velocities:

$$\frac{v_f + v_i}{2} = \frac{d}{t} \tag{1 \& 2}$$

We also have another handle on the initial and final velocities. We can find one or the other from the definition of acceleration:

$$a \equiv \frac{v_f - v_i}{t} \tag{3}$$

---

[2] This process done with infinitesimally small chunks is the key to integral calculus. It is, in effect, the opposite of taking the derivative.

It now appears that we have two equations in five unknowns. That might seem pretty bad, but there is a ray of hope. We can always reduce the problem to one equation in four unknowns. This turns out to be better than it sounds, because it means that we can solve for either the distance or the final velocity in terms of the initial conditions. If we want to know how far something will travel under constant acceleration, for example, we could find the final velocity from the definition of acceleration, which is equation (3), and use it to find distance from the combined equation (1) & (2). Problems of this nature are so common, however, that many people prefer to work the problem out in general and remember the result. The same is true, as well, for the general solution of final velocity in terms of initial velocity, acceleration, and distance.

The algebra necessary to get these two general solutions looks a little simpler if we express the combined equations (1) & (2) as well as the definition (3) in terms of the sum and difference of the initial and final velocity:

$$v_f + v_i = \frac{2d}{t} \qquad \text{(1) & (2)}$$

$$v_f - v_i = at \qquad \text{(3)}$$

It is now clear that we can eliminate the final velocity v_f by subtracting these two expressions. (We could instead get rid of the initial velocity v_i by adding them, but most problems are given in terms of initial conditions.)

$$0 - 2v_i = at - \frac{2d}{t}$$

This relationship can be expressed in terms of distance d. Multiply both sides by $t/2$ and rearrange:

$$\boxed{d = \tfrac{1}{2}at^2 + v_i t}$$

This formula is worth remembering. Before going on to derive the other formula, let us consider a couple of special cases as examples. If acceleration happens to be equal to zero, this formula reduces to rate × time.

$$d = 0 + v_i t$$

This is the result we would get directly from the definition of velocity. If initial velocity, on the other hand, is equal to zero, the formula tells us how far something will move from a position at rest:

$$d = \tfrac{1}{2}at^2 + 0$$

We can see, for example, that an object falling with the acceleration due to gravity—at 32 ft/s²—will fall only 16 ft in the first second of free fall:

$$d = \tfrac{1}{2}(32 \text{ ft/s}^2)(1\text{s})^2 = 16 \text{ ft}$$

We could have figured out the same thing without the formula by using the idea of an average velocity. At the instant the object starts to fall, it is going at 0 ft/s; after 1 s of free fall, it has accelerated up to an instantaneous velocity of 32 ft/s. The average of 0 and 32 ft/s is 16 ft/s. After having traveled 1 s with this average velocity, the object will therefore have fallen only 16 ft.

This result, by the way, comes as a surprise to some people. They think that an object falling at 32 ft/s² should fall 32 ft in the first second. This mistake results from a failure to distinguish between velocity and acceleration. An object in free fall does not fall at constant velocity, it falls at constant acceleration.

We have just seen how the formula for distance under constant velocity can be applied to situations in which either the acceleration or initial velocity is zero. We have also seen how we could get the same result using the idea of an average velocity instead of the formula. This should come as no surprise, since that is how we got the formula in the first place. It is only a general solution of the problem as to the distance traveled under a constant velocity. The only reason for remembering it is to save time. If you are going to use it, however, you should know where it came from and be sure you know how to solve the same problem with the idea of an average velocity.

The other formula worthy of being remembered is the general solution for the final velocity in terms of the initial conditions. This formula may be found by eliminating time, t, between equations (1) & (2) and (3). The easy way to do that is to multiply them together in the form in which they were last expressed on page 159. (in terms of the sum and difference of the velocities). This eliminates time, t, since it is in the numerator of the right-hand side of one and the denominator of the right-hand side of the other:

$$(v_f + v_i)(v_f - v_i) = \left(\frac{2d}{t}\right)(at)$$

The product on the left-hand side of this expression is the difference of the squares of the velocities, since, as you should remember from algebra, $(a + b)(a - b) = a^2 - b^2$ (*see* Appendix I).

$$v_f{}^2 - v_i{}^2 = 2ad$$

Subtracting the initial velocity squared from both sides gives the general solution of the final velocity, or at least its square, in terms of the initial velocity, acceleration, and the distance:

$$\boxed{v_f{}^2 = v_i{}^2 + 2ad}$$

The square of the final velocity is therefore equal to the square of the initial velocity plus two times the acceleration times the distance. This formula is also worthy of remembering because it will save a great deal of time in solving problems.

As an example using the second formula, suppose we want to know how fast something will be moving after falling 20 ft in free fall. Assuming it falls from rest, so that $v_i = 0$,

$$v_f{}^2 = 0 + 2(32 \text{ ft/s}^2)(20 \text{ ft})$$
$$= 1{,}280 \text{ ft}^2/\text{s}^2$$

Or taking the square root of both sides,

$$v_f = \sqrt{1{,}280 \text{ ft}^2/\text{s}^2}$$
$$= 11.3 \text{ ft/s}$$

CHECK QUESTIONS

1. An automobile enters a freeway on-ramp with an initial velocity of 50 ft/s and accelerates at 3.0 ft/s² for 10 s. How far does the automobile travel in this amount of time?

    Answer: 650 ft

2. A sports car accelerates at 7.0 ft/s² for a distance of 150 ft, starting with an initial velocity of 20 ft/s. What is the sports car's final velocity at the end of this distance?

    Answer: 50 ft/s

▶ Acceleration Due to Gravity

The acceleration of an object in free fall has been used several times in this chapter as an example, because this type of motion is very common. Surprisingly, it was discovered for the first time by Galileo in the seventeenth century. Before Galileo, scholars believed that light objects fall more slowly than heavy objects. Aristotle taught that an object ten times as heavy as another object would fall ten times as fast. His authority was so strong that nobody ever bothered to experiment and see if he was right until Galileo's day. It is generally believed that Galileo disproved Aristotle's theory by dropping a light and a heavy stone at the same time from the tower of Pisa, Italy. No one knows if he actually did this experiment, but he did write a book in which two characters did and then argued afterward about the meaning of the results. The story pointed out that experience is more valuable than relying on authority.

You have enough experience with falling objects to know that an object in free fall does not start falling suddenly at a constant velocity. You immediately sense the exaggeration in a cartoon character who walks off a ledge, looks around, and then suddenly plummets downward. Your sense of reality tells you that he should not acquire speed suddenly. He should speed up gradually.

You would probably also agree that a coin dropped from your hand picks up more and more speed as it falls. You know you hit the ground harder when you fall further because you are going faster. Many other examples of ever-increasing velocity of falling objects are common to our experience. There still remains a lingering belief in some people's minds, however, that everything falls with the same constant speed. Some authority from out of their past seems to tell

FIGURE 5-10 *A cartoon may exaggerate reality by showing a change in velocity as nearly instantaneous.*

them that all falling objects go at 32 ft/s. Again we should point out that this mistaken belief usually results from a failure to distinguish between velocity and acceleration.

There is, however, a grain of truth in the belief in a constant velocity for freely falling objects. Objects in free fall do stop accelerating as they reach a *terminal* velocity when the force of air friction balances their weight. This constant velocity is, however, far from being the same constant for all objects or even for the same object under different circumstances. A piece of paper floating downward at one terminal velocity can be made to fall nearly as fast as a book by crumpling it into a ball. But this we will discuss at the end of Chapter 6.

For now, we should note that objects do fall with the same acceleration as long as air friction is negligible. If two objects of different weights are dropped at the same time in a vacuum, their velocities will constantly increase together. At the surface of the earth the rate of change of velocity of a freely falling object is

$$g = 32 \text{ ft/s}^2 = 9.8 \text{ m/s}^2$$

The letter g is used to denote this particular value of acceleration.

The value of g has been experimentally measured. Various objects have been dropped and their velocities measured after certain time intervals. The value of g is found to vary (in the third significant figure) at different places on the earth's surface, but it is always found to be the same at any one place for any two different objects. (The reason is based on an exact proportionality between mass and weight, which will also be discussed in Chapter 6.)

Once the value of g is determined, it may be used as a perfectly good unit of acceleration. A sports car accelerating at 8.0 ft/s² might be said to have an acceleration of 0.25 g:

$$a = 8.0 \text{ ft/s}^2 \left(\frac{1\,g}{32 \text{ ft/s}^2} \right) = 0.25\,g$$

That is, the car is accelerating at one-quarter the rate that it would if it were dropped. An airplane pilot can take 10 g of acceleration if she is wearing a special tight-fitting g suit to keep her blood from rushing from her head.

Figure 5-11 is a full-scale diagram showing the position of a falling penny at intervals of 0.01 s after it is released from rest. This drawing, or a copy of it, can be used to test someone's reaction time. Hold the drawing vertically upside down and have the other person hold his fingers at the zero-second mark. Drop the drawing between his fingers and let him catch the first image he can. A good reaction time is close to 0.15 s. A fair reaction time is closer to 0.20 s.

Some people cut out the picture to perform this experiment. Other people use the whole page. Still others like to drop the whole book—for some reason or other.

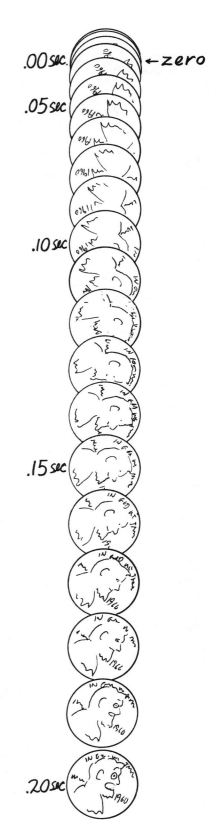

FIGURE 5-11 *A full-scale drawing of a coin at equal time intervals after falling may be used to test reaction time.*

PROBLEM SET 5

The problems in this chapter are concerned with motion under constant velocity or under constant acceleration in a straight line. The definitions of **velocity** and **acceleration** are

$$v \equiv \frac{\Delta d}{\Delta t} = \frac{d_f - d_i}{t_f - t_i}$$

$$a \equiv \frac{\Delta v}{\Delta t} = \frac{v_f - v_i}{t_f - t_i}$$

If the time interval Δt is very small, an instant of time, the preceding definitions refer to instantaneous velocity and instantaneous acceleration. Otherwise they refer to average values of velocity and acceleration. (A time interval is short enough to be considered an instant if it is small compared to time over which velocity or acceleration fluctuates or changes.)

If velocity is changing, but at a constant rate of acceleration, the average velocity v_{ave} turns out to be the average of the initial and final velocities over a given time interval:

$$v_{ave} = \frac{v_i + v_f}{2}$$

This average velocity may be used to find the distance traveled under constant acceleration:

$$d = \tfrac{1}{2}at^2 + v_i t$$

The final velocity may also be found as a function of acceleration and the distance over which the acceleration was constant:

$$v_f{}^2 = v_i{}^2 + 2ad$$

One common constant acceleration situation is a body in free fall. The acceleration due to gravity g is a constant near the surface of the earth:

$$g = 9.8 \text{ m/s}^2 = 32 \text{ ft/s}^2$$

This value is only constant to two significant figures for a few kilometers from the earth's surface and in the absence of air friction. Under those conditions the distance traveled in free fall is

$$d = \tfrac{1}{2}gt^2 + v_i t$$

assuming the positive direction to be downward.

5-1   In the story of the race between a rabbit and a turtle, little is said of the officials who covered the start and finish lines. Actually, Mother Nature started the race with a flash of lightning near the starting line, and a bear watched the finish. The bear determined the distance from the start to the finish by counting the time between the lightning and the thunder, using the speed of sound (350 m/s).

Given: The rabbit covers the first mile (1.61 km) in 4.00 min. The bear, who noticed a time lapse of 8.30 s between lightning and thunder, watches for the rabbit.

Find: (a) What is the velocity of the rabbit in meters per second?
     (b) How far is it from the start to the finish line?
     (c) How long will it take the rabbit from start to finish (no stops)?

Your Solution

Discussion

People frequently say that you can estimate the distance between yourself and a flash of lightning by counting the seconds between a flash of lightning and the thunder. This is true, because it takes almost no time for the light to get to you, whereas the sound travels relatively slowly. This problem gives you a chance to measure distance using that technique and to use that distance to predict the time it will take for a rabbit to run it at an even slower constant velocity.

Sample Solution 5-1

(a) $v \equiv \dfrac{\Delta d}{\Delta t}$ conversion factors

$= \dfrac{1.61\ km}{4.00\ min}\left(\dfrac{1\ min}{60s}\right)\left(\dfrac{10^3\ m}{1\ km}\right)$

$= 6.708\ meter/s \cong \boxed{6.71\ m/s}$

(b) $d = vt$

$= (350\ m/s)(8.30s)$

$= 2{,}905\ meter \cong \boxed{2.90\ km}$

(c) $t = \dfrac{d}{v}$

$= \dfrac{2{,}905\ m}{6.708\ m/s}$

$= 433s\ \left(\dfrac{1\ min}{60s}\right)$

$= \boxed{7.22\ min}$

Discussion

The velocity of the rabbit is found directly from the definition of velocity. A three-line equality sign may be used to mean "is defined as." Velocity is more than equal to change of distance over time, it is *defined* as change of distance over change in time.

The distance and time in parts (b) and (c) are found by solving the definition stated in part (a) for the appropriate variable.

While answers should only be reported to three significant figures, the limit of accuracy of the data, four digits may be retained in intermediate results to avoid a round-off error. Otherwise you shouldn't be overly concerned if your answer disagrees in the last significant figure with that in the book.

5-2 A 300-lb box falls from a flat-bed truck into the path of your car. You are wondering how far it will slide before it comes to rest.

Given: The initial velocity of the box is 5.0 m/s the instant it falls. It slides to a stop in 3.0 s with some constant negative acceleration.

Find: (a) What is the acceleration of the box?
(b) What is the average velocity of the box?
(c) How far does the box slide?

Your Solution

Discussion

This problem is an example of motion under constant acceleration. The box is assumed to decelerate at a constant rate until it stops. Please note the distance traveled by the box is not as great as if it continued moving at a constant velocity.

Sample Solution 5-2

(a) $a \equiv \dfrac{\Delta V}{\Delta t} = \dfrac{V_f - V_i}{t}$

$= \dfrac{(0\,m/s) - (5.0\,m/s)}{(3.0s)} = \boxed{-1.7\,m/s^2}$

(b) $V_{ave} = \dfrac{V_f + V_i}{2}$

$= \dfrac{(0\,m/s) + (5.0\,m/s)}{2} = \boxed{2.5\,m/s}$

(c) $d = V_{ave}\,t$

$= (2.5\,m/s)(3s) = \boxed{7.5\,m}$

or $\begin{cases} d = \frac{1}{2}\,at^2 + V_i\,t \\ = \frac{1}{2}(-1.7\,m/s^2)(3.s)^2 + (5.0\,m/s)(3.0s) \end{cases}$

Discussion

The acceleration can be found from the definition of acceleration, since the change in velocity and the time are given. The negative result means that the acceleration is in the opposite direction from that in which the box is traveling.

Average velocity is the same thing as instantaneous velocity, as long as the acceleration is constant. Distance can either be found from this average velocity or from the equation giving distance traveled under constant acceleration.

5-3 One measure of a sports car's acceleration is the time necessary for it to get from 0 to 60 mph. Another common measure is the amount of time necessary for it to go a ¼ mile, starting from rest.

Given: A Triumph TR-3 in good tune can accelerate from 0 to 26.8 m/s in 11.0 s. (Assume that the acceleration remains constant.)

Find: (a) What is the acceleration in meters per second squared?
(b) What is the distance traveled in 11.0 seconds?
(c) What is the time necessary to travel 400 m (¼ mile)?

Your Solution

Discussion

This problem is another example of motion under constant acceleration. The assumption that the sports car accelerates at a constant rate is really only a first-order approximation, however, since much greater acceleration is available in lower gears. Vintage sports cars just don't come with fluid-o-matic transmissions.

Sample Solution 5-3

(a) $a \equiv \dfrac{\Delta v}{\Delta t} = \dfrac{v_f - v_i}{t}$

$$= \dfrac{26.8 \text{ m/sec} - 0 \text{ m/sec}}{11.0 \text{ sec}}$$

$$= \boxed{2.44 \text{ m/sec}^2}$$

(b) $d = v_{ave}\, t$

$$= \left(\dfrac{v_f + v_i}{2}\right) t$$

$$= \left(\dfrac{26.8 \text{ m/sec} + 0}{2}\right)(11.0 \text{ sec}) = \boxed{147\,m}$$

$$\left(\text{or } d = \tfrac{1}{2}\, at^2 + v_i t = \tfrac{1}{2}\,(2.44 \text{ m/sec}^2)(11 \text{sec})^2 + 0\right)$$

(c) from $d = \tfrac{1}{2}\, at^2$

$$t = \sqrt{\dfrac{2d}{a}}$$

$$t = \sqrt{\dfrac{2\,(400 \text{ m})}{(2.44 \text{ m/sec}^2)}} = \sqrt{327 \text{ sec}^2} = \boxed{18.1 \text{ sec}}$$

Discussion

The distance in part (b) can be found either from the average velocity or from the equation for distance traveled under constant acceleration.

The time in part (c) can be found by solving this equation for time. The same average velocity cannot be used, because the final velocity turns out to be 44 m/s or about 90 mph, which is not the same as in part (b).

5-4 Some primitive societies were known to throw young maidens into "bottomless" pits. This fact is sometimes used as an example of patriarchy in primitive societies.

Given: One lovely young maiden hits the bottom of a bottomless pit after 3.5 s in free fall.

Find: (a) How deep is the bottomless pit?
(b) What is the velocity of the lovely young maiden as she strikes the bottom.

Your Solution

Discussion

This is an example of motion under constant acceleration in that special case where the acceleration is the acceleration due to gravity. It is useful, by the way, to know the value of the acceleration due to gravity in both English and metric units (9.8 m/s^2 = 32 ft/s^2).

All of the well-known bottomless pits used by primitive societies have been found to actually have bottoms. Otherwise, they might not have been so bad.

Sample Solution 5-4

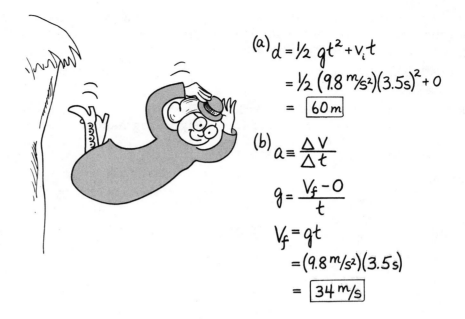

(a) $d = \frac{1}{2}gt^2 + v_i t$

$= \frac{1}{2}(9.8\,^m/s^2)(3.5s)^2 + 0$

$= \boxed{60\,m}$

(b) $a \equiv \dfrac{\Delta V}{\Delta t}$

$g = \dfrac{V_f - 0}{t}$

$V_f = gt$

$= (9.8\,^m/s^2)(3.5s)$

$= \boxed{34\,^m/s}$

Discussion

The distance is found from the formula for distance under constant acceleration, where the acceleration is g and the initial velocity is zero. The final velocity can be found from the definition of acceleration, since we know the time.

5-5 A foolish archer is one who shoots an arrow directly upward and then waits for the arrow to return without moving. Neglecting air resistance, the arrow will return to earth with the same velocity as that with which it left the bow.

Given: An arrow is shot directly upward with an initial velocity of 50 m/s.

Find: (a) How long does the arrow travel upward before it stops?
(b) How high does it go?
(c) When will it return?

Your Solution

Discussion

This is another example of motion under constant acceleration. The initial and final velocity ($v_f = 0$) are given, and you are asked to find the time. From the time it takes to go up and stop and the initial velocity, you should be able to find how high the arrow goes before returning. You should also be able to figure out how long it takes the arrow to make the total trip, going up and coming back down.

Sample Solution 5-5

(a) $a = \dfrac{V_f - V_i}{t}$

$t = \dfrac{V_f - V_i}{a} = \dfrac{0 - (V_\uparrow)}{(-g)}$

$= \dfrac{-(50\,m/s)}{(-9.8\,m/s^2)}$

$= \boxed{5.1\,s}$

(b) $V_{ave} = \dfrac{d}{t}$

$d = (V_{ave})(t)$

$= \left(\dfrac{V_f + V_i}{2}\right)(t)$

$= \left(\dfrac{0 + 50\,m/s}{2}\right)(5.1\,s) = \boxed{128\,m}$

(c) $V_{ave} = \dfrac{d}{t}$

$t = \dfrac{d}{V_{ave}} = \dfrac{d}{\frac{1}{2}(V_i + V_f)}$

$t_\downarrow = t_\uparrow$ since $(V_i + V_f) =$ same thing in both up and down cases

$t_{total} = t_\uparrow + t_\downarrow$

$= 2\,t_\uparrow = 2(5.1\,s) = \boxed{10.2\,s}$

Discussion

The time necessary to reach the top of the arc can be found from the acceleration and the change of velocity. The acceleration is just the acceleration due to gravity. The negative sign for acceleration, $-g$, is used, because the acceleration is in the direction opposite to that of the initial velocity. It is a deceleration. While the arrow is going up, it is falling to a stop.

The distance in part (b) can either be found from the average velocity or from the formula for velocity under constant acceleration.

$$(v_f)^2 = (v_i)^2 + 2ad$$

You might find it interesting to solve this equation for d and try it.

The total time can be found in part (c) by considering that the time necessary for it to fall up is equal to the time necessary for it to fall the same distance down. An alternative approach would be to set the distance equal to zero in the equation for distance traveled under constant acceleration. This approach is discussed in Chapter 7.

5-6 A professional baseball player develops the ability to judge quickly the velocity of the ball in both magnitude and direction, so that she can predict its future behavior.

Given: One-quarter second after a baseball is struck, an outfielder starts running to where the ball will land, having noticed that the ball moved forward a distance of 6 m and upward 8 m in this first moment of its flight.

Find: (a) What is the magnitude of the ball's velocity?
(b) What angle does the ball's velocity make with the horizontal?

Your Solution

Discussion

Velocity is a vector quantity, having both magnitude and direction. This problem is an exercise in finding both parts of the vector. You can assume that in such a short period of time, the ball will travel in essentially a straight line. For our present purposes, we will assume that it is deflected from a straight line only a negligible amount by gravity.

Sample Solution 5-6

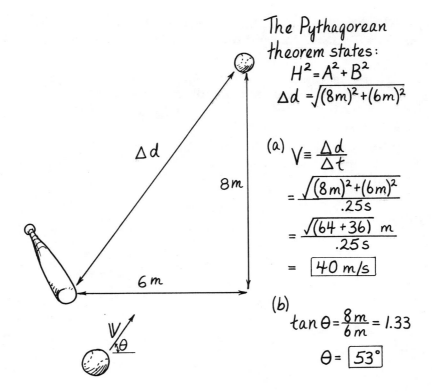

The Pythagorean theorem states:

$$H^2 = A^2 + B^2$$

$$\Delta d = \sqrt{(8m)^2 + (6m)^2}$$

(a)
$$V \equiv \frac{\Delta d}{\Delta t}$$

$$= \frac{\sqrt{(8m)^2 + (6m)^2}}{.25s}$$

$$= \frac{\sqrt{(64 + 36)} \; m}{.25s}$$

$$= \boxed{40 \; m/s}$$

(b)
$$\tan \theta = \frac{8m}{6m} = 1.33$$

$$\theta = \boxed{53°}$$

Discussion

The distance traveled is found from the Pythagorean theorem, and the time is given. The magnitude of the velocity can therefore be found from the definition. The angle can be found from trigonometry, assuming that the path of the ball is essentially a straight line making the hypotenuse of a right triangle. Since the sides of this triangle are given, the angle is found from the tangent.

5-7 A boat going directly across the Mississippi River must head somewhat upstream to avoid being taken downstream by the water current.

Given: A boat is traveling at 3.0 m/s relative to the water. The water is flowing downstream at 1.0 m/s. The river is 300 m wide.

Find: (a) What is the magnitude of the resultant velocity if the boat heads upstream to compensate for the current?

(b) At what angle must the boat head to compensate for the current if the boat captain wants to get to a point directly across the river from where he started?

(c) How far downstream will the boat land if the boat captain doesn't compensate?

Your Solution

Discussion

The vector nature of velocity becomes important when you add two velocities together. Problems of this nature happen when you have something moving in a medium that is itself moving. This situation is common in navigation.

Sample Solution 5-7

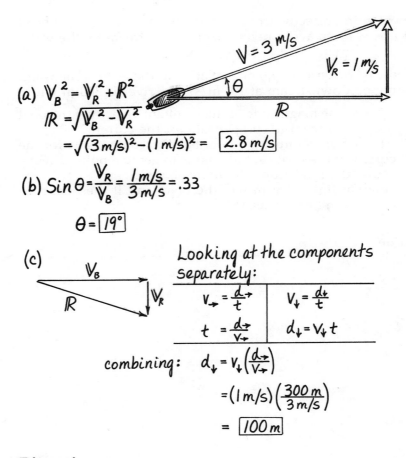

(a) $V_B^2 = V_R^2 + R^2$

$R = \sqrt{V_B^2 - V_R^2}$

$= \sqrt{(3 \, m/s)^2 - (1 \, m/s)^2} = \boxed{2.8 \, m/s}$

(b) $\sin \theta = \dfrac{V_R}{V_B} = \dfrac{1 \, m/s}{3 \, m/s} = .33$

$\theta = \boxed{19°}$

(c)

Looking at the components separately:

$V_\rightarrow = \dfrac{d_\rightarrow}{t}$	$V_\downarrow = \dfrac{d_\downarrow}{t}$
$t = \dfrac{d_\rightarrow}{V_\rightarrow}$	$d_\downarrow = V_\downarrow t$

combining: $d_\downarrow = V_\downarrow \left(\dfrac{d_\rightarrow}{V_\rightarrow} \right)$

$= (1 \, m/s) \left(\dfrac{300 \, m}{3 \, m/s} \right)$

$= \boxed{100 \, m}$

Discussion

This problem differs from some vector addition problems in that the resultant is one leg of a right triangle, instead of the hypotenuse, in parts (a) and (b). You know the direction you want the resultant to be in, and you can find its magnitude by solving the Pythagorean theorem for one of the sides. The direction you need to point the boat to produce this resultant can be found from the sine, since you know the hypotenuse and the side opposite the angle.

Part (c) assumes that the given velocities are the sides of a right triangle in velocity space and that the resultant is along the hypotenuse. Since the boat is traveling downstream at the same time that it is traveling across the river, the time necessary to find the distance downstream can be found from those components of its velocity and distance across stream.

5-8 An ocean liner going from San Francisco to Tokyo stops for a few days in Honolulu. You are on board and wish to estimate the time it will take you to get to Tokyo.

Given: It is 3,700 km from San Francisco to Honolulu and 6,200 km from Honolulu to Tokyo. It took you 5.1 days to reach Honolulu from San Francisco.

Find: (a) What is the average velocity of the ocean liner in km/hr?
(b) How long will it take to reach Tokyo from Honolulu?

5-9 The deceleration (negative acceleration) of an automobile in a full skid on dry pavement is a little greater than one-half "*g*," or half the acceleration you would experience in free fall. Suppose you are going down the highway and you suddenly decide you need to stop your car.

Given: Your maximum acceleration is -5.5 m/s² (assumed constant) and you are traveling at 26.7 m/s.

Find: (a) What is the minimum time it will take to stop?
(b) What will your average velocity be while stopping?
(c) How much distance will it take you to stop?

5-10 The brakes on a parked car fail and it accelerates at a constant rate, rolling into a lamp post down the hill.

Given: The car reaches a velocity of 4.3 m/s by the time it smashes into the lamp post. The time between the brakes failing and the smashing of the car is 3.7 s.

Find: (a) What is the car's acceleration?
(b) What was the average velocity of the car while it was rolling?
(c) How far did the car roll before the collision?

5-11 Windows are placed at the top of the Washington Monument to give tourists a view of their nation's capital and a gasp of fresh air. The Park Department has had to put guard screens over these windows to keep people from throwing objects out and injuring other tourists below.

Given: The Washington Monument is 167 m high. A tourist manages to squeeze a nickel through the guard screen to see if it will dent the sidewalk below.

Find: (a) How much time will the nickel take to fall?
(b) How fast will it be traveling when it strikes?

5-12 A juggler tosses a ball upward and waits for its return. During the first half of the time that the ball is in the air, it is falling to a stop. After reaching its highest point, the ball returns to the juggler's hand.

Given: The initial upward velocity of the ball is 15 m/s.

Find: (a) How long does it take the ball to reach its highest point?
(b) How high does it go?
(c) What is the ball's downward velocity upon hitting the juggler's hand after it falls back from this highest point?

5-13   An orangutan accidentally fires a rifle at a hunter in a tree some distance away. Fortunately, the hunter sees the bullet coming and jumps out of the tree.

Given: The muzzle velocity of the bullet is 600 m/s at an angle of 40.0° above the horizontal. The hunter notes the position of the bullet 0.100 s after it is fired.

Find: (a) What are the horizontal and vertical components of the muzzle velocity?
(b) How far has the bullet traveled in the horizontal and vertical directions (to the proper number of significant figures) by the time the hunter notes its position?

5-14   Crewmen aboard the USS *Enterprise* are playing shuffleboard on the flight deck. While the aircraft carrier steams north, a crewman makes a starboard shot (toward the east).

Given: While the *Enterprise* is traveling north at 13.0 m/s, the shuffleboard puck is traveling east, relative to the deck, at a velocity, on the average, of 7.0 m/s.

Find: (a) What is the magnitude of the resultant velocity relative to the water (assumed stationary)?
(b) What is the direction of the resultant velocity measured from the north?
(c) How far north does the puck travel while traveling 7.0 m east?

MORE INTERESTING PROBLEMS

5-15   A man jumps off a bridge to see how high it is. He measures his time of descent.

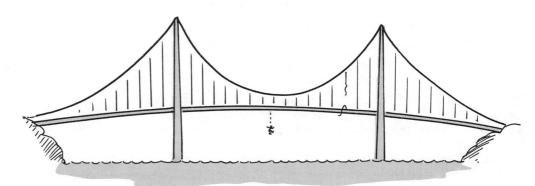

Given: He is in the air for 3.0 s before splashing into the water below.

Find: (a) How high is the bridge above the water?
(b) How fast is the man going as he hits the water?

5-16 A freight train requires a great deal of time and distance to stop. The engineer sees an amber signal and applies the brakes, stopping the train at a constant, but small, deceleration.

Given: After noting that the train has traveled 1.6 km in 64 s, starting from an initial velocity of 30 m/s, the engineer sees the reason for the warning signal. She wishes to predict how much further the train will travel before stopping.

Find: (a) What is the average velocity between the time the engineer first applies the brakes and the time she sees the reason for the warning signal?
(b) What is the deceleration (negative acceleration) of the train?
(c) How much will the train travel after the engineer sees the reason for the warning?
Hint: Parts (b) and (c) might be easier to visualize if you make a velocity-time graph.

5-17 A rock takes longer to fall from a certain height above the surface of the moon than it does to fall from the same distance above the surface of the earth.

Given: The acceleration due to gravity on the moon is one-sixth of that on the earth. It requires a rock 2.5 s to fall from a certain height above the surface of the earth.

Find: (a) How long would it require the same rock to fall this distance on the moon?
(b) How high would that be above the surface of the moon?

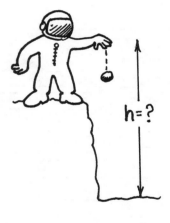

5-18 A woman gets on an elevator with an armload of packages. As the elevator accelerates upward, she drops a package.

Given: The package is initially 1.2 m from the floor of the elevator. The elevator accelerates upward at a constant rate of 3.0 m/s^2 while the package is falling. (Assume the initial velocity of both the package and elevator start out at zero.)

Find: (a) What is the time required for the package to hit the floor? (The floor and the package actually come together.)
(b) How fast is the package going relative to the floor when they meet?
(c) How fast is the floor going relative to the earth at the meeting point?

5-19 An ion propulsion rocket would drive a spacecraft forward by ejecting electrically charged particles out the back. Such a method of propulsion would not give large accelerations under existing technology, but the little acceleration produced would last for long periods of time.

Given: An ion propulsion unit accelerates a spaceship at 0.6 cm/s². The distance from the earth to the moon is 390,000 km.

Find: (a) How long would it take to go a distance equal to that from the earth to the moon, starting from rest and traveling at a constant acceleration?

(b) How fast would you be going when you got there?

5-20   Relativistic effects do not begin to become noticeable below about one-tenth the velocity of light. Above that velocity, men on a spacecraft will experience a time dilation that will shorten their journey. The rules we have learned apply only below that velocity.

Given: A spacecraft accelerates up to 3×10^7 m/s (which is about a tenth the velocity of light) with a constant acceleration of 9.8 m/s².

Find: (a) How many years will it take to reach this velocity?

(b) How far will the spacecraft have traveled in the meanwhile?

5-21   Drivers' manuals published by the highway departments of most states point out the dangers of fast driving by showing, usually graphically, the disproportionately large stopping distances for high velocities.

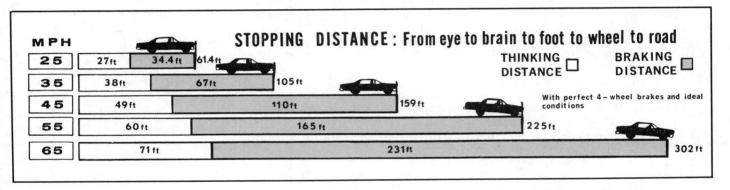

Given: The 1976 *California Driver's Handbook* shows the above chart.

Find: (a) What is the reaction time associated with the "thinking distance"?

(b) What is the deceleration associated with the "braking distance" used in calculating these stopping distances?

5-22   BART trains (Bay Area Rapid Transit—modern mass transportation in the San Francisco Bay area) are computer-controlled to accelerate up to a given velocity, determined by a signal received from a particular track section.

Given: A BART train accelerates at the rate of 4.8 (km/hr)/s.

Find: (a) What is the acceleration in m/s²?

(b) How long will it take for a BART train to move 50 m in pulling out of the station?

(c) What will be the average velocity over the first 50 m?

5-23 A blunderbuss is an obsolete short firearm with a flaring muzzle for use at close quarters. It was usually loaded with a number of balls, like a sawed-off shotgun.

Given: A blunderbuss ball is fired directly upward from the ground and returns to its starting point in 8.8 s.

Find: (a) How much time did the ball take to reach its maximum height?
(b) What was the muzzle velocity of the ball?
(c) How high does the ball go?

ANSWERS *(The answers to parts (b) and (c) may depend on how you round off in previous parts. Your best bet is to keep one extra digit in an intermediate result to avoid a round-off error in the final answer.)*

5-8   (a) 30.2 km/hr $\cong$ 30 km/hr; (b) 205 hr $\cong$ 8.5 days.
5-9   (a) 4.85 s $\cong$ 4.9 s; (b) 13.3 m/s (or 13.4 m/s) $\cong$ 13 m/s;
(c) 64.8 m $\cong$ 65 m.
5-10 (a) 1.16 m/s²; (b) 2.15 m/s $\cong$ 2.2 m/s; (c) 8.0 m.
5-11 (a) 5.84 s; (b) 57.2 m/s.
5-12 (a) 1.5 s; (b) 11 m; (c) -15 m/s.
5-13 (a) 460 m/s horizontal, 386 m/s $\cong$ 390 m/s vertical; (b) 46.0 m horizontal, 38.5 m vertical.
5-14 (a) 14.8 $\cong$ 15 m/s; (b) 28° east of north; (c) 13 m.
5-15 (a) 44 m; (b) 29 m/s.
5-16 (a) 25 m/s; (b) 0.156 m/s² $\cong$ 0.16 m/s²; (c) 1.28 km (or 1.25 km).
5-17 (a) 6.1 s; (b) 31 m.
5-18 (a) 0.43 s; (b) 5.5 m/s; (c) 1.3 m/s.
5-19 (a) 3.6 × 10⁵ s; (b) 2.2 × 10³ m/s.
5-20 (a) 3.1 × 10⁶ s $\cong$ 0.1 yr; (b) 4.6 × 10¹⁰ km.
5-21 (a) 0.74 s; (b) 20 ft/s² = 0.6 *g*.
5-22 (a) 1.33 m/s²; (b) 8.66 s or 8.67 s; (c) 5.8 m/s.
5-23 (a) 4.4 s; (b) 43 m/s; (c) 95 m.

▶ The First and Second Laws of Motion

In a science fiction movie of the early 1950s, *Destination Moon,* a crew of six climbed aboard a single-stage rocket ship and shot off into outer space. Once the large initial acceleration was over, all the crew put on their space suits and magnetic boots to go out and clomp around on the outside of the space ship.

The audience gasped as one member of the crew seemed to get too close to the rocket engines. The script had emphasized the fact that the ship was going thousands of miles per hour, and the audience just assumed that the engines had to be running to keep the ship going that fast. This audience had grown up reading Flash Gordon, who always had fire coming out of his rockets as he zipped through space. As the crew member in this new movie leaned over and looked into one of the engines, however, *nothing happened.* The crew member was not blown away, because the engines were not in operation. The new movie was more sophisticated than Flash Gordon. An object in motion does not need a force to keep it in motion. The rocket ship was coasting toward the moon, and it did not need power.

This idea that force is not necessary to keep an object moving at a constant velocity has been with us since Galileo. Before him people had assumed that a force was necessary to keep an object in motion. They couldn't see the frictional forces, so they naturally assumed that the force they had to apply was the only force acting. They had a hard time, however, explaining the fact that a ball, once thrown, continues moving. They thought that the air parted in front of the ball and rushed around to push it from the rear. Galileo was the first to point out that if constant velocity was considered the natural state of things, complicated mechanisms required to maintain motion were unnecessary. An external force was only needed, he said, to balance the invisible forces of friction. If friction could be reduced to nothing, then there would be no need of an external force to keep an object moving.

FIGURE 6-1 *Galileo was thinking of a frictionless surface.*

His closest approximation to frictionless motion was a hard round object rolling on a very smooth surface. If he had been able to reduce friction still further, he would have found that his theory worked even better. An air track would have done beautifully. It is a hollow track with little air holes spaced along its surface. Air from the pressure side of a vacuum cleaner is forced through these holes, making a glider placed on the track float on a thin layer of air in nearly frictionless contact. Galileo would have loved to have owned an air track. His theories of motion might not have been improved, but he would have had an easier time convincing his contemporaries that his ideas were correct.

Galileo's contemporaries were not ready to accept his theories of motion, which were contrary to the theology of his day. The idea that an object in motion tends to remain in motion lent credence to the Copernican theory, which said that the same thing is going on in the heavens. The idea that the same sort of motion occurs in the heavens as on earth was just too much. The authorities showed Galileo the instruments of torture and asked that he take back his teachings. He was no fool. He took them back.

Culture is always evolving. Movie audiences in the early 1950s still believed that Flash Gordon's rocket ship on its way to the moon had to have its engines operating all the way to maintain its speed. Most school children today know better than that. They can tell you, from what they have seen on television, that a space ship will coast along under its own inertia until ground control orders a "burn."

A generation or so after Galileo, Newton said nearly the same thing and got away with it. People were willing to accept the heresy of Galileo as the gospel of Newton. With little change, therefore, Galileo's idea is now called Newton's first law of motion.

▶ Newton's First Law, The Law of Inertia

Every object in the universe has the property of resisting change in motion in the absence of a force acting upon it. This fact is called Newton's first law. Galileo said it first, but Newton improved it a little. He gave the name **inertia** to this property of resisting change in motion, and he defined what is meant by a change in motion. A change in the magnitude of velocity is clearly a change in motion, but so is a change in direction. As stated by Newton, the first law says, **An object at rest will, in the absence of a force, remain at rest, and an object in motion will remain in motion at constant velocity along a straight line.** Since we have already discussed velocity as a vector quantity, we can summarize Newton's law by saying that the velocity of an object will remain constant (in both magnitude and direction) as long as the force applied to it is zero. Since a change in velocity is defined as acceleration, we could also say that *if the force is zero, the acceleration will be zero.*

$$\text{If } \mathbf{F} = 0, \text{ then } \mathbf{a} = 0 \qquad \text{(Newton's first law)}$$

You can perform several experiments that will confirm the first law. One of the most dramatic is to jerk the tablecloth out from under a full service of china, crystal wine goblets, and dinnerware. This experiment works best when the plates are filled with food and the goblets with wine. The more stuff there is, the more inertia it has. Before the experiment, tell your guests, "An object at rest will remain at rest." If you perform the experiment just right, you can also demonstrate that objects in motion will remain in motion.

Like all physical concepts, the first law is only a model of reality. There is no real object in the universe for which the sum of all the forces acting on it is exactly zero. This condition is never exactly satisfied. There is therefore no real object in the universe for which the first law really works. It *nearly* works, however, when the force is very small. Also, the acceleration is hardly noticeable when the force acts only for a very short time. The smaller the force or the shorter the time, the better the first law works. We say that the first law holds true *in the limit where the force is zero,* since it works better the closer we get to that limit. The force acting on the place settings in the tablecloth trick, for example, is not really zero; but if the cloth is whipped out quickly enough, the small force exerted will not last long enough to disturb things—much.

Another dramatic experiment demonstrating inertia is done by putting an anvil on someone's stomach and striking it with a sledge hammer. This has been an effective, although dangerous, lecture demonstration. One excellent physics teacher successfully performed this demonstration in four out of five of his classes one semester, using student volunteers to strike the anvil. He found that some students are more accurate than others. Amazingly, he still performs this dramatic demonstration every semester.

FIGURE 6-2 *The anvil's inertia protects the instructor's stomach.*

▶ Newton's Second Law of Motion

Both the anvil and the tablecloth experiments involve unbalanced forces, but the effects of these forces are small because they are applied for such a short period of time that only a small change in motion is produced. The anvil rebounds a little from the sledge hammer blow, but its motion is small enough to be absorbed by even a skinny person's stomach. The wine goblets in the tablecloth trick do have some velocity after the tablecloth is whipped out from underneath them. For this reason, it is best to use a waxed table, so that the goblets slide gently to a stop without tipping over.

Thus, experience tells us that the converse of the first law of motion is also true. If the force is not zero, the acceleration is not zero:

$$\text{If } F \neq 0, \text{ then } a \neq 0$$

Further, the amount of acceleration will be proportional to, and in the same direction as, the force applied:

$$\mathbf{F} \propto \mathbf{A}$$

This proportionality is Newton's second law of motion. The proportionality depends upon something called the object's "inertia." An anvil has more inertia than a wine glass; it requires greater force to give it the same acceleration.

The second law can be written with an equality sign when using a proportionality constant, the value of the constant being a measure of the object's inertia.

$$\boxed{\mathbf{F} = m\mathbf{a}}$$

The proportionality constant m is the **mass** of the object.

Mass is the best measure of how much of an object you have. The same force will accelerate a particular object the same amount anywhere in the universe. In the old British system, things were measured by weight. When you bought 10 lb of sugar, 10 lb was the gravitational force between the earth and the bag of sugar. On the moon, the same bag of sugar would weigh one-sixth as much. If you buy 10 *kilograms* of sugar, however, it will be 10 kilograms anywhere in the universe, including the moon. This is one of the many advantages of the modern metric system.

Mass and acceleration are defined in the metric system in terms of fundamental quantities. The mass of an object is compared to the mass of a standard kilogram reference block kept in France. Acceleration is measured in meters and seconds, both of which are defined in terms of absolute standards.

Force in the metric system is defined in terms of mass and acceleration. One newton of force is that amount which will accelerate a 1-kg mass at the rate of 1 m/s². Thus, a newton is defined in terms of Newton's second law

$$\mathbf{F} = m\mathbf{a}$$

where the mass and acceleration are unity (1) as:

$$1 \text{ N} \equiv (1 \text{ kg})(1 \text{ m/s}^2)$$

This same force will accelerate the same object at the same rate anywhere in the universe.

The force referred to in Newton's second law is the **net,** or unbalanced, force. Many forces may be acting upon an object, some canceling the effect of others. It is the vector sum of all the forces that produces acceleration. If a large weight is suspended by a single rope from the ceiling, the rope exerts whatever force is necessary to balance the weight. As long as the net force on it is zero, a stationary weight will remain at rest. If you push horizontally on it while it is in

FIGURE 6-3 *Newton was thinking of an unbalanced force.*

its equilibrium position, and there is nothing to counter this force, the weight will accelerate. If you repeated this experiment on the moon, it would take only one-sixth as much force in the rope balancing one-sixth as much weight; but the same horizontal force would result in the same acceleration. Since gravity is balanced in both cases, it should have no effect on our thought experiment. If the vector sum of all the forces is the same, the same acceleration results anywhere you happen to be.

Of course, once the weight swings any distance at all from its equilibrium position, the situation becomes altogether different. The force in the rope would then have a vertical component to balance its weight and a horizontal component to oppose the force you applied. Any force applied to the weight would effect its acceleration. Your hand, the weight, components of the rope force, air friction, and everything else would have to be taken into account. The force in

FIGURE 6-4 *The same force produces the same acceleration anywhere in the universe.*

$$\Sigma\ \mathbf{F} = m\mathbf{a}$$

is the resultant, or vector sum, of *all* of the forces acting. The upper case Greek letter sigma, Σ, is used to denote "the sum of."

CHECK QUESTION

An object of mass 5.0 kg is found to accelerate at 2.0 m/s² when a certain unbalanced force acts upon it. (a) How large is this unbalanced force? (b) How much will the same object accelerate if an unbalanced force of 20 N acts upon it?

Answer: (a) 10 N, (b) 4.0 m/s²

► Weight

If a force of 1.0 N will produce an acceleration of 1.0 m/s² when applied to a 1.0-kg mass, a force of 2.0 N will accelerate the same mass at 2.0 m/s². That is what is meant by a proportionality.

This proportionality holds the other way around as well. If a 1.0-kg mass is observed to accelerate at a rate of 4.0 m/s², we may assume that an unbalanced force of 4.0 N is being applied to it. If a complex set of forces causes this object to accelerate at 8.0 m/s², you know that the resultant of all those forces is 8.0 N and that its direction is the same as that of the acceleration.

Drop the 1.0-kg object and you will observe that gravity accelerates it downward at the rate of 9.8 m/s². Since its mass is 1.0 kg, the force of gravity necessary to cause this acceleration must be 9.8 N.

The same reasoning can be applied to any object, large or small. A 10.0-kg stone will require 10.0 N to accelerate it at 1.0 m/s². If you observe a 10-kg stone accelerating at 10 m/s², you may assume that the resultant force acting on it must be 100 N. If it accelerates downward at 9.8 m/s², the gravitational force acting on it, its weight must be 98 N.

$$F = ma$$
$$= (10.0 \text{ kg})(9.8 \text{ m/sec}^2)$$
$$= 98 \text{ N} \qquad\qquad \text{since } 1 \text{ N} = (1 \text{ kg})(1 \text{ m/sec}^2)$$

The situation in which the force of gravity produces acceleration amounts to a special case of Newton's second law.

$$F_{\text{gravity}} = ma_{\text{gravity}}$$
$$\text{wt} = mg$$

Weight is defined as the force due to gravity. We can use this relationship to determine mass or weight when one or the other is known.

$$\boxed{\text{wt} = mg}$$

▶ British Units

In the British system of units substances are generally measured by weight instead of mass. Weight is a perfectly good measure of a substance as long as we are confined to the earth's surface, give or take a few miles, and as long as we are only concerned with precision to two or three significant figures. To that extent, the acceleration due to gravity g may be considered constant, so that weight is proportional to mass:

$$\text{wt} = mg$$

One might think that it makes absolutely no difference whether one measures by weight or by mass as long as this proportionality does hold. As a matter of fact, most difficulties arise only in working problems that involve Newton's second law (of which the relationship between mass and weight is a just special case).

$$F = ma$$

When using this law, it is important to keep in mind that m is mass and not weight. Since the amount of stuff you have is usually measured in terms of its weight, we usually need to go through an additional step to get its mass. Solving the weight-mass relationship for mass,

$$m = \frac{\text{wt}}{g}$$

The unit of mass usually used in the British system is called the **slug,** probably because it is a measure of the sluggishness of the object being accelerated.

$$1 \text{ slug} \equiv \frac{1 \text{ lb}}{1 \text{ ft/s}^2}$$

A slug is defined as the amount of mass that will be accelerated at 1 ft/s² by a force of 1 lb.

Since many people have not heard of this unit of mass, it might be useful to consider an example involving a slug of whiskey. Assume that we start with a full jug of whiskey and that we remove it a little at a time until we find that 1 lb of force applied in a sideward direction causes it to accelerate in that direction at 1 ft/s². The amount of mass that we would have left would then be 1 slug. We would then be able to expect that 2 lb of force would accelerate it at 2 ft/s², that 10 lb of force would accelerate it at 10 ft/sec², and so on.

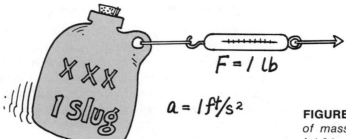

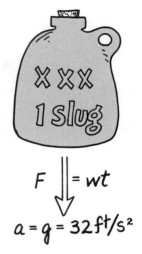

FIGURE 6-5 *A slug is the amount of mass that is accelerated at 1 ft/s² by a force of 1 lb.*

How much does this much whiskey weigh? Some people might say, "One pound, of course." Not so! One pound is the force that accelerates it sidewards at 1 ft/s²; it is not the weight. The purpose of talking about a sideward force and a sideward acceleration was to avoid the effects of weight (gravity). The weight of a falling object always turns out to accelerate it at 32 ft/s², at least on the surface of the earth. We have been talking about much smaller accelerations than that so far. But let us escalate. If 10 lb of force makes the jug accelerate at 10 ft/s², how much force is necessary to make the same jug accelerate at 32 ft/s²? If you dropped the jug for some reason, that would in fact be the acceleration you would observe.

We should therefore conclude that the force of gravity that produces this acceleration is in fact 32 lb. The special case of Newton's second law that happens when the unbalanced force is the whole weight

$$F = ma$$

of the object is that the gravitational force produces the gravitational acceleration:

$$\text{wt} = mg$$
$$= (1 \text{ slug})(32 \text{ ft/s}^2) = 32 \text{ lb}$$

Of course, this does not mean that all jugs of whiskey weigh the same 32 lb just because they would all accelerate at the same 32 ft/s² if dropped. If someone drank half of the original slug of whiskey before dropping the other half, the remaining half slug would only weigh 16 lb. This 16 lb will do to a half slug, however, what 32 lb will do to a whole slug.

$$F = ma$$

$$a = \frac{F}{m} = \frac{16 \text{ lb}}{\frac{1}{2} \text{ slug}} = 32 \text{ ft/s}^2$$

FIGURE 6-6 *If an object is dropped, the unbalanced force is the weight of the object, and the gravitational force produces the acceleration due to gravity.*

The same argument will hold for a quarter slug, which weighs only 8 lb, or an eighth slug, which weighs only 4 lb, or any other amount of whiskey. The resulting acceleration is always the same, because the ratio of the mass to the weight is always the same.

CHECK QUESTION

A mid-sized automobile weighs 3,200 lb. (a) What is its mass? (b) How much will it accelerate if acted upon by an unbalanced force of 200 lb?

Answer: (a) 100 slug, (b) 2.0 ft/s²

▶ The Principle of Equivalence

We have been assuming that the mass of something is what gives it weight. We could draw a distinction between the inertial mass of an object, which is a measure of its resistance to acceleration, and the gravitational mass, which would be the thing that gives it weight. There is really no reason why the inertial mass has to be the same as the gravitational mass, when you stop to think about it. There is no reason, that is, except for the experimental fact that we have never found anything that had even a bit more of the one without having exactly the same amount more of the other.

What would we do if we found something that had more inertial mass than gravitational mass? We would use it to make hammers. Carpenters need to carry around several different sizes of hammers for different sizes of nails. A 20-oz hammer, which is good for framing a house, is too large for small work. It has so much mass that it tends to knock cabinets apart rather than knock them together. A 16-oz hammer, which is good for general work, does not have enough mass to efficiently drive the big nails used in structural lumber. Carpenters would welcome a set of light hammers that would be easy to carry around but still had the required mass. Unfortunately for carpenters, there is no such material. They don't make hammers out of aluminum, for example, because aluminum lacks mass in exactly the same proportion as it lacks weight.

The misfortune of the carpenter turns out to be good fortune for theoretical physicists. The fact that the inertial mass of an object is always equivalent to the gravitational mass is called the **principle of equivalence,** and the principle of equivalence is the foundation of the general theory of relativity. That is heavy physics, so to speak. The principle of equivalence, however, is well within most people's understanding of reality. Hand a penny and a dime to some unsuspecting person and ask which is heavier. Explain that the dime is a tiny bit smaller but is made from a different metal, which might be more dense. (A real silver dime would be great for this experiment if you could lay your hands on one.) The result is not important. The point is to watch what that person does in trying to decide which is heavier. People almost always toss small objects in their hand when trying to resolve small weight differences. When they hold the coin statically in their hand, they are testing for gravitational mass. When they toss the coin gently about, they are testing for inertial mass. They just naturally assume that if it has more of one, it will have more of the other.

▶ "Weighing" the Earth

We can use the principle of equivalence in an experiment to determine the mass of the earth. We will first need to develop the idea of Newton's law of universal gravitation. Next, we will consider an experiment that provides the value of the proportionality constant in this universal law. We will then be in a position to calculate the mass of the earth.

Newton's law of universal gravitation states that every object in the universe pulls on every other object with a force proportional to the product of their masses and inversely proportional to the square of the distance between their centers. You can see why this might be true once we assume that some kind of attraction between any two

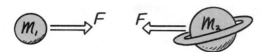

FIGURE 6-7 *There is a gravitational force of attraction between any two masses.*

objects exists. We might expect the strength of this pull to be directly proportional to the mass of the object since, if we were to replace one of the objects with two identical objects of the same size, we would

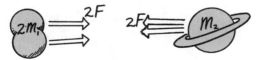

FIGURE 6-8 *The gravitational force is directly proportional to the mass.*

expect the force to double. Tripling the objects would triple the force, and so on. Therefore, the force might reasonably be proportional to the mass of object number one.

$$F \propto m_1$$

Similarly, the force might reasonably be proportional to the mass of object number two.

$$F \propto m_2$$

Any time a single quantity is independently proportional to two independent quantities, it is also proportional to their product; doubling either of the quantities will double their product.

$$F \propto m_1 \cdot m_2$$

If either m_1 or m_2 is doubled, F will double. Therefore, the force between any two objects may be expected to be proportional to the product of their masses.

The force between any two objects gets weaker as the objects get further apart. The fact that the force is *inversely* proportional to the *square* of the distance has been observed experimentally. This relationship,

$$F \propto \frac{1}{d^2}$$

just seems to work, but it works in a reasonable fashion. When the objects are twice as far apart, the force is only one-fourth as great as before. When they are three times as far apart, the force is only one-ninth as great, and so on. Combining the inverse square law with the proportionality for the product of the masses:

$$F \propto \frac{m_1 m_2}{d^2}$$

CHECK QUESTION

A small object a certain distance from the center of Planet X is found to experience a gravitational attraction of 4.0 N toward the center of that planet. (a) How much force of attraction would the same object experience if it were the same distance from the center of Planet Y, assuming Planet Y has 10 times the mass of Planet X? (b) How much force of attraction would the object experience toward the center of Planet X if it were 2.0 times as far from the center of that planet?

Answer: (a) 40 N, (b) 1.0 N

We note that doubling either of the masses doubles the force. If we double both of the masses simultaneously, the force is doubled squared, or four times as great. Doubling the distance then puts you right back where you started.

$$F \propto \frac{(2\,m_1)(2\,m_2)}{(2d)^2} = \frac{m_1 \cdot m_2}{d^2}$$

Newton's law of universal gravitation can be expressed in terms of an equality sign and a proportionality constant:

$$F = G\,\frac{m_1\,m_2}{d^2}$$

The value of this proportionality constant G is determined by experiment. Just hold two objects a certain distance apart and see how hard they pull on each other.

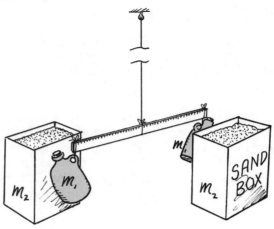

FIGURE 6-9 *A torsion balance may be used for measuring gravitational forces between room-sized objects.*

One way of measuring this force is to hang a pair of masses, say milk bottles, on the end of a stick, supported at the middle by a thread. Pulling two other masses, say boxes of sand, up close to the two masses hanging from the stick will cause tiny gravitational forces to act, tending to turn the stick and twist the thread. It is just a question of using a weak enough string and making it long enough so that this tiny force makes a measurable difference. You can then measure the amount of force it takes to twist a shorter piece of the same string, and you will have a way of getting an experimental value for the force of attraction, F, that exists between room-sized objects. If you measure the masses of the milk bottles, m_1, and the boxes of sand, m_2, as well as the distance d between the centers of the bottles and sand boxes, you will have measured all the parameters in Newton's law of universal gravitation. The only unknown is therefore the value of the proportionality constant G.

$$F = G\,\frac{m_1 \cdot m_2}{d^2}$$

Solving the gravitational law for this constant, we get

$$G = \frac{Fd^2}{m_1 \cdot m_2}$$

It turns out that you always get the same value for G whenever you do the experiment or wherever you do it. Whenever you plug in the experimental values for everything on the right-hand side of this equation, you get the same result:

$$G = 6.67 \times 10^{-11} \text{ N} \cdot \text{m}^2/\text{kg}^2$$

People sometimes confuse this gravitational constant, capital G, with the g used for the acceleration due to gravity. People sometimes use the word *gravity* to refer to both constants. They might say that the universal law of gravitation states, "The force between any two objects in the universe is equal to gravity times the product of their masses divided by the square of the distance between them." They might then say that the relationship between mass and weight

$$\text{wt} = mg$$

is that "weight equals mass times gravity." This question of whether *gravity* best refers to G or g is further complicated by the fact that most people would say that gravity is the thing that holds us down on the earth's surface. In this latter case, they clearly mean the *force* of gravity, or weight, wt. It is therefore wise always to specify the phenomenon you are thinking of when you say gravity. Speak either of the gravitational constant (G), the acceleration due to gravity (g), or the force of gravity (wt), but never of "gravity" alone.

Experiments to measure the universal gravitational constant, such as the one we have just described, are sometimes called weighing the earth. The reason is that there is a close relationship between G and g. If one of the masses in the law of universal gravitation is the mass of the earth,

$$m_1 = m_e$$

and the distance is the radius of the earth,

$$d = r_e$$

the force given by the law of universal gravitation is simply the weight on the earth's surface:

$$\text{wt} = G \frac{m_e \, m_2}{r_e^2}$$

This weight is the same thing as is given by the relationship between weight and mass:

$$\text{wt} = m_2 g$$

We can get the relationship between G and g by equating these two expressions for weight:

$$m_2 g = G \frac{m_e \, m_2}{r_e^2}$$

Dividing both sides by the mass m_2, which is the same quantity in both expressions, we see that g is related to G by the mass of the earth and the radius of the earth.

$$g = G \frac{m_e}{r_e^2}$$

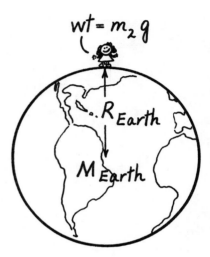

FIGURE 6-10 *Weight is force due to gravity at the earth's surface.*

Once we know the value of G, g, and r_e, we can solve this equation for the mass of the earth:

$$m_e = \frac{gr_e^2}{G} = \frac{(9.8 \text{ m/s}^2)(6 \times 10^6 \text{ m})^2}{6.7 \times 10^{-11} \text{ N} \cdot \text{m}^2/\text{kg}^2}$$

All we need now to find the mass of the earth is to measure g by dropping something and to measure the radius of the earth r_e by taking a long walk. People who do such things always seem to come out with the same result:

$$m_e = 6 \times 10^{24} \text{ kg}$$

▶ The Acceleration Due to Gravity

We have said that everything falls with the same acceleration. This experimental fact is directly related to the principle of equivalence. We use the preceding argument on weighing the earth, but change our nomenclature just a bit. Assuming that the only force acting on an object is the gravitational force of attraction exerted on the object by the earth, we can show that the resulting acceleration is the same without regard to the mass of the object. First, we can express this gravitational force in terms of the mass of the earth, m_e, the radius of the earth, r_e (which is also the distance between the center of the object and the center of the earth), and the mass m of the object.

$$F = G \frac{m_e \, m}{r_e^2} \qquad \text{Gravitational}$$

Still assuming that this is the only force acting on the object, the resulting acceleration a is given by Newton's second law of motion, as usual:

$$F = ma \qquad \text{Inertial}$$

The principle of equivalence assumes that the gravitational mass in the law of universal gravitation is the same quantity as the inertial mass in the second law of motion. The same symbol m represents mass in both equations. If the gravitational force in the first equation is the force used to overcome inertia of the object in the second equation, the right-hand sides of the equations are also equal:

$$ma = G \frac{m_e}{r_e^2} \cdot m$$

Assuming that the masses are the same on both sides of the equation, in agreement with the principle of equivalence, we can simplify this expression by dividing both sides by m.

$$a = G \frac{m_e}{r_e^2}$$

The acceleration therefore does not depend on the mass of the object but only upon the universal gravitational constant, G, the mass of the earth, m_e, and the radius of the earth, r_e. These quantities are all the same for any object on the earth's surface.

You might argue that r_e would be different on the top of a mountain. This change is, however, negligible compared to 4,000 miles. All life as we know it is confined to a bubble-thin shell, called the biosphere. Every point within this shell is so nearly the same distance from the center of the earth that almost all objects within our direct experience fall with the same acceleration.

Of course, the sameness of the acceleration due to gravity for all objects in this bubble-thin shell holds only when no forces other than weight act on the objects. This assumption is valid only when frictional forces, such as air friction, can be neglected. Eliminate air friction, and even a feather will fall like a stone. This can be shown by placing both a feather and a stone together in a glass tube from which the air can be removed. With air in the tube, the stone will fall like a stone and the feather will float down like a feather. Evacuate the tube, however, and the feather will fall from one end of the tube to the other with the same acceleration as the stone.

The feather and stone experiment, which has been a classic demonstration for hundreds of years, can be done almost as well by dropping your physics book at the same time as you drop a sheet of paper. The paper will accelerate for only a tiny distance before it reaches its **terminal velocity,** the velocity at which air friction balances downward force. With the paper crumpled into a ball to minimize air friction, however, the paper will be found to strike the floor at about the same time as the book. But if you repeat the experiment once more, throwing both objects out of a high window, you can observe that the force of air friction will again become important. The book will eventually reach a higher terminal velocity than the paper.

Skydivers make use of air friction. They do not open their parachutes for a long time after jumping out of the airplane. Much of the time a skydiver is falling, however, she is not accelerating much at all. Acceleration takes place only until the skydiver gets up to a speed of about 100 mph. Air friction then balances the weight and the skydiver falls with a constant velocity. This is not really free fall, because there is no unbalanced force to cause an acceleration. The skydiver can control this terminal velocity, however, by holding her hands and arms in different positions. If she wishes to join another skydiver below, she has only to pull her arms in close to her body to reduce air friction. Suppose, because of this, that the force of air friction drops to only three-quarters of her weight; the unbalanced force acting on her will be one-fourth of her weight, and she will accelerate downward at $0.25\,g$. As she reaches a new terminal velocity, the air friction increases to balance her weight and acceleration drops to zero.

If Galileo could only have jumped out of an airplane, he could have once more shown that an object in motion tends to stay in motion at a constant velocity if all the forces are balanced to zero. Then imagine Newton tucking his arms in and saying, "But if the forces are unbalanced. . . ."

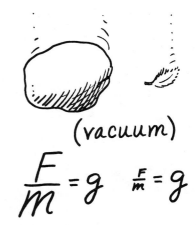

FIGURE 6-11 *All objects fall with the same acceleration in the absence of air friction.*

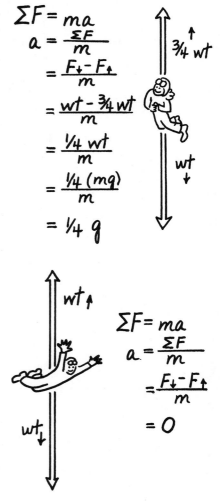

FIGURE 6-12 *The skydiver falls with terminal velocity only after air friction becomes large enough to balance her weight. If she pulls her arms in, air friction will decrease and she will accelerate downwards.*

The problems in this chapter deal with situations in which an unbalanced force acting on an object produces acceleration. The amount of acceleration, according to **Newton's second law** of motion, is proportional to the vector sum of all the forces acting on the object:

$$\Sigma \ \mathbf{F} = m\mathbf{a}$$

The proportionality constant m is called the mass of the object. Force in the metric system of units is defined in terms of mass and acceleration. One **newton** of force is defined as the amount of force that will produce 1 m/s² of acceleration when acting on a mass of 1 kg:

$$(1 \ N) \equiv (1 \ kg)(1 \ m/s^2)$$

In the British system of units, however, mass is the dependent quantity and is defined in terms of force and acceleration. One **slug** of mass is defined as the amount of mass that will be accelerated at 1 ft/s² by a force of 1 lb.

$$(1 \ lb) = (1 \ slug)(1 \ ft/s^2)$$

A special case of Newton's second law is the relationship between mass and weight. If the only force acting on an object is its weight (the apparent gravitational force between it and the earth), the same acceleration will always result. This particular acceleration is called the **acceleration due to gravity,** g. The relationship between mass and weight is that weight equals mass times the acceleration due to gravity:

$$wt = mg$$

The acceleration due to gravity at the earth's surface has a value of 9.8 m/s², or, in British units, 32 ft/s².

$$g = 9.8 \ m/s^2 = 32 \ ft/s^2$$

The fact that all objects fall with the same acceleration shows that the same property of an object that gives it inertia, that is, mass in Newton's second law, also gives it weight, or gravitational force. Newton's **law of universal gravitation** states that every body in the universe pulls on every other body with a force proportional to the product of their masses and inversely proportional to the square of the distance between them:

$$F = G \ \frac{m_1 m_2}{d^2}$$

The proportionality constant G is called the **universal gravitational constant** and has a value of $6.67 \times 10^{-11} \ N \cdot m^2/kg^2$.

6-1 One measure of a sports car's performance is its ability to accelerate from rest up to a given speed in some amount of time. The better the engine is running, the shorter will be the time.

Given: A 1960 Triumph TR-3 with driver has a mass of 1,000 kg. In good tune, it is able to accelerate from 0 to sixty mph (26.8 m/s) in 11.0 s.

Find: (a) What is the acceleration in meters per second squared?
(b) What is the unbalanced force acting on the car, in newtons?

Your Solution

Discussion

This problem illustrates Newton's second law of motion. The greater the unbalanced force action on an object, the greater will be its acceleration. The unbalanced force necessary to produce a reasonable acceleration of an automobile can be calculated by figuring out its acceleration and its mass.

Sample Solution 6-1

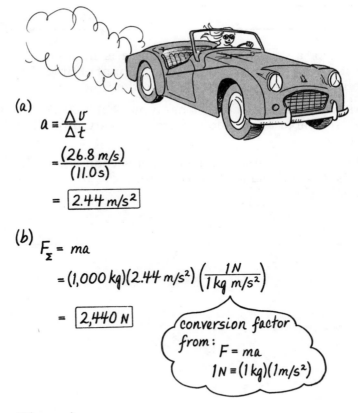

(a)
$$a \equiv \frac{\Delta v}{\Delta t}$$
$$= \frac{(26.8 \text{ m/s})}{(11.0 \text{ s})}$$
$$= \boxed{2.44 \text{ m/s}^2}$$

(b)
$$F_\Sigma = ma$$
$$= (1,000 \text{ kg})(2.44 \text{ m/s}^2)\left(\frac{1 \text{ N}}{1 \text{ kg m/s}^2}\right)$$
$$= \boxed{2,440 \text{ N}}$$

conversion factor from:
$$F = ma$$
$$1 \text{ N} \equiv (1 \text{ kg})(1 \text{ m/s}^2)$$

Discussion

The acceleration can be calculated directly from the definition. This is an average acceleration, since the acceleration of an automobile is greater in lower gears. The result turns out to be about 25% the acceleration due to gravity.

The force is calculated directly from Newton's second law. The combination of units, a kilogram times a meter divided by a second squared, is called a newton, since a newton of force is defined as the amount of force that will accelerate 1 kg of mass at the rate of 1 m/s².

6-2 Space travelers of the future will conduct commercial transactions in terms of the mass of a commodity rather than its weight, mass being independent of a gravitational field.

Given: A sack of dried leaves has a mass of 1.0 kg and is acted upon by an unbalanced force of 200 N for 1.0 s.

Find: (a) What is the acceleration of the sack?
   (b) How far will the sack have gone at the end of the time period?

Your Solution

Discussion

Mass is a measure of the stuff involved, whereas weight is a result of the gravitational interaction between the stuff and some nearby planet. The mass is therefore the same anywhere, whereas the weight depends on where you are. If you wanted to measure the mass of something in an unknown gravitational field, one way would be to apply a known force and measure the acceleration.

Sample Solution 6-2

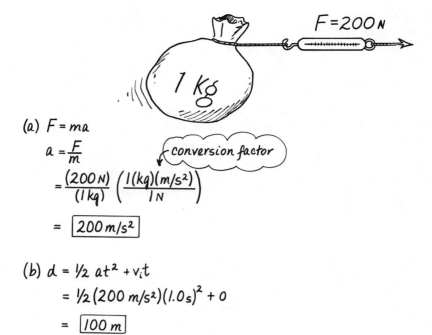

$F = 200\,N$

(a) $F = ma$

$a = \dfrac{F}{m}$

conversion factor

$= \dfrac{(200\,N)}{(1\,kg)} \left(\dfrac{1(kg)(m/s^2)}{1\,N} \right)$

$= \boxed{200\,m/s^2}$

(b) $d = \frac{1}{2}\,at^2 + v_i t$

$= \frac{1}{2}(200\,m/s^2)(1.0\,s)^2 + 0$

$= \boxed{100\,m}$

Discussion

The acceleration is found by solving Newton's second law for acceleration. The distance can be found from this result by using the formula for distance traveled under constant acceleration, assuming the initial velocity is zero.

6-3 A nice mother is taking her baby for a stroll on level ground. To give the baby a little thrill, she gives the baby carriage a shove with a force greater than that necessary to overcome friction. She then lets go and allows the carriage to roll to a stop.

Given: The carriage and baby have a mass of 45 kg. She shoves with a force of 90 N and finds that the carriage accelerates at 1.8 m/s².

Find: (a) How much unbalanced force is necessary to produce this acceleration?

(b) How much of the mother's shove is used in overcoming friction?

(c) What will be the deceleration of the carriage when friction alone acts?

Your Solution

Discussion

This problem is an example of a force situation where more than one force is acting on an object. The force of friction is in the opposite direction from the driving force and partly diminishes its effect. If you know the driving force, you should be able to figure out the friction by its effect.

Sample Solution 6-3

(a) $\Sigma F = ma$

$\qquad = (45\,kg)(1.8\,m/s^2)\left(\dfrac{1\,N}{1\,kg\,m/s^2}\right)$

$\qquad = \boxed{81\,N}$

(b) $\Sigma F = F_{shove} - F_{friction}$

$\quad F_{friction} = F_{shove} - \Sigma F$

$\qquad = (90\,N) - (81\,N) = \boxed{9\,N}$

(c) $F = ma$

$\quad a = \dfrac{F}{m}$

$\qquad = \dfrac{(9\,N)}{(45\,kg)}\left(\dfrac{1\,kg\,m/s^2}{1\,N}\right)$

$\qquad = \boxed{0.2\,m/s^2}$

Discussion

The total force can be found by a direct application of Newton's second law. Since this total force is the difference (or vector sum) of the force with which the mother shoves and the force of friction, the frictional force is the amount by which the total force is smaller than the mother's shove. If we know the size of the frictional force, Newton's second law tells us the rate at which deceleration of the baby carriage occurs when friction alone acts.

6-4 All pizzas, small, medium and large, accelerate downward at the same rate when dropped. In New York City the acceleration is 9.802 m/s².

Given: A pizza cook throws a 1.5-kg pizza pie upward into the air with an acceleration of 35 m/s².

Find: (a) How much does this pizza pie weigh?
(b) What is the unbalanced upward force acting while the pizza is thrown?
(c) How hard must the cook throw the pizza to overcome the weight and still produce enough unbalanced force to cause the given upward acceleration?

Your Solution

Discussion

This problem is similar to the last one, except that one of the forces is the force of gravity. You can find the weight by remembering that it is just the gravitational force that results in the gravitational acceleration when it is unbalanced.

Sample Solution 6-4

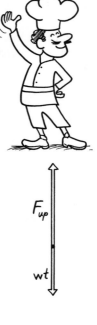

(a) $wt = mg$

$\qquad = (1.5\ kg)(9.802\ m/s^2)$

$\qquad = 14.7\ N$

$\qquad \cong \boxed{15\ N}$

(b) $\Sigma F = ma$

$\qquad = (1.5\ kg)(35\ m/s^2)$

$\qquad = \boxed{52\ N}$

(c) $\Sigma F = F_{up} - wt$

$\quad F_{up} = \Sigma F + wt$

$\qquad = (52\ N) + (15\ N)$

$\qquad = \boxed{67\ N}$

Discussion

The relationship between mass and weight, which was developed as a proportionality in Chapter 1, may now be looked upon as a special case of Newton's second law:

$$F = ma$$

If the unbalanced force is the weight, $F = wt$, the resulting acceleration will be the acceleration due to gravity, $a = g$:

$$wt = mg$$

You can figure out the unbalanced force acting on the pizza when it is tossed into the air from its mass and acceleration. Since you know that this unbalanced force is the difference between the upward force applied by the pizza cook and the weight (or vector sum, since they are in opposite directions), you can find the upward force by solving this equality.

6-5   One way to measure the acceleration of your car is to see how far a pendulum swings from the vertical. That is why so many highly intelligent drivers have objects of art hanging from their rear-view mirrors.

Given: When the car is accelerating, an object of art having a mass of 0.2 kg is found to swing 15° from the vertical.

Find: (a) What is the weight of the object?
  (b) What is the force of acceleration necessary to make it swing that much?
  (c) What is the acceleration of the car?

Your Solution

Discussion

This is another example of an acceleration produced by the combined effect of two forces acting on an object. The forces are neither in the same nor opposite direction, however, and must be combined by the rules of vector addition.

Sample Solution 6-5

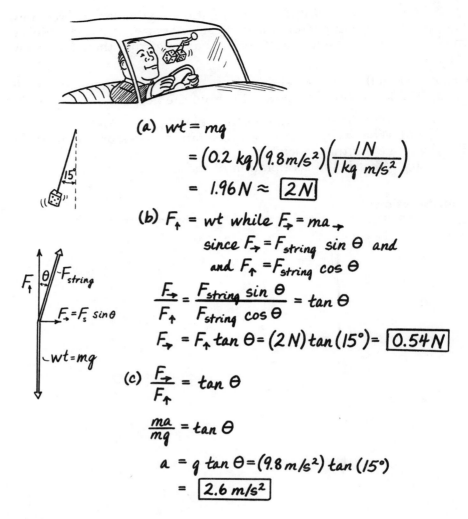

(a) $wt = mg$

$$= (0.2 \text{ kg})(9.8 \text{ m/s}^2)\left(\frac{1N}{1 \text{kg m/s}^2}\right)$$

$$= 1.96 N \approx \boxed{2N}$$

(b) $F_\uparrow = wt$ while $F_\rightarrow = ma_\rightarrow$

since $F_\rightarrow = F_{string} \sin \theta$ and

and $F_\uparrow = F_{string} \cos \theta$

$$\frac{F_\rightarrow}{F_\uparrow} = \frac{F_{string} \sin \theta}{F_{string} \cos \theta} = \tan \theta$$

$$F_\rightarrow = F_\uparrow \tan \theta = (2 N) \tan (15°) = \boxed{0.54 N}$$

(c) $\dfrac{F_\rightarrow}{F_\uparrow} = \tan \theta$

$$\frac{ma}{mg} = \tan \theta$$

$$a = g \tan \theta = (9.8 \text{ m/s}^2) \tan (15°)$$

$$= \boxed{2.6 \text{ m/s}^2}$$

Discussion

The force acting in the string has two components. The upward component balances the weight. By figuring out the weight in part (a), we can use this component to figure out the force in the string. We could then find the sideward component and the acceleration it produces to get the answers to parts (b) and (c). The solution shown here, however, shows a slightly different approach. The upward and sideward components of the force in the string are expressed in terms of the sine and cosine of the angle to the vertical, and these components are divided, one by the other, to produce the tangent of that angle. One component can then be solved in terms of the other. A similar approach is shown for the accelerations. The answers in parts (a) and (b) are limited to one significant figure by the accuracy to which we know the mass. The answer to part (c), however, can be reported to two significant figures, since the mass divides out of the result.

6-6 Both the sun and the moon cause tides in the earth's oceans by virtue of their gravitational forces acting on water on the earth's surface.

Given: The sun has a mass of 2.0×10^{30} kg and is at an average distance of 1.5×10^{11} m from the earth. The moon has a mass of 7.5×10^{22} kg and is at an average distance of 3.8×10^{8} m from the earth. The universal gravitational constant has a value of 6.67×10^{-11} N $\cdot$ m^2/kg^2.

Find: (a) What is the gravitational force of attraction between the sun and 1.0 kg of water on the earth's surface?
   (b) What is the force of gravity between the moon and 1.0 kg of water on the surface of the earth?

Your Solution

Discussion

Most people think of the moon alone as causing the tides. That is because the tides caused by the moon are several times as large as those caused by the sun. When these tides get together, however, the result is an extra large tide, called a spring tide.

This problem is an application of Newton's law of universal gravitation. It will give you a chance to compare the gravitational force on earthly objects due to the sun and the moon.

Sample Solution 6-6

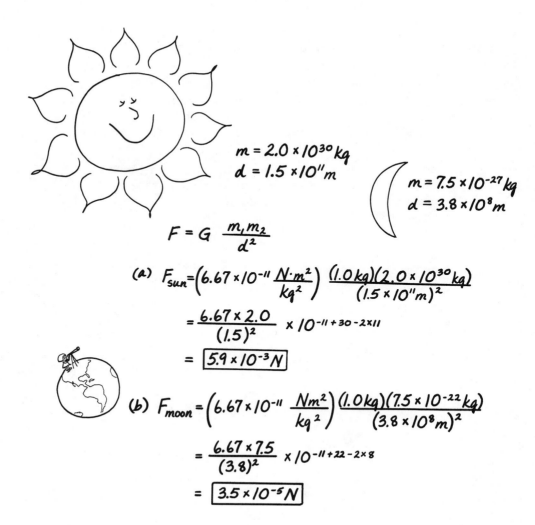

$m = 2.0 \times 10^{30}\,kg$
$d = 1.5 \times 10^{11}\,m$

$m = 7.5 \times 10^{-27}\,kg$
$d = 3.8 \times 10^{8}\,m$

$$F = G\,\frac{m_1 m_2}{d^2}$$

(a) $F_{sun} = \left(6.67 \times 10^{-11}\,\dfrac{N \cdot m^2}{kg^2}\right)\dfrac{(1.0\,kg)(2.0 \times 10^{30}\,kg)}{(1.5 \times 10^{11}\,m)^2}$

$= \dfrac{6.67 \times 2.0}{(1.5)^2} \times 10^{-11 + 30 - 2 \times 11}$

$= \boxed{5.9 \times 10^{-3}\,N}$

(b) $F_{moon} = \left(6.67 \times 10^{-11}\,\dfrac{Nm^2}{kg^2}\right)\dfrac{(1.0\,kg)(7.5 \times 10^{-22}\,kg)}{(3.8 \times 10^{8}\,m)^2}$

$= \dfrac{6.67 \times 7.5}{(3.8)^2} \times 10^{-11 + 22 - 2 \times 8}$

$= \boxed{3.5 \times 10^{-5}\,N}$

Discussion

Both parts (a) and (b) are solved by inserting the given values into Newton's law of universal gravitation. The results, as you can see, are quite small. Some people feel, however, that these tiny forces are responsible for all sorts of human events. Astrologers make lengthy calculations based, some say, upon these forces.

The forces exerted by the sun on our 1-kg sample of water turn out to be over two orders of magnitude greater than the force exerted by the moon. Can you think of a reason why the resulting tide is one order of magnitude smaller? Can you explain why the sun and the moon each cause two tidal bulges in the earth's oceans?

6-7   An object on Mars weighs less than the same object on earth, because Mars has much less mass. The smaller mass is, however, somewhat compensated for by the shorter distance from the surface of that planet to its center as compared to the earth.

Given: The mass of Mars is 0.11 times the mass of the earth, and the radius of Mars is 0.53 times that of the earth. A certain physics instructor weighs 900 N on the earth.

Find: (a) How much would he weigh at the earth's radius from the center of Mars?

(b) How much would he weigh on the surface of Mars?

Your Solution

Discussion

This problem is an example of Newton's law of universal gravitation. This law combines the effects of mass and distance to produce a complex proportionality. One way to see the combined effect is to look at a change in mass and distance at separate steps. To see what your weight would be on Mars, for example, imagine that someone were to sneak in at night when you were not looking and replace the earth with Mars. Your weight would decrease because of the change in mass of the planet attracting you, but you would find yourself considerably above the surface of the smaller planet. Your weight would then start to increase somewhat as you fell to the surface of Mars.

Sample Solution 6-7

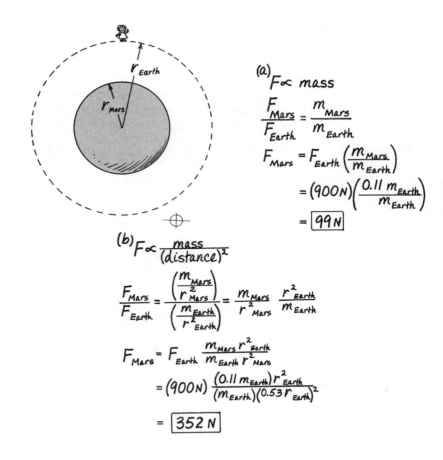

(a) $F \propto mass$

$$\frac{F_{Mars}}{F_{Earth}} = \frac{m_{Mars}}{m_{Earth}}$$

$$F_{Mars} = F_{Earth}\left(\frac{m_{Mars}}{m_{Earth}}\right)$$

$$= (900N)\left(\frac{0.11\, m_{Earth}}{m_{Earth}}\right)$$

$$= \boxed{99N}$$

(b) $F \propto \dfrac{mass}{(distance)^2}$

$$\frac{F_{Mars}}{F_{Earth}} = \frac{\left(\dfrac{m_{Mars}}{r^2_{Mars}}\right)}{\left(\dfrac{m_{Earth}}{r^2_{Earth}}\right)} = \frac{m_{Mars}}{r^2_{Mars}}\,\frac{r^2_{Earth}}{m_{Earth}}$$

$$F_{Mars} = F_{Earth}\,\frac{m_{Mars}\, r^2_{Earth}}{m_{Earth}\, r^2_{Mars}}$$

$$= (900N)\,\frac{(0.11\, m_{Earth})\, r^2_{Earth}}{(m_{Earth})(0.53\, r_{Earth})^2}$$

$$= \boxed{352\,N}$$

Discussion

The law of universal gravitation says that the force is directly proportional to the product of masses and inversely proportional to the distance squared:

$$F = G\,\frac{m_1 \cdot m_2}{d^2}$$

Part (a) just looks at the proportionality between force and mass. A direct proportion can be expressed as a ratio, which can then be solved for one of the unknowns. Part (b) uses the same approach but includes the inverse proportion with the square of the distance.

6-8 A space shuttle ignites its rockets for a burn to correct its course in accordance with calculations made by ground control.

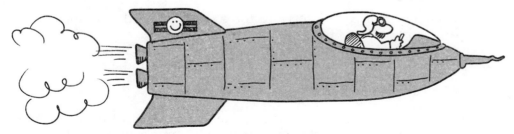

Given: The space shuttle has a mass of 17,000 kg. Ground control orders a burn of 4.5 s to change the velocity by 150 m/s.

Find: (a) What is the average acceleration in meters per second squared?
    (b) What is the thrust of the rockets necessary to provide this acceleration?

6-9 One of the problems encountered in the Skylab program was measuring without the aid of gravity the mass of waste products generated by biological experiments. The technique actually used involved applying a given force and measuring the acceleration.

Given: An unknown mass of waste product experiences an acceleration of 8.3 m/s^2 when a force of 5.0 N is applied to it. This amount of force is applied for 0.25 s.

Find: (a) What is the mass of the waste product?
    (b) How far did it go while the force was applied?

6-10 A jet aircraft takes off after accelerating down the runway under full throttle.

Given: The mass of the plane is 65,000 kg. It reaches a takeoff velocity of 50 m/s in 35 s, using an engine thrust of 100,000 N.

Find: (a) What is the actual acceleration (assumed constant) of the aircraft?
    (b) How much unbalanced force produces this acceleration?
    (c) What was the force of friction that opposed this acceleration?

$V_t = Constant$

6-11 An elephant gets on an elevator. He presses a floor button and notices that the force with which the elevator presses up on his feet changes.

Given: The elephant has a mass of 2,000 kg. Just after the button is pressed, the force against his feet is 15,000 N.

Find: (a) What is the elephant's weight?
    (b) How much unbalanced force acts on the elephant?
    (c) What is the elephant's acceleration? Up or down?

6-12 A person standing on a bus during rush hour finds it necessary to lean forward when the bus starts moving from the bus stop into traffic.

Given: The bus accelerates forward with an acceleration of 1.5 m/s². The person has a mass of 55 kg.

Find: (a) How much forward force must act on the person?
(b) What is the person's weight?
(c) At what angle to the vertical must the person lean to remain balanced if it is impossible to find a polite place to hold on?

6-13 The *Queen Elizabeth* and the *Queen Mary* were two famous ocean liners once operated by the same shipping company. They were sister ships and both had about the same mass.

Given: Both ocean liners had a mass of about 75,000 tonnes (7.5×10^7 kg). Once they were docked at the same port with a distance of only 306 m between their centers. ($G = 6.67 \times 10^{-11}$ N m²/kg²)

Find: (a) What was the gravitational force of attraction between these two sister ships?
(b) How far apart would they need to be if the force of attraction were to be 1.0 N?

6-14 A 1-kg bag has a weight of 9.8 N on the surface of the earth, but the same amount of mass would be much heavier on the surface of Jupiter.

Given: Jupiter has a mass of 1.9×10^{27} kg and a radius of 7.0×10^4 km. The earth has a mass of 6.0×10^{24} kg and a radius of 6.4×10^3 km.

Find: (a) How much would the 1-kg bag weigh if it were to move away from the center of the earth to a distance equal to Jupiter's radius?
(b) How much would the bag weigh if someone were to replace the earth with Jupiter when no one else was looking (which would be the same as moving the bag to the surface of Jupiter)?
(c) What is the acceleration due to gravity on Jupiter?

MORE INTERESTING PROBLEMS

6-15 A skydiver falling at a constant velocity with her arms outstretched wishes to fall faster to execute a kiss pass. She pulls her arms in close to her body to reduce air friction.

Given: Her mass is 60 kg. With her arms pulled in, she accelerated downward at 2 m/s².

  Find: (a) What was the force of air friction while her arms were still outstretched?
   (b) What was the force of air friction after she pulled her arms in?

6-16 A bullet in a rifle barrel is propelled by expanding gases. People who load their own ammunition select gunpowders with different burning rates depending on the length of the gun barrel and the mass of the bullet.

Given: A 30-caliber bullet typical for sporting use has a mass of 11.7 g. It is accelerated up to a velocity of 620 m/s over a distance of 0.51 m.

 Find: (a) What is the acceleration of the bullet?
   (b) How much force is exerted on the bullet by the expanding gases?
   (c) How long is the bullet in the rifle barrel?

6-17 A bricklayer inadvertently rediscovers the Atwood machine, a device sometimes used to measure acceleration due to gravity. This arrangement is composed of two unequal masses supported by a cord passing over a pulley. (Assume the cord does not stretch.)

Given: The bricklayer has a mass of 75 kg, and the bricks together with the barrel have a mass of 110 kg, which, as he notices, is more. The bricks start out 12 m from the ground. He decides to hold on.

 Find: (a) What is the acceleration of the bricklayer (same magnitude but opposite direction as the bricks)?
   (b) How fast is the bricklayer traveling when he reaches the top, providing that he misses the bricks as they pass?

6-18 Your body seems heavier than usual when you are in a jet airliner that is accelerating down the runway for a takeoff. You could actually measure the increased weight of a mass with a spring fish scale. Of course, the best mass to use for this experiment to avoid suspicious looks from fellow passengers would be a fish.

Given: The fish normally weighs 40 N. During takeoff the weight of the fish seems to increase to 47 N as it swings from the vertical.

 Find: (a) To what angle does the fish swing from the vertical?
   (b) What is the acceleration of the airliner?
   Note: Unlike the example, you know the hypotenuse and one side in force space.

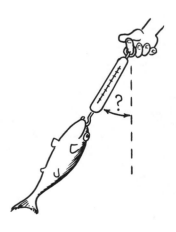

6-19 One common experiment used to verify Newton's second law of motion involves a cart on a horizontal track. The cart is accelerated by a weight, which is attached to a string running over a pulley. Just as in the Atwood machine (*see* Problem 6-17), the unbalanced force accelerates both masses.

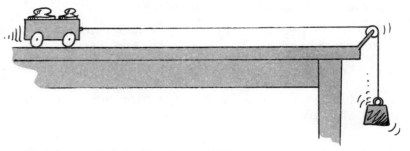

Given: The cart has a mass of 1.00 kg. The weight has a mass of 0.25 kg in one experiment and 2.50 kg in a second experiment.

Find: (a) What is the total mass being accelerated in the first experiment?
(b) What is the acceleration of the cart in the first experiment?
(c) What is the acceleration in the second experiment?

6-20 A block of ice slides down a wooden plank with very little friction. The acceleration of the ice is less than g, because only the component of the weight parallel to the plank is unbalanced.

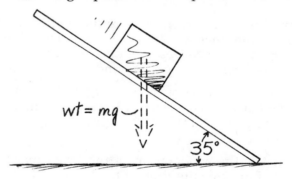

Given: The plank makes an angle of 35° with the horizontal. The block of ice has a mass of 11.5 kg.

Find: (a) How much force is acting on the block in the direction parallel to the plank?
(b) What is the acceleration of the block?
(c) What would be the acceleration of the ice block if it had twice as much mass?

ANSWERS (*These answers were calculated by keeping an extra digit in intermediate results to avoid round-off errors.*)

6-8 (a) 33 m/s² = 3.4 g; (b) 5.67 × 10⁵ N ≅ 5.7 × 10⁵ N.

6-9 (a) 0.60 kg; (b) 0.26 m.

6-10 (a) 1.43 m/s², (b) 9.3 × 10⁴ N; (c) 7 × 10³ N.

6-11 (a) 1.96 × 10⁴ N; (b) 4.6 × 10³ N; (c) 2.3 m/s² down.

6-12 (a) 83 N; (b) 539 N ≅ 5.4 × 10² N; (c) 8.7°.

6-13 (a) 4.0 N; (b) 612 m.

6-14 (a) 0.0819 N ≅ 8.2 × 10⁻² N; (b) 25.9 N ≅ 26 N; (c) 26 m/s² = 2.6 g.

6-15 (a) 588 N ≅ 5.9 × 10² N; (b) 468 N ≅ 4.7 × 10² N.

6-16 (a) 3.77 × 10⁵ m/s² ≅ 3.8 × 10⁵ m/s²; (b) 4.4 × 10³ N; (c) 1.6 × 10⁻³ s.

6-17 (a) 1.85 m/s² ≅ 1.9 m/s²; (b) 6.7 m/s.

6-18 (a) 32°; (b) 6 m/s² = 0.6 g.

6-19 (a) 1.25 kg; (b) 1.96 m/s² ≅ 2.0 m/s²; (c) 7.0 m/s.

6-20 (a) 65 N; (b) 5.6 m/s² = 0.57 g; (c) 5.6 m/s² = 0.57 g (same as part (b)).

Motion Along a Curved Path

7

▶ Projectile Motion

In Chapter 6 we discussed acceleration produced by a force that was either acting in the same or opposite direction from that which the object was originally moving. The resulting motion was always confined to a straight line. For an object to vary from a straight-line path, it must experience a component of force perpendicular to its original direction. The result is a curved path—our subject for this chapter.

Suppose you are being served coffee in an airliner. Your coffee cup is traveling at 500 mph in a horizontal path. As your coffee is being poured, it accelerates under its own weight downward towards the cup.

$$F = ma$$
$$wt = mg$$

The acceleration of the coffee depends only upon its mass and weight. Neither one of these are affected by the fact that the coffee is traveling horizontally. You would be surprised, in fact, if the steward were to compensate for the velocity of the plane by pouring the coffee ahead of your cup.

The idea that motion in one direction is independent of motion in another is the **principle of independence of motion.** This principle is quite general and useful for understanding any motion along a

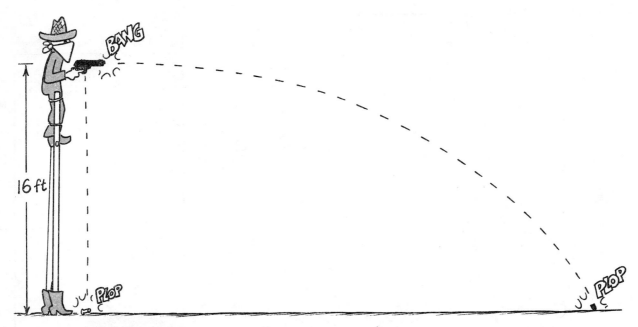

FIGURE 7-1 *The shell casing ejected and the bullet fired horizontally from a 16-ft height will both require 1 s to hit the ground.*

curved path. Let us consider the motion of a bullet fired horizontally from an automatic pistol. As the bullet leaves horizontally the spent shell casing drops vertically. You might think that the bullet would hit the ground first, since it is going faster. Or you might argue that the shell casing would hit the ground first, since it does not go as far.

Neither of these contentions is correct. Let us neglect air friction and assume that the ground is level. The bullet and the shell casing fall the same vertical distance in the same amount of time, just as coffee pours the same in a jet plane traveling 500 mph as it does when the plane is standing on the runway. Both the bullet and casing hit the ground at the same time, because they both start out with zero vertical velocity. The horizontal velocity of the bullet does not affect its motion in the vertical direction.

Let us study an example with the automatic pistol, assuming a convenient height and a reasonable distance for the bullet to go before hitting the ground, and we will see how to calculate the muzzle velocity of the bullet as it leaves the gun. We will assume that the bullet and shell casing both leave the gun without any vertical velocity. The large horizontal velocity of the bullet carries the bullet a considerable horizontal distance, say, 600 ft, while the shell casing falls right beside the shoe or boot of the person firing the gun.

A nice height to deal with would be 16 ft. Although we would need a tall Texan to hold the gun, we already know that anything dropped from rest falls 16 ft in the first second. We therefore know that the shell casing takes exactly 1.0 s to hit the ground.

If you didn't know that, you could figure it out from the equation for distance traveled under constant acceleration, which we developed in Chapter 5:

$$d = \tfrac{1}{2}at^2 + v_i t$$

For our case, the distance is downward and the acceleration is downward, g, and the initial velocity is zero—at least in the downward direction.

$$d_\downarrow = \tfrac{1}{2}gt^2 + 0$$

This equation can be solved for time by multiplying both sides of the equation by 2, dividing both sides by g,

$$\frac{2d_\downarrow}{g} = t^2$$

and taking the square root of both sides,

$$t = \sqrt{\frac{2d_\downarrow}{g}}$$

Inserting 16 ft for the height $d_\downarrow$ and the known value for g in English units, we get the time of fall, which you probably already knew:

$$t = \sqrt{\frac{2(16 \text{ ft})}{32 \text{ ft/s}^2}}$$

$$\boxed{= 1.0 \text{ s}}$$

The only new part of this discussion is that the same argument is good for the bullet as for the shell casing. Neither is being acted upon by any force other than its weight after it leaves the gun. Both start with zero initial *vertical* velocity, and so both have the same vertical height at all times. The bullet takes 1.0 s to fall 16 ft, just as the shell casing does.

In the meantime, the bullet travels a large horizontal distance, 600 ft, while it is falling. This is motion at a constant velocity, since there is no force in the horizontal direction, neglecting air friction. We can therefore use the definition of velocity to calculate the horizontal velocity with which the bullet left the gun:

$$v_\rightarrow = \frac{d_\rightarrow}{t}$$

$$= \frac{600 \text{ ft}}{1.0 \text{ s}}$$

$$\boxed{= 600 \text{ ft/s}}$$

This is also the horizontal component of velocity that the bullet has as it strikes the ground. The only thing that happens to its velocity is that it picks up a vertical component as it falls.

▶ Trajectories

The path of a projectile is called the projectile's **trajectory.** People frequently anticipate the trajectory of a projectile in order to predict its **range** (how far it will go), whether the projectile happens to be a wad of waste paper they are tossing into a waste basket or a package they are dropping from an airplane.

Perhaps you recall old newsreel films showing packages being released from airplanes. The packages were often long and cylindrical, with fins on the back so that they would line themselves up with wind to reduce air friction. The movies showed the packages dropping away from the plane but staying underneath the plane all

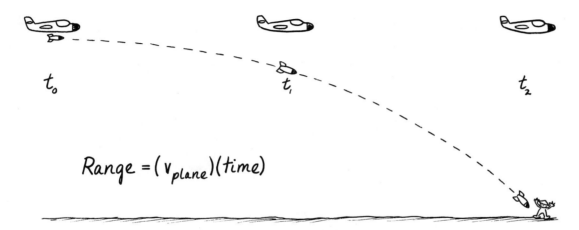

$$\text{Range} = (v_{plane})(\text{time})$$

FIGURE 7-2 *The range of an object dropped from an airplane flying horizontally is the velocity of the plane times the time it takes to get over the landing site, since the object always stays under the plane.*

the way down to the ground. The physics demonstrated by these old newsreels is clear, whether we agree with the application or not. The airplane is always directly over the packages, because of the law of independence of motion. The plane stays over the packages, because they are both going at the same horizontal velocity. The packages are falling, but they are only falling down. They are coasting in the sideward direction, since their shape makes the sideward force of air friction small.

Pretend, if you like, that these packages contain food and medical supplies; and let us calculate the range or horizontal distance from the drop point to the place where the packages hit the ground. Knowing that the plane is overhead when our package hits the ground (since they are both traveling at the same horizontal velocity, $v_{\rightarrow}$), we can compute the range, $d_{\rightarrow}$, from the definition of velocity:

$$v_{\rightarrow} \equiv \frac{d_{\rightarrow}}{t}$$

We can get the time, t, in this expression from the vertical motion of the package, since it is moving in both directions at once. The time required for an object to drop a vertical height, $d\downarrow$, from the plane is given by the formula for the distance traveled under constant acceleration:

$$d\downarrow = \tfrac{1}{2}gt^2$$

As in the example with the automatic pistol, we can solve this equation for the time of flight in terms of this downward distance and downward acceleration g. First multiply both sides by 2, to get rid of the $\frac{1}{2}$, then solve for time and take the square root:

$$2d\downarrow = gt^2$$
$$t = \sqrt{\frac{2d\downarrow}{g}}$$

The time thus calculated from the accelerated motion in the downward direction can then be used to determine $d_{\rightarrow}$, the horizontal distance, from the constant horizontal velocity $v_{\rightarrow}$, since both motions are going on at the same time:

$$d_{\rightarrow} = v_{\rightarrow}t$$
$$= v_{\rightarrow}\sqrt{\frac{2d\downarrow}{g}}$$

We can calculate the range to the target area for a plane flying at 600 mph (880 ft/s) at a height of 10,000 ft by first calculating the time necessary for the package to fall:

$$t = \sqrt{\frac{2d_\downarrow}{g}}$$

$$= \sqrt{\frac{2(10{,}000 \text{ ft})}{32 \text{ ft/s}^2}}$$

$$= 25 \text{ s}$$

The distance to the target area can be found from the horizontal velocity of the plane, assuming that components of the package retain the horizontal component of velocity (880 ft/sec) while falling:

$$d_\rightarrow = v_\rightarrow t$$

$$= (880 \text{ ft/s})(25 \text{ s})$$

$$= 22{,}000 \text{ ft}$$

Thus the package must be dropped a horizontal distance of 22,000 ft, or a bit more than four miles, before the target.

A frequent error in reasoning is to try to put the acceleration due to gravity $g_\downarrow$ and the horizontal velocity $v_\rightarrow$ into the same equation for distance under a constant acceleration.

$$d_\nearrow = \tfrac{1}{2}g_\downarrow t^2 + v_\rightarrow t \qquad \text{WRONG!}$$

The resulting distance $d_\nearrow$ is neither in the horizontal nor vertical direction. This is wrong. Don't do it. The whole idea of independence of motion is that you have to treat velocity and distance in one direction independently from velocity and distance in the other. Only time links the two equations of motion for the two different directions.

CHECK QUESTION

An object is thrown horizontally with a velocity of 10 m/s from the top of a building that is 19.6 m tall. (a) How long does it take the object to fall the vertical distance of 19.6 m? (b) How far does the object travel horizontally while it is in free fall?

Answer: (a) 2.0 s, (b) 20 m

▶ Vertical Components of Velocity, $v_\uparrow$

We have so far only considered cases where the initial motion of a projectile is entirely horizontal, so that there has been no initial vertical component to worry about. Under these conditions we found the formula for vertical displacement is

$$d_\uparrow = \tfrac{1}{2}g t^2$$

If there is an initial vertical component of velocity $v_\uparrow$ as well, the vertical displacement is

$$d_\uparrow = \tfrac{1}{2}gt^2 + v_\uparrow t$$

If, for example, a juggler throws a ball upward with a velocity of 64 ft/s, the height to which the ball rises at time t is given by

$$d_\uparrow = \tfrac{1}{2}(-32 \text{ ft/s}^2)t^2 + (64 \text{ ft/s})t$$

(If we consider up the positive direction, acceleration due to gravity is negative.) At the end of the first second $t = 1$ s and the height is given by

$$\begin{aligned} d_\uparrow &= \tfrac{1}{2}(-32 \text{ ft/s}^2)(1^2 \text{ s}^2) + (64 \text{ ft/s})(1 \text{ s}) \\ &= -16 \text{ ft} + 64 \text{ ft} \\ &= 48 \text{ ft} \end{aligned}$$

At the end of the second second $t = 2$ s and the height is given by

$$\begin{aligned} d &= \tfrac{1}{2}(-32 \text{ ft/s}^2)(4 \text{ s}^2) + (64 \text{ ft/s})(2 \text{ s}) \\ &= -64 \text{ ft} + 128 \text{ ft} \\ &= 64 \text{ ft} \end{aligned}$$

At the end of the third second $t = 3$ s. The acceleration term is beginning to gain on the velocity term, and the height is back to

$$\begin{aligned} d_\uparrow &= \tfrac{1}{2}(-32 \text{ ft/s}^2)(9 \text{ s}^2) + (64 \text{ ft/s})(3 \text{ s}) \\ &= -144 \text{ ft} + 192 \text{ ft} \\ &= 48 \text{ ft} \end{aligned}$$

At the end of the fourth second, however, we find that $d_\uparrow = 0$ ft. The acceleration term is equal to the velocity term, and the ball is back in the juggler's hand.

$$\begin{aligned} d_\uparrow &= \tfrac{1}{2}(-32 \text{ ft/s}^2)(16 \text{ s}^2) + (64 \text{ ft/s})(4 \text{ s}) \\ &= -256 \text{ ft} + 256 \text{ ft} \\ &= 0 \text{ ft} \end{aligned}$$

At any time after 4 s have elapsed, the negative acceleration term is even larger than the positive velocity term. The resulting negative distance just means that the ball has gone below the original starting position.

Thus, we see that the equation for distance under a constant acceleration can be used to predict the height of the ball, or any other object for that matter, at the end of any given time. We just insert the initial velocity and time into the equation

$$d_\uparrow = \tfrac{1}{2}(-g)t^2 + v_\uparrow t$$

We could also turn the equation around to find the time at which the ball will be at any particular height. It is just a quadratic equation in time, and you probably learned how to solve it when you took algebra. We could always haul out the old quadratic equation and turn the crank to find the time for any height we like.

There is one particular height of frequent interest, however, for which we don't even need the quadratic equation. That is the height zero. We often want to know the time that something thrown up in the air will come down again. As we saw from our numerical example, $d_\uparrow = 0$ when the ball returns to the juggler's hand after 4 s. We

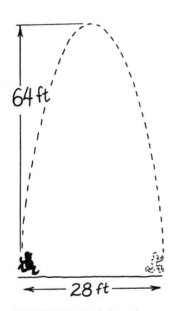

FIGURE 7-3 *A juggler sees the path of an object moving along with him as motion in a vertical path alone. From the vantage point of the audience it is seen to travel a curved path called a parabola.*

could find the time necessary for the ball to return to this height for any initial velocity if we solve the equation for values of time with the height equal to zero.

$$0 = -\tfrac{1}{2}gt^2 + v_\uparrow t$$

One way to find the values for time that will produce this result is to factor out time from the two terms in the preceding equation. We will then have the product of two things that are equal to zero.

$$0 = (-\tfrac{1}{2}gt + v_\uparrow) \cdot t$$

This expression is of the form $a \cdot b = 0$. We know that if that is true, either a must be equal to zero or b must be equal to zero (or they may both be equal to zero). Then we see that either

$$t = 0$$

or

$$-\tfrac{1}{2}gt + v_\uparrow = 0$$

will satisfy the equation. The first solution, $t = 0$, we knew already. That was when the ball was first thrown. The other solution gives the condition that must be satisfied when the ball returns. The minus term in the parentheses must equal the plus term.

$$\tfrac{1}{2}gt = v_\uparrow$$

When that happens, all the stuff inside the parentheses of the preceding equation, are equal to zero, and the result will produce a zero value in the distance formula. Solve this expression for time, and we have a general solution to the question of how long it takes something thrown upward to return to the hand.

$$t = \frac{2\,v_\uparrow}{g}$$

For the juggler's ball in our example, the initial velocity $v_\uparrow$ was 64 ft/s. The acceleration due to gravity, g, is just 32 ft/sec^2.

$$t = \frac{2(64 \text{ ft/s})}{32 \text{ ft/s}^2} = 4 \text{ s}$$

We already found this result by the trial-and-error method. We started at time zero and put in larger and larger values for time until the height of the ball returned to zero.

The trial-and-error method could be used to find the time of return for any initial upward velocity. This is called calculation by successive approximation. It is useful if one has either a programable computer or lots of time. The algebraic method is more powerful, however, because it brings the force of mathematical logic to bear on the problem and produces a general solution that is true for a similar situation.

Suppose that the juggler in the preceding example was walking across the stage. He has a certain horizontal velocity, say, 7 ft/s. The juggler would see the ball going up and right back down into his hand. The audience, however, would see the ball going up and back down in a curved path, as it is moving horizontally at a constant velocity. The ball would meet the juggler after he had walked 28 ft.

$$v \equiv \frac{d}{t}$$

$$d_\rightarrow = v_\rightarrow t$$
$$= (7 \text{ ft/s})(4 \text{ s})$$
$$= 28 \text{ ft}$$

CHECK QUESTION

An object is thrown with a velocity having an upward component of 9.8 m/s and a sideward component of 30 m/s. (a) How much time does it take the object to go up and come back down to the same level? (b) What horizontal distance does the object travel in this amount of time?

Answer: (a) 2.0 s, (b) 60 m

The shape of the path taken by the ball, or any other projectile, for that matter, is a parabola.

Let us examine the parabolic path of a projectile to see how we can control the range by changing the direction of the initial velocity, usually called muzzle velocity. Assume that the magnitude of the muzzle velocity has been determined already but that we can point the muzzle at whatever direction we like to hit any target within some maximum range. As we shall see, this maximum range happens when we point the muzzle upward at 45°.

It should come as no surprise to find that you can hit a closer target by pointing the muzzle upwards at a sharper angle. Increasing the angle with the horizontal decreases the horizontal component of the muzzle velocity while increasing the vertical component, as shown in Figure 7-4, as well as Figure 7-5. A smaller horizontal component of velocity should make for a shorter range, since the horizontal distance, $d_\rightarrow$, that something travels while in the air a certain amount of time, t, can be found from the definition of velocity (as we just saw in our discussion of the juggler):

$$d_\rightarrow = v_\rightarrow t$$

Thus, making $v_\rightarrow$ smaller also makes the range smaller.

However, you can do the same thing by decreasing the angle, thereby increasing $v_\rightarrow$. The reason is that a smaller angle to the hori-

FIGURE 7-4 *A parabolic trajectory is calculated by resolving the muzzle velocity into horizontal and vertical components.*

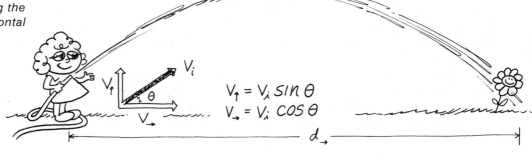

$$V_\uparrow = V_i \sin \theta$$
$$V_\rightarrow = V_i \cos \theta$$

zontal has the effect of decreasing the time, t, that whatever we shot up in the air takes to go up and come down.

As in our discussion of the juggler, we can use algebra to figure out the time it takes something to go up and come back down to the same level. We just use the formula for distance traveled under constant acceleration,

$$d_\uparrow = \tfrac{1}{2}(-g)t^2 + v_\uparrow t$$

and solve it for when the object is back down on the ground again, that is, when $d_\uparrow = 0$. As before, the two solutions for time that satisfy this condition are $t = 0$ when the projectile left the ground, and t equals twice the velocity divided by the acceleration due to gravity, that is, when the object got back:

$$t = \frac{2v_\uparrow}{g}$$

We therefore have an algebraic expression for the range $d_\rightarrow$ in terms of the horizontal component of the velocity and the time, and another expression for the time in terms of the vertical component of velocity and the acceleration due to gravity. Putting these two formulas together, we have a general solution for range:

$$d_\rightarrow = v_\rightarrow \left(\frac{2v_\uparrow}{g}\right)$$

Expressing these components in terms of the angle θ that the initial velocity makes with the horizon, we must use trigonometry to determine where the projectile will land:

$$d_\rightarrow = v_i \cos\theta \, \frac{2(v_i \sin\theta)}{g}$$

Rearranging, we get

$$d_\rightarrow = \frac{2v_i^2 \sin\theta \cos\theta}{g}$$

This expression is symmetrical with respect to sine and cosine. Since the sine of an angle is the cosine of the complement of that angle, we can replace the angle in the expression with its complement and get the same range. This means that we will hit the same target aiming at θ above the horizontal as we will aiming at $(90° - \theta)$. The maximum range for any given velocity is achieved by shooting

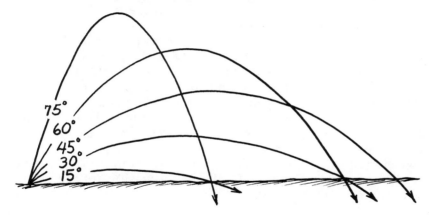

FIGURE 7-5 *Any target within the maximum range can be hit by shooting high or by shooting low.*

at 45° above the horizontal. At that angle, the sine and the cosine are equal.

CHECK QUESTION

A stone is thrown from a slingshot with a velocity of 32 ft/s with a certain upward angle, starting from ground level. (a) How far does the stone go before hitting the ground if the certain angle is 45°? Remember, $\sin 45° = \cos 45° = \sqrt{2}/2$. (b) How far does the stone go if the certain angle is 37°? The sine of 37° is 0.60; the cosine of 37° is 0.80.

Answer: (a) 32 ft, (b) 31 ft

► Circular Motion

The man in the moon
Came tumbling down,
And ask'd his way to Norwich.
He went by the south,
And burnt his mouth
With supping hot pease porridge.

Ritson's *Gammer Gurton's Garland* (1784)

Have you ever wondered what holds the moon up? The gravitational force of the earth certainly extends as far as the moon. In fact the gravitational force acting in the other direction is what causes our ocean tides. If the force of attraction is there, why is it that the moon does not come tumbling down?

The answer is that nothing holds the moon up. The moon is falling. The gravitational attraction of the earth causes the moon to accelerate continually toward the earth. Fortunately, it misses. Instead of falling into, it falls around, the earth. Newton's first law of motion tells us that the moon would travel at constant velocity along a straight line if there were no force acting upon it. The gravitational force acting on the moon causes it to fall *away from a straight line* and into a circle.

We started this chapter talking about the motion of coffee as it is poured in a moving airplane. In all our discussions of the motion of different objects in a gravitational field, we assumed that we were

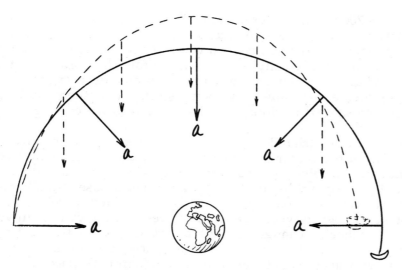

FIGURE 7-6 *Acceleration in circular motion is always perpendicular to the path of the object and toward the center, whereas the acceleration in parabolic motion, shown in dotted lines for comparison, is always in the same (downward) direction.*

close to the surface of the earth and that the force on the object was always in the same direction, namely down. If, for some reason, you were to throw that cup of coffee up into the air, its path would be a parabola. The force of gravity acting on it would be constant in magnitude and constant in direction in the sense that the force at any one point would be parallel to the force at any other point. If an object is acted upon by a force that is constant in magnitude but directed always perpendicular to its direction of motion, the object will travel in a circle. Both situations are shown in Figure 7-6. The path of the coffee cup, shown in dotted lines, is a parabola. The path of the moon, which travels in a nearly circular path, is shown with the direction of its acceleration always pointing toward the center of the earth and always perpendicular to its direction of motion.

This perpendicular acceleration produces a change in the moon's velocity—in direction, not magnitude. The moon is 380,000 km from earth today, and it is traveling at 3,700 km/hr. Tomorrow it will still be 380,000 km away and (assuming a circular path) will still be trav-

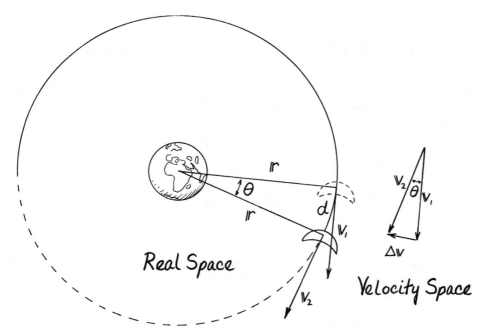

FIGURE 7-7 *The velocity in circular motion changes in direction but not in magnitude. Two successive velocities and their change in velocity form an isosceles triangle similar to a triangle formed by two successive positions (measured from the center) and the change in position.*

eling at 3,700 km/hr. After having fallen for a whole day, the moon's velocity will have changed, but only in direction. This change in velocity may be represented as a velocity vector $\Delta \mathbf{v}$, which must be added to the first velocity, $\mathbf{v}_1$, in order to form the new velocity, $\mathbf{v}_2$, as a resultant.[1] This change in velocity can be most easily visualized by drawing a diagram in velocity space, a mathematical space having the same directions as real space but in which length represents speed.

An isosceles triangle is formed in real space by the two radii drawn from the center of the earth and by the distance traveled by the moon. (Actually the base of this "triangle" is the curved part of a circle, but we can think of it as straight if we choose a small enough time interval.) An isosceles triangle is also formed in velocity space by the two velocity vectors and the corresponding change in velocity. The velocity triangle cannot be congruent to the real space triangle—they aren't even in the same space—but it is similar. Each of the two velocities is tangent to the circle in real space and therefore perpendicular to the corresponding radius. Since two lines that are perpendicular to two other lines must meet at the same angle, the apex angle of both isosceles triangles must be the same. That makes them similar.

We can now use these two similar triangles to get the acceleration from the ratio of their sides. It is proper to drop the vector notation at this point, because we are interested in taking the ratio of their magnitudes. Using the fact that the ratio of the sides of two similar triangles are always equal, we can say that the base of the velocity space triangle is to one side as the base of the real space triangle is to one side. We can eliminate the distance, Δd, by expressing it in terms of

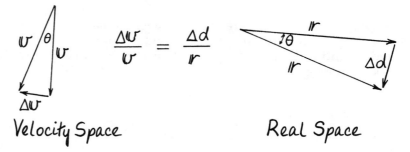

$$\frac{\Delta v}{v} = \frac{\Delta d}{r}$$

FIGURE 7-8 *Similar triangles have proportional sides.*

Velocity Space   Real Space

the velocity, v, of the moon and the time, Δt, during which the change of velocity takes place. From the definition of velocity, we have

$$\Delta d = v \cdot \Delta t$$

Combining expressions, we get

$$\frac{\Delta v}{v} = \frac{v \cdot \Delta t}{r}$$

[1] The triangle, the Greek letter delta, is a mathematical operator that stands for "the difference in". In this case, we use vector notation:

$$\Delta \mathbf{v} \equiv \mathbf{v}_2 - \mathbf{v}_1$$

$\Delta \mathbf{v}$ is the vector difference between the velocities.

Multiplying both sides by v and dividing both sides by Δt, we get

$$\frac{\Delta v}{\Delta t} = \frac{v^2}{r}$$

This is the moon's acceleration toward the center, since the definition of acceleration is the change in velocity over the change in time:

$$a_c = \frac{v^2}{r}$$

This is usually referred to as the **centripetal** acceleration, from the Latin words meaning *to move toward the center.*

The force that causes this acceleration is, from Newton's second law,

$$F = ma_c$$

which is given as

$$F_c = \frac{mv^2}{r}$$

Since this force is toward the center, it is called **centripetal force.**

People often say mistakenly that an earth satellite is held up by something called **centrifugal** force. That cannot be, because *centrifugal* means *to flee the center.* It is true that there is a centrifugal force, but it acts on the earth, not on the satellite. If there were such a thing as centrifugal force acting on the moon in such a direction as to balance the force of gravity, the net force acting on the moon would be zero and the moon would travel in a straight line.[2] It is because the only force acting on the moon is *toward the center* of its path that the moon travels in a circle, being deflected away from straight motion. Therefore, nothing holds the moon up. It is falling.

CHECK QUESTION

An object is traveling at a velocity of 3.0 m/s around a circle having a radius of 1.0 m. (a) What is the centripetal acceleration of the object? (b) If the object has a mass of 2.0 kg, how much centripetal force is necessary to keep the object in its circular path?

Answer: (a) 9.0 m/s², (b) 18 N

[2] Centrifugal force can also refer to an imaginary force, called an **inertial force,** caused by our frame of reference accelerating. Something will seem to accelerate even though it doesn't if it is in a frame of reference where everything around it accelerates in the opposite direction. Such an object would require a force to keep it from accelerating in such a frame of reference. Examples are common. Passengers in a car feel a centrifugal force when the car goes around a curve. The car is really pushing them toward the center to deflect their bodies from a straight line. They might interpret this as a mysterious force pulling them away from the center, just as they might feel a mysterious attraction for the front of the car as it stops, but that is only because their frame of reference is accelerating.

The problems in this chapter are related to motion along a curved path. Motion of this type is produced by a force having a component perpendicular to the velocity of the object.

When an external force acting on an object remains in the same direction, such as the force of gravity on the earth's surface pulling always downward on a projectile, the resulting path is a **parabola.** The horizontal motion of a projectile is independent of the vertical motion. A projectile moves at a constant horizontal velocity while experiencing a constant vertical acceleration under the influence of its weight. The horizontal component of its motion is described by

$$d_{\rightarrow} = v_{i\rightarrow}t$$

while the vertical component is described by

$$d_{\downarrow} = \tfrac{1}{2}gt^2 + v_{i\downarrow}t$$

When an external force acting on an object remains always perpendicular to the direction in which the object is going, the resulting path is **circular.** The acceleration changes the direction, but not the magnitude, of the velocity. The acceleration is called **centripetal acceleration,** a_c, since it is always directed toward the center of the circle. Centripetal acceleration is related to the velocity v of the object and the radius r of the circle by

$$a_c = \frac{v^2}{r}$$

The force that produces this acceleration is called **centripetal force.** In accordance with Newton's second law, $F = ma_c$, the force acting toward the center in circular motion is given by

$$F_c = m\,\frac{v^2}{r}$$

EXAMPLE PROBLEMS

7-1 David Fall is rigging a hoist on the flat roof of his house and manages to knock himself over the edge. Fortunately, the flat roof of his garage some distance away is lower than the roof of his house. He makes a mighty jump in the horizontal direction.

Given: The roof of the garage is 1.35 m lower than that of the house. The garage is 2.4 m from the house.

Find: (a) How much time does David take to fall to the level of his garage?

(b) What horizontal velocity does David need to reach the roof of his garage in time?

Your Solution

Discussion

This problem is an example of projectile motion of the simple kind, where the initial velocity is all in the horizontal direction. At the same time David is falling, he must be moving sideways at a certain rate if he is going to make it to the roof. The motions in the two directions may be thought of as independent. His downward motion is a constant acceleration with a zero initial velocity. His sideward motion is a constant velocity, which must be figured out from the data.

Sample Solution 7-1

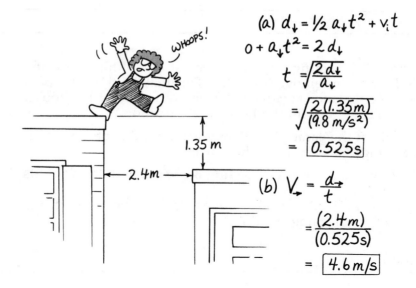

(a) $d_\downarrow = \frac{1}{2} a_\downarrow t^2 + v_i t$

$0 + a_\downarrow t^2 = 2 d_\downarrow$

$t = \sqrt{\dfrac{2 d_\downarrow}{a_\downarrow}}$

$= \sqrt{\dfrac{2(1.35m)}{(9.8 \, m/s^2)}}$

$= \boxed{0.525s}$

(b) $V_\rightarrow = \dfrac{d_\rightarrow}{t}$

$= \dfrac{(2.4m)}{(0.525s)}$

$= \boxed{4.6 \, m/s}$

Discussion

The downward motion follows the rule for distance traveled under constant acceleration where the initial velocity is zero. The initial velocity is zero in the downward direction, since it is all sidewards and has no downward component. The formula for distance under constant acceleration is therefore solved for time and the result is then used in part (b). This time from part (a) can be used in part (b), since both vertical and horizontal motions, while independent, take place literally at the same time.

7-2 Daniel is admiring the view from a cliff and wishes to know how high it is above the ocean. Having taken physics, Dan casts a stone horizontally from the cliff and measures the time necessary for it to splash into the sea.

Given: The stone is in the air for 3.0 s and strikes the water 75 m horizontally from where it is thrown.

Find: (a) How high is the cliff above the ocean?
(b) How fast did Dan throw the stone horizontally?

Your Solution

Discussion

This problem is another example of projectile motion of the kind where the initial velocity is all in the horizontal direction. In this case, however, both of the distances in the horizontal and vertical directions are unknown and must be found. You must know which is constant velocity and which is constant acceleration.

Sample Solution 7-2

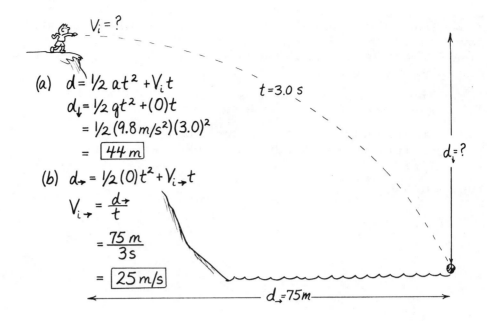

(a) $\quad d = \frac{1}{2}at^2 + V_i t$

$\quad d_\downarrow = \frac{1}{2}gt^2 + (0)t$

$\quad = \frac{1}{2}(9.8\,m/s^2)(3.0)^2$

$\quad = \boxed{44\,m}$

(b) $\quad d_\rightarrow = \frac{1}{2}(0)t^2 + V_{i\rightarrow}t$

$\quad V_{i\rightarrow} = \dfrac{d_\rightarrow}{t}$

$\quad = \dfrac{75\,m}{3s}$

$\quad = \boxed{25\,m/s}$

Discussion

The downward distance is found from the formula for motion under constant acceleration. The initial velocity is taken as zero in this part, because only the downward component counts, and the initial velocity is all in the horizontal direction as the problem is stated.

Part (b) may be thought of as being solved with the same formula, but the acceleration in the sideward direction is zero. From this point of view, this formula results in the definition of velocity in the case where velocity is constant.

7-3 A handball player must meet the ball on the first bounce and return it with the same motion so that it strikes the front wall before hitting the floor again.

Given: A handball leaves the player's hand with a velocity of 20 m/s at 25° above the horizontal from a point just above the floor and 7.2 m from the front wall.

Find: (a) What are the horizontal and vertical components of the velocity?
(b) How much time will the ball take to reach the front wall, in seconds?
(c) How high up on the wall will it hit?

Your Solution

Discussion

This problem is an example of trajectory motion in which the projectile has an initial velocity with both horizontal and vertical components.

The ball may still have a vertical component as it hits the front wall. A good handball shot is a bounce off the ceiling after the ball has hit the front wall.

Sample Solution 7-3

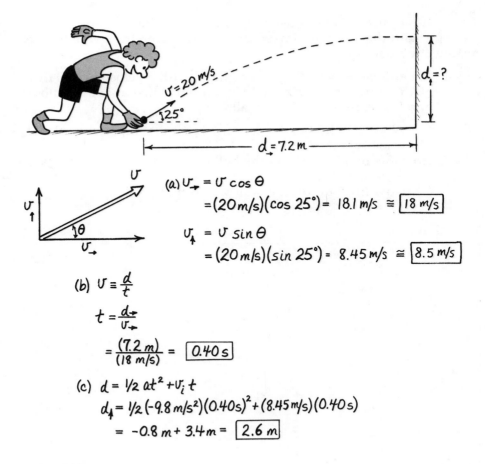

(a) $v_\rightarrow = v \cos \theta$

$\qquad = (20 \text{ m/s})(\cos 25°) = 18.1 \text{ m/s} \cong \boxed{18 \text{ m/s}}$

$v_\uparrow = v \sin \theta$

$\qquad = (20 \text{ m/s})(\sin 25°) = 8.45 \text{ m/s} \cong \boxed{8.5 \text{ m/s}}$

(b) $v \equiv \dfrac{d}{t}$

$t = \dfrac{d_\rightarrow}{v_\rightarrow}$

$\qquad = \dfrac{(7.2 \text{ m})}{(18 \text{ m/s})} = \boxed{0.40 \text{ s}}$

(c) $d = \frac{1}{2} a t^2 + v_i t$

$d_\uparrow = \frac{1}{2}(-9.8 \text{ m/s}^2)(0.40 \text{ s})^2 + (8.45 \text{ m/s})(0.40 \text{ s})$

$\qquad = -0.8 \text{ m} + 3.4 \text{ m} = \boxed{2.6 \text{ m}}$

Discussion

The components of the initial velocity are found using the sine and cosine of the angle to the vertical. The time necessary to reach the front wall can be found from the definition of velocity, since the horizontal component of the motion is at a constant velocity.

The vertical motion follows the rule for distance under constant acceleration. The acceleration due to gravity is considered negative in this instance, because it is in a direction opposite to that of the initial velocity. The vertical component of the initial velocity must be used in this equation, since we cannot assume that the ball is at the top of its trajectory as it hits the front wall.

7-4   A golf player uses a nine iron when he is close to the green. He then wants to hit the ball with a high trajectory so that it will not roll much after landing.

Given: A golf ball leaves the nine iron with a velocity of 35 m/s at an angle of 65° above the horizontal.

Find: (a) What are the horizontal and vertical components of the velocity?
(b) How far does the ball go horizontally before landing?
(c) What is the maximum height of the ball while it is in the air?

Your Solution

Discussion

This problem is an example of a full trajectory, in which a projectile goes up and comes back to its original level while it is moving sidewards at a constant velocity. This special case assumes level ground.

Sample Solution 7-4

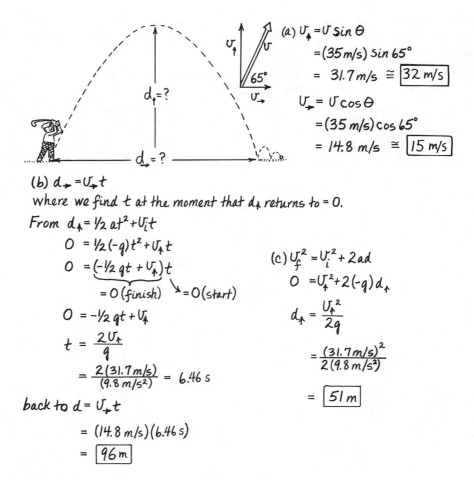

(a) $U_\uparrow = U \sin \theta$

$= (35\,m/s) \sin 65°$

$= 31.7\,m/s \cong \boxed{32\,m/s}$

$U_\rightarrow = U \cos \theta$

$= (35\,m/s) \cos 65°$

$= 14.8\,m/s \cong \boxed{15\,m/s}$

(b) $d_\rightarrow = U_\rightarrow t$

where we find t at the moment that $d_\uparrow$ returns to $= 0$.

From $d_\uparrow = \frac{1}{2} a t^2 + U_i t$

$0 = \frac{1}{2}(-g) t^2 + U_\uparrow t$

$0 = \underbrace{(-\frac{1}{2} g t + U_\uparrow)}_{= 0 \,(finish)} \underbrace{t}_{= 0 \,(start)}$

$0 = -\frac{1}{2} g t + U_\uparrow$

$t = \frac{2 U_\uparrow}{g}$

$= \frac{2(31.7\,m/s)}{(9.8\,m/s^2)} = 6.46\,s$

back to $d = U_\rightarrow t$

$= (14.8\,m/s)(6.46\,s)$

$= \boxed{96\,m}$

(c) $U_f^2 = U_i^2 + 2 a d$

$0 = U_\uparrow^2 + 2(-g) d_\uparrow$

$d_\uparrow = \frac{U_\uparrow^2}{2g}$

$= \frac{(31.7\,m/s)^2}{2(9.8\,m/s^2)}$

$= \boxed{51\,m}$

Discussion

The time necessary for the ball to return to its original level is found by solving the formula for distance traveled under constant acceleration for the case where the distance is zero. This quadratic equation has two solutions. The trivial solution is when $t = 0$ at the start of the trajectory. The next time that d equals zero is the time that the ball has fallen back to its original level.

An alternative approach to part (b) would be to use the general solution derived for this problem in the text of this chapter. That approach is not very good, however, as it involves memorizing an equation that holds true for only the special case of level ground. It would be better to derive that equation when you need it.

7-5 A pilot can withstand ten times the acceleration due to gravity, 10 g, if she is wearing a special tight-fitting g-suit designed to keep the blood from rushing from her brain.

Given: An 85-kg pilot who is traveling east at 900 km/hr decides she wants to go west. She makes a circle with 10 g acceleration.

Find: (a) What is the radius of the smallest circle in which the plane can turn around?

(b) How much centripetal force acts on the pilot while she is in the turn?

Your Solution

Discussion

This problem illustrates centripetal acceleration and force in circular motion. An object traveling in a circular path must experience a certain acceleration toward the center to deflect it from a straight line, the amount of the acceleration depending on the velocity of its motion and the radius of the circle. This force must result from an unbalanced force, in accordance with Newton's second law.

Sample Solution 7-5

(a) $a_c = \dfrac{v^2}{r}$

$r = \dfrac{v^2}{a_c}$

$= \dfrac{(900 \text{ km/hr})^2}{(10g)} \left(\dfrac{1.0g}{9.8 \text{ m/s}^2}\right)\left(\dfrac{10^3 \text{ m}}{1 \text{ km}}\right)^2\left(\dfrac{1 \text{ hr}}{3600 \text{ s}}\right)^2$

$= 638 \text{ m} \cong \boxed{0.64 \text{ km}}$

(b) $F = ma$

$= (85 \text{ kg})(10g)\left(\dfrac{9.8 \text{ m/s}^2}{1g}\right)\left(\dfrac{1 \text{ N}}{1 \text{ kg m/s}^2}\right)$

$= \boxed{8.3 \times 10^3 \text{ N}}$

Discussion

The radius of the circle can be found by solving the centripetal acceleration formula. The acceleration is given in terms of the acceleration due to gravity and must be changed into meters per second squared using a conversion factor. The units of velocity must also be changed into meters per second to make things match.

The force necessary to produce this acceleration can be found from Newton's second law.

7-6 A little girl riding a merry-go-round thinks she feels a force pulling her away from the center. Actually, the merry-go-round horse is pushing her toward the center to keep her from going in a straight line.

Given: The merry-go-round takes her around in a circle of radius 3.0 m every 7.5 s. Her mass is 25 kg.

Find: (a) What is the magnitude of her velocity at any moment?
(b) How much centripetal force is acting on the little girl?

Your Solution

Discussion

This is another problem illustrating centripetal acceleration in circular motion. The velocity of the object, however, is given in terms of the radius of the path and the time necessary to travel one circumference.

Sample Solution 7-6

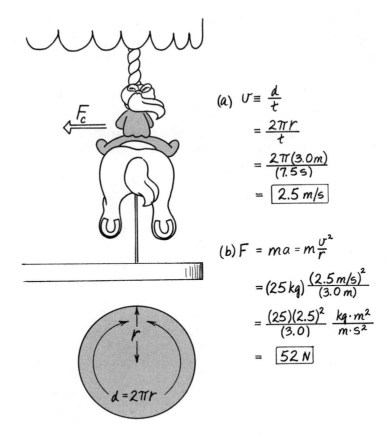

(a) $v \equiv \dfrac{d}{t}$

$ = \dfrac{2\pi r}{t}$

$ = \dfrac{2\pi(3.0m)}{(7.5s)}$

$ = \boxed{2.5 \ m/s}$

(b) $F = ma = m\dfrac{v^2}{r}$

$ = (25kg)\dfrac{(2.5 \, m/s)^2}{(3.0 \, m)}$

$ = \dfrac{(25)(2.5)^2}{(3.0)}\dfrac{kg \cdot m^2}{m \cdot s^2}$

$ = \boxed{52 \ N}$

Discussion

Velocity is defined as distance over time. Since the time of one complete revolution is given, the proper distance to use is one circumference of the circle.

Part (b) is found from Newton's second law, where the acceleration is given by the centripetal acceleration formula.

ESSENTIAL PROBLEMS

7-7 A hammerhead flies off the handle and sails horizontally from a tall building. The hammerhead breaks a window 20 stories below on a neighboring building some distance away.

Given: The window is 50 m below and 35 m horizontally from the worker.

Find: (a) How much time elapses between the loss of the hammerhead and the breaking of the window?
(b) What is the initial horizontal velocity of the hammerhead?

7-8 A trapeze artist lets go at the bottom of his swing when his velocity is all in the horizontal direction. A trampoline has been placed on the floor at the correct horizontal distance to catch him.

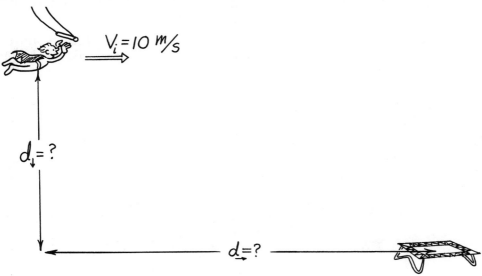

Given: His initial velocity is 10 m/s in the horizontal direction. He is in the air for 1.1 s.

Find: (a) What vertical distance does he fall in that amount of time?
(b) How far must the trampoline be positioned in the horizontal direction to catch him?

7-9 A short basketball player shoots the ball upward with the correct horizontal and vertical velocity components to bounce it from the backboard into the net.

Given: The ball has a horizontal velocity of 8.4 m/s and an initial upward velocity of 5.3 m/s. The ball is in the air for 0.40 s before striking the backboard.

Find: (a) What is the magnitude and direction of the resultant initial velocity?
(b) How far does the ball go in the horizontal direction?
(c) What is the vertical velocity of the ball just as it hits the backboard?

7-10   A football player starts the game by kicking the ball into the air from the ground. It is caught on the first bounce by a player on the other team, who was clever enough to judge the ball's initial velocity and forecast its landing place.

Given: The ball has an initial velocity of 25.0 m/s at an angle of 52.0° above the horizontal.

Find: (a) What are the initial horizontal and vertical components of velocity?
(b) How far does the ball go before landing?
(c) What is the maximum height of the ball while in the air?

7-11   An automobile is rounding a turn of constant radius of curvature. A passenger, who happens to be wearing a blindfold at the time, notices that the armrest is pushing her toward the center of the turn with a constant force.

Given: The passenger has a mass of 60 kg, and the force with which the armrest is pushing against her is 150 N. The forward velocity of the automobile is 26 m/s.

Find: (a) If the armrest is exerting the only horizontal force on the passenger, how much acceleration is produced?
(b) What is the radius of curvature of the turn?

7-12   A bug is sitting on a long-playing record going around at a constant speed.

Given: The record is turning at a constant rate of 33.3 revolutions per minute. The bug, whose mass is 1.4 g, is sitting 12 cm from the center of the record.

Find: (a) How long does it take the bug to go around once?
(b) How fast is the bug going? (*Hint:* Figure out the distance associated with part (a).)
(c) How much centripetal force acts on it?

MORE INTERESTING PROBLEMS

7-13   An automatic pistol fires a bullet horizontally at the same time that it kicks the brass shell casing out the side of the gun, also with a horizontal velocity. Both bullet and shell casing hit the ground at the same time.

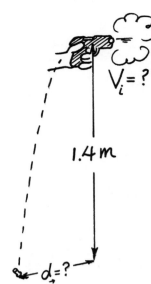

Given: The pistol is 1.4 m above level ground. The bullet hits the ground 250 m downrange. The shell casing has an initial velocity of 0.8 m/s in the horizontal direction.

Find: (a) What is the time that both the bullet and shell casing are in the air?

(b) What is the muzzle velocity of the bullet?

(c) How far does the shell casing hit the ground from a point directly beneath the gun?

7-14   A juggler tosses a ball in the air while she is walking across the stage. The horizontal component of the ball's velocity is the same as that of the juggler, so that, to the juggler, the ball seems to simply go up and come back down.

Given: The initial upward component of the ball's velocity is 15 m/s. Both the ball and juggler have a horizontal velocity of 0.85 m/s.

Find: (a) How long does it take the ball to reach the top of its trajectory?

(b) How long does it take for the ball to go up and come back down?

(c) How far does the juggler walk between the time she tosses the ball and the time she catches it?

7-15   A cannon fires a cannon ball on level ground. Once the muzzle velocity has been determined by the gunpowder charge, the range may be adjusted by changing the elevation or angle above the horizontal at which the cannon is fired.

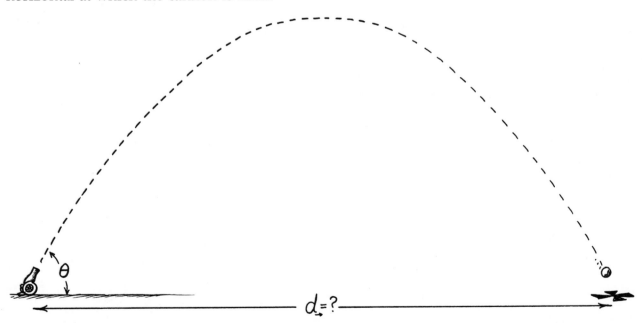

Given: The cannon ball has a muzzle velocity of 200 m/s at an angle of 27° above the horizontal.

Find: (a) How far does the cannon ball go horizontally before hitting the ground?

(b) What other elevation will give the same range?

ANSWERS

7-7	(a) 3.2 s; (b) 11 m/s.
7-8	(a) 5.9 m; (b) 11 m.
7-9	(a) 9.9 m/s at 32° to horizontal; (b) 3.4 m; (c) 1.4 m/s.
7-10	(a) 15.4 m/s, 19.7 m/s; (b) 61.9 m; (c) 19.8 m.
7-11	(a) 2.5 m/s^2; (b) 270 m.
7-12	(a) 1.8 s; (b) 0.418 m/s $\cong$ 42 cm/s; (c) 2.0 × 10^{-3} N.
7-13	(a) 0.53 s; (b) 468 m/s; (c) 0.43 m.
7-14	(a) 1.53 s; (b) 3.06 s $\cong$ 3.1 s; (c) 2.6 m.
7-15	(a) 3.3 km; (b) 63°.

Momentum 8

▶ Newton's Third Law

The Zen Buddhist contemplates the sound of one hand clapping. The physics student contemplates the concept of a single force. Both are only one-half of an interaction.

Nothing can exert a force on an object without that object exerting an equal force back. We cannot touch another person without that person touching us. We cannot nudge something without that thing nudging us back. Forces always come in pairs. Newton's third law of motion expresses this by saying: **For every force of action, there is an equal but opposite force of reaction** on another object.

It does not matter which is the **action** force and which is the **reaction,** as long as we realize that neither exists without the other. If the force exerted by a bat on a ball is the action, the force exerted by the ball on the bat is the reaction. Which is which depends on your point of view. You may wish to think of the bat as pushing the ball in the positive direction. Then the ball pushes the bat in the negative direction. If action is positive, then reaction is negative in direction, but equal in magnitude:

$$F_a = -F_r$$

The reaction force acting on the bat has a noticeable effect, which can be felt by the batter. Applying Newton's second law to both the

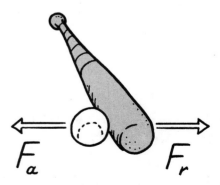

FIGURE 8-1 *For every force of action there is an equal and opposite force of reaction.*

action and reaction forces, we see that the backward acceleration of the bat is proportional to the forward acceleration of the ball.

$$F_{ball} = -F_{bat}$$
$$m_{ball}\, a_{ball} = -m_{bat}\, a_{bat}$$

We can see that the accelerations so produced are inversely proportional to the ratio of their masses. The more massive the bat, the less it will accelerate backwards relative to the forward acceleration of the ball.

$$a_{bat} = -\left(\frac{m_{ball}}{m_{bat}}\right) a_{ball}$$

The reaction force is always present, even when its effect is not noticeable. A 120-lb student in free fall will be accelerated downward by her unbalanced weight. The unbalanced reaction force is 120 lb pulling the earth upward. The acceleration of the earth is not noticed, because the earth is vastly more massive than the student.

$$m_{student}\, a_{student} = -m_{earth}\, a_{earth}$$
$$a_{earth} = -\left(\frac{m_{student}}{m_{earth}}\right) a_{student}$$
$$= -\left(\frac{10^2 \text{ kg}}{10^{25} \text{ kg}}\right) g$$
$$= -10^{-23} g$$
$$\cong -10^{-22} \text{ m/s}^2$$

Since the mass of the earth is on the order of 10^{25} kg and the mass of the student is something like 60 kg, which is on the order of 10^2 kg, the acceleration of the student will be about 10^{23} times as great as that of the earth. The earth therefore accelerates upward at $10^{-23}\, g$. Although this acceleration is far too weak to measure with even the most delicate instrument, it nonetheless exists—at least in theory.

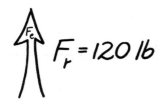

FIGURE 8-2 *The unbalanced force pulling down on the student is equal to the unbalanced force pulling up on the earth.*

▶ Momentum

The mass of an object is of great importance in determining the effect of its interaction with other objects. Intuitively this is to be expected. When a massive truck interacts with a small car, we expect the larger vehicle to win the argument, so to speak, but we also expect the car to have at least some effect on the larger vehicle. Suppose they approach each other, each traveling with a velocity of 10 m/s; the smaller vehicle has a mass of 1,000 kg and the truck a mass of 5,000 kg. Further, suppose the wreckage fuses together after the collision and moves off with a single final velocity, v_f. Let us see how we can predict this velocity.

Newton's third law tells us that the force of action F_{car} on the small car is equal and opposite to the force of reaction F_{truck} on the truck:

$$F_{car} = -F_{truck}$$

Applying Newton's second law, $F = ma$, to this expression, we obtain the following relationship between the accelerations:

$$m_{car}\, a_{car} = -m_{truck}\, a_{truck}$$

$$v_{car} = 10 \ m/s \qquad\qquad v_{truck} = -10 \ m/s$$

We will call one velocity positive and the other negative. (This is just the same trick we have been using all along to deal with vector quantities that are in the same or opposite directions. Since vectors in the same direction add numerically, and those in the opposite direction subtract, we can treat them just like scalars as long as we choose a positive direction.) Let us say that right is positive and left is negative. That makes the initial velocity of the car v_{car} positive and the velocity of the truck v_{truck} negative. The acceleration of each vehicle can be expressed in terms of the change of velocity to its final value, v_f, from its initial value, v.

FIGURE 8-3 *Two massive objects approaching each other each have a certain amount of momentum.*

$$a = \frac{v_f - v}{\text{time}}$$

Thus, the average acceleration of the car and the truck can be stated as follows:

$$m_{car}\frac{(v_f - v_{car})}{t} = -m_{truck}\frac{(v_f - v_{truck})}{t}$$

The time of acceleration, t, is the same for the car as for the truck, since the force of action and the force of reaction both start and stop at the same instant. We can therefore eliminate the time term by multiplying both sides of the equation by t:

$$m_{car}(v_f - v_{car}) = -m_{truck}(v_f - v_{truck})$$

Thus, the mass times the change in velocity is the same for the car as it is for the truck. This happens to be a general result that holds for balls and bats or any other two things that interact. Always the mass times the change of velocity of one is equal to the mass times the change of velocity of the other. You might expect at this point that there is something special about the product of mass and velocity. If so, you are quite right—as we can see by continuing our example to get everything in terms of initial and final values.

$$v_f = ?$$

FIGURE 8-4 *The velocity at time of collision can be found from the fact that the momentum after the collision is equal to the vector sum of the momentums before.*

We can manipulate the preceding expression to express everything as the products of masses and velocities:

$$m_{car} \, v_f - m_{car} \, v_{car} = -m_{truck} \, v_f + m_{truck} \, v_{truck}$$

Moving those terms containing v_f to one side of the equation,

$$m_{car} \, v_f + m_{truck} \, v_f = m_{car} \, v_{car} + m_{truck} \, v_{truck}$$

If we factor out v_f,

$$(m_{car} + m_{truck})v_f = m_{car} \, v_{car} + m_{truck} \, v_{truck}$$

we see that the total mass of the car and truck, the final mass of the wreck, times its final velocity turns out to be the sum of the products of the initial masses and velocities of the car and the truck. This is an important expression. It is true in general. The final product of mass and velocity is always equal to the sum of the initial products of mass and velocity.

This expression allows us to predict, among other things, the final velocity of two things after they collide. We have been looking at the simple case in which they stick together after collision, but it can be extended to any case. Before doing so, however, let us finish our example. At this point we need to remember to use a negative value for the truck's velocity. (We assumed that it was negative to start with, but we only used the sign convention to work out the algebra. The algebraic result is therefore quite general, and we need to go back to the sign convention when we insert particular values.)

$$(1{,}000 \text{ kg} + 5{,}000 \text{ kg})v_f = (1{,}000 \text{ kg})(10 \text{ m/s})$$
$$+ (5{,}000 \text{ kg})(- 10 \text{ m/s})$$
$$(6{,}000 \text{ kg})v_f = 10{,}000 \text{ kg} \cdot \text{m/s} - 50{,}000 \text{ kg} \cdot \text{m/s}$$
$$v_f = \frac{(10{,}000 - 50{,}000) \text{ kg} \cdot \text{m/s}}{6{,}000 \text{ kg}}$$
$$v_f = - 6.7 \text{ m/s}$$

The minus sign means that the final velocity of the wreck is in the original direction of the truck, since we had decided that the truck's initial velocity was negative. We could have made the truck's velocity positive and the car's negative, but the result would have been the same. The velocity of the wreck would still have been in the truck's direction.

The wreck does not have to go in the direction of the more massive object. The small car could have made up for the fact that it only has one-fifth the mass of the truck by going five times as fast. Then we would substitute those data into the expression:

$$[(1{,}000 \text{ kg}) + (5{,}000 \text{ kg})]v_f = (1{,}000 \text{ kg})(50 \text{ m/s})$$
$$+ (5{,}000 \text{ kg})(- 10 \text{ m/s})$$
$$(6{,}000 \text{ kg})v_f = 50{,}000 \text{ kg} \cdot \text{m/s} - 50{,}000 \text{ kg} \cdot \text{m/s}$$
$$= 0$$

In this case, the velocity of the wreck would turn out to be zero:

$$v_f = 0$$

If the car were going faster than five times the velocity of the truck, the velocity of the wreck would be in the direction of the car. Perhaps this is the reason people drive small cars so fast.

We have seen that the ultimate result depends upon not only the mass of the colliding objects but upon the products of their masses and velocities. This product is called the **momentum** of the object.

$$\text{Momentum} \equiv (mv)$$

One commonly used convention is to let the letter p stand for momentum (since m is already taken for mass). Rather than introduce a new symbol in this book, however, we will either spell it out or use (mv) to stand for momentum.

CHECK QUESTION

A woman having a mass of 60 kg is running with a velocity of 2.0 m/s. What is her momentum?

Answer: 120 kg-m/s

Momentum is important, because it is absolutely conserved in all interactions of nature—at least all that we have examined so far. We always find the same amount of momentum after an interaction as before. We can see why this conservation law is true, or at least consistent with Newton's laws of motion, by starting with the third law. If the force of action is considered to be in the positive direction, it will be equal to a negative force of reaction:

$$F_a = -F_r$$

This is the third law of motion. We can combine it with the second law to prove the law of conservation of momentum. First we express Newton's second law in terms of momentum. Acceleration, a, as it appears in the usual expression of the second law

$$F = ma$$

is the same thing as the rate of change of velocity. We should write it out as such, because momentum is expressed in terms of velocity rather than acceleration. The change in velocity, $\Delta v = (v_f - v_i)$, may be used to express the acceleration in Newton's second law in terms of a change in momentum:

$$
\begin{aligned}
F &= m\frac{\Delta v}{\Delta t} \\
&= \frac{m(v_f - v_i)}{\Delta t} \\
&= \frac{mv_f - mv_i}{\Delta t}
\end{aligned}
$$

Then, multiplying both sides of the equation by Δt, we have: **Force times the time interval over which the force acts equals the change of momentum.**

$$F\,\Delta t = (mv_f - mv_i)$$
$$\text{Impulse} = \Delta(mv)$$

The term *impulse* is frequently used for the product of force times time. In this form Newton's second law says: **Impulse equals change in momentum.**

Newton's third law tells us that every action impulse $F_a\Delta t$ is accompanied by a reaction impulse $F_r\Delta t$ acting on some other body. The forces of action and reaction are always equal and act for exactly the same amount of time since the forces are equal at every instant. Therefore, every impulse that changes the momentum of any object is accompanied by an equal and opposite reaction impulse, which brings about an opposite change of momentum of some other body in the system. Thus, the total amount of momentum in any system obeying Newton's laws of motion remains constant.

▶ Impulse

Golfers and batters learn that following through on a swing improves the amount of momentum transferred to a ball. Following through does not increase the amount of force, but it has the effect of

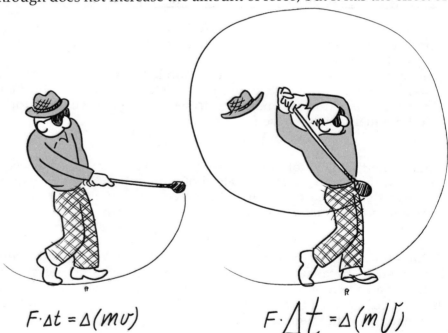

FIGURE 8-5 *A golfer improves the momentum transference with a smooth follow-through, as in (b).*

keeping the force applied for a longer period of time. The position of the club or bat long after the ball has left can reveal what was going on in the short period when they were in contact.

We can visualize the effect of follow-through with a graph of force versus time. The areas under the curves in Figure 8-6 repre-

FIGURE 8-6 *Although the maximum force is the same, the longer the club is held in contact with the ball, the greater the impulse.*

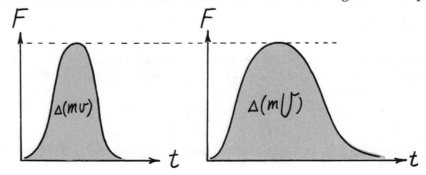

sent the change in momentum $\Delta\,(mv)$ resulting from a force applied over a period of time. In the simplest approximation, we can think of the impulse area as being a rectangle bounded by time, Δt, and average force, F_{ave}. This approximation is adequate for calculating the average force in a situation. For example, suppose a full golf swing takes 1.3 s and results in a 0.1-kg ball traveling 200 yd in about 5 s. Let us see if we can use these reasonable figures to tell how much force had to be applied to the golf ball in the short time it was in contact with the club head. We can figure out the momentum of the ball from the mass and velocity. Its mass is given, and its velocity is such that it travels 200 yards, or about 200 m, in 5 s. Thus, its velocity must be roughly 40 m/s:

$$v \equiv \frac{d}{t} = \frac{200 \text{ m}}{5 \text{ s}} = 40 \text{ m/s}$$

Now that we know the momentum, at least in principle, we can estimate the time, Δt, that the club head is in contact with the ball. If it is in contact while the club swings through about 10°, and if the full

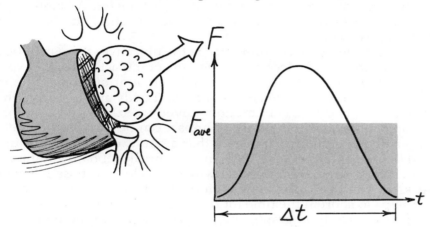

FIGURE 8-7 *The impulse applied to a golf ball can be approximated by a rectangle on a force-time graph.*

swing goes through about $1\frac{1}{4}$ of a full circle—or about 450°—remembering our assumption that a full swing takes 1.3 s, the time of contact is approximately 0.03 s:

$$\Delta t = \left(\frac{10°}{450°}\right) 1.3 \text{ s} \cong 0.03 \text{ s}$$

We showed in the last section that one way to state Newton's second law is to say that impulse equals change in momentum:

$$F \,\Delta t = \Delta(mv)$$

We can solve this expression for force, F, since we have figured out everything else:

$$F = \frac{\Delta(mv)}{\Delta t}$$

$$= \frac{(0.1 \text{ kg})(40 \text{ m/s})}{0.03 \text{ s}} = 1.3 \times 10^2 \text{ N}$$

This force is the average force with which the club pushes on the ball. The actual force would first increase to some peak value and then decrease to zero at the end of the time of contact. Although the actual force is a complicated function of time, we can approximate the momentum transfer produced by it with a rectangle having the same area in force-time space. (Since impulse equals force times time, area in force-time space represents momentum transfer.) We should recognize that the actual peak value of the force might be something like twice the average value. Since the average force is about 130 N, which would be about 30 lb, the peak force would be on the order of twice this much.

We therefore see that a golfer, whose strength lies in force applied to the club, can achieve a maximum momentum transfer by prolonging the time of contact as much as possible through skill. This same skill becomes important in other ball-and-stick games such as baseball and tennis.

A related skill is involved in sports where a force is applied with the bare, or nearly bare, hand. A karate expert or boxer has learned to "think through" the point of impact when delivering a blow. Part of his skill is increasing the momentum transfer by keeping up the force as long as possible, but another part of his skill is in increasing the amount of momentum available for transfer. A karate expert or boxer thus learns to put as much of the arm and body behind the blow as possible.

In sports where maximum force is the objective, there may be an advantage in actually decreasing the time of impact. A karate expert, for example, can generate just a certain amount of momentum depending on the mass of his hand, arm, and body and his strength in setting those body parts in motion. Given a certain fixed amount of available momentum, the force of impact can be increased by delivering a sharp blow that transfers that amount of momentum in as short a time as possible. If he wants to break a board, he has the board solidly supported so that it will give as little as possible before breaking. Delivering the blow with the hard part of the hand or foot also helps to shorten the time for the momentum transfer. Viewed in terms of a force-time graph, the idea is to squeeze the momentum area along the time axis to produce a high peak force.

The same idea can be used in reverse by the recipient of the blow. A boxer, for example, frequently wishes to reduce the average force of a blow. Not his own, of course, but that of the other fellow. A professional boxer learns to put a good deal of momentum of both his arm and body behind a punch he is delivering, just as does the karate expert. The average street fighter might be seriously injured, or even killed, by such a blow. A professional boxer's fist is therefore legally considered a deadly weapon outside the ring. Inside the ring, it is not so deadly. Not only do both contestants wear padded gloves but they are both trained to take punches with minimum injury. Both fighters know from considerable experience that they will be hurt more if they get hit flat-footed than if they roll with the punch.

Moving the head a few feet per second one way or the other might seem to make little difference compared to the hundred or so feet per second that a fist is traveling. By moving his head away from the punch, however, the boxer stretches his muscles and joints in the proper direction to give with the blow and spread the momentum transfer over a maximum amount of time.

If you have ever gone skating at a roller rink, you probably have had direct experience with the transfer of momentum. It does not

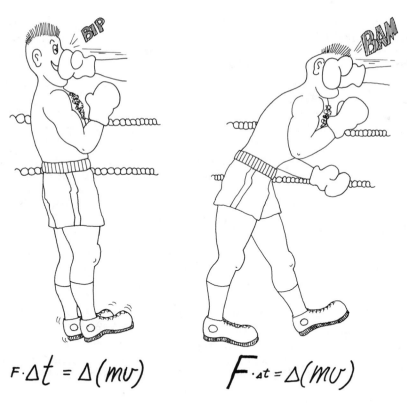

$$F \cdot \Delta t = \Delta (mv)$$

$$F \cdot_{\Delta t} = \Delta (mv)$$

FIGURE 8-8 *Force can be reduced for momentum transfer by prolonging the time of contact.*

seem to hurt as much to fall on an oak floor at a roller rink as it does to fall on a concrete sidewalk. The wooden floor may seem very hard to the touch, but it has considerably more give than concrete. If you happened to be resting in the middle of such a floor and another person tripped over you, you would feel the floor shake as he landed. The floor's small amount of give allows momentum to be spread out over more time, thus reducing the average force.

CHECK QUESTION

A roller skater falls and experiences a force of 600 N for 0.40 s on the first bounce. (a) How much momentum is transferred to the floor by this impulse? (b) If the skater has a mass of 60 kg, what is her change of velocity associated with this change of momentum?

Answer: (a) 240 kg-m/s, (b) 4.0 m/s

▶ Impulse and Momentum Conservation

Falling is an art. The first thing you learn when you take judo lessons is how to fall. That is because you will be doing a lot of it, and you do not want to get hurt. You are taught to land flat on your side so that the force is spread out over as much area as possible. This reduces the pressure at any one point. This is not surprising to most people, but you are also taught to do something that does surprise many people. You are taught to slap the mat as hard as you can with your arm when you fall. A few people might think this is to make a loud sound to scare your opponent. Of course this isn't true. The arm slap has the effect of reducing the force on the rest of your body.

As you approach the mat, you have a certain amount of momentum. You cannot lessen this amount of momentum, but you can transfer as much as possible into your arm. The more momentum transferred to the mat by your arm, the less is transferred by the rest of your body.

Your arm has a less complicated internal structure than your rib cage and can better take the blow. A judo expert has toughened arms to take a mighty blow with little pain. A white-belt student, however, must live with a sore arm for many days. Many a neophyte has given up this martial art after the first time he slipped and fell on the sidewalk after having developed the proper reaction for falling on the mat. Concrete has much less give than those thin little mats used in judo schools, and this seemingly slight difference becomes painfully apparent as the student slaps hard with his arm as he hits the ground.

Still another example of impulse and the conservation of momentum is a common physics demonstration used to demonstrate inertia. It consists of a massive ball attached to a support with string. A downward force is applied to the same kind of string attached to the bottom of the ball. The question is, "Which string will break first?" This turns out to be a trick question. Either string may be made to break at the will of the string puller.

The top string may obviously be made to break by pulling with a slowly increasing force. It supports both the weight of the ball and the force in the bottom string. The top string breaks, however, only if the bottom string is not jerked.

Having thus established that forces in the same direction add up to a larger force, the string puller ties the ball back up and gives a hard jerk on the bottom string. The bottom string then breaks, leaving the top string intact. A short explanation for this phenomenon is that the inertia of the ball protects the top string. A fuller explanation involves the concepts of impulse and momentum.

When the bottom string is jerked, the force builds up suddenly to the point where the string breaks. This impulse does in fact transmit some momentum to the ball, and the ball does move downward. If the impulse applied by the bottom string is sharp enough, however, the momentum transfer will be small and the ball will move downward only slowly. The force in the top string then increases as the ball stretches it, and this increased force tends to stop the ball by producing an impulse in the opposite direction. If the ball is moving slowly enough, however, the impulse applied by the top string will be spread out over enough time so that the increased force will never reach the point of breaking the string. The ball will return to the equilibrium position, bouncing up and down slightly due to the springiness of the string, and eventually come to rest.

Momentum is conserved, although the forces are different. A sudden jerk of the hand transfers momentum to the ball through the bottom string before it breaks, and this same momentum is transferred over a longer period of time through the top string to the hook in the ceiling. The graphs of force versus time in Figure 8-10 represent this momentum as the area under the force curve and above the

FIGURE 8-9 *Either the top or bottom string can break, depending on how the bottom string is pulled.*

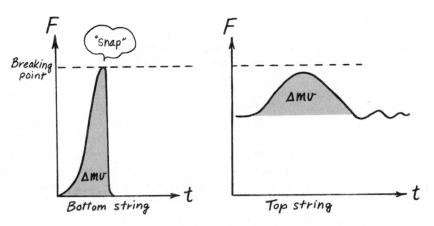

FIGURE 8-10 *The bottom string breaks if it is jerked suddenly; the same impulse is spread out over a longer time in the top string.*

equilibrium position. Equilibrium for the lower string is zero when no force is applied. Equilibrium for the top string is the force of the weight of the ball. The areas are the same, because momentum is conserved. All the momentum gained by the ball is lost through the top string by the time the ball returns to its equilibrium position. Even a modest increase in time will keep the force in the top string below the breaking point, even though it has a head start—in terms of force—on the bottom string.

▶ Velocity Before and After an Interaction

We have seen how the law of conservation of momentum can be derived from Newton's second and third laws of motion. Every bit of momentum that something receives had to be given up by something else, since every force of action must be accompanied by an equal and opposite force of reaction and since both forces act for exactly the same period of time.

Thus, the total momentum of a system remains unchanged. Different parts of the system can trade the momentum around by pushing on each other with equal and opposite forces of action and reaction, but the total momentum before an interaction is the same amount as after.

This assumes that we choose a large enough system to include everything that is in the act, so to speak. That includes everything pushing or pulling on anything else in the system with no unbalanced external forces that we would need to worry about. The little car and big truck were considered to be a single system, because we were concerned only with the forces of action and reaction between these two objects. We did not need to worry about the weight of these two vehicles, which would involve forces of action and reaction between the objects and the earth, because this force was balanced by their wheels acting against the ground. In any real case there would also be a force of friction between the wheels and the ground, but we ignored this little problem in our simple example, because its effect would be small during the short time that the wreck takes place compared to the vastly greater forces of collision. Thus, the car and the truck were taken together as an isolated system, because the only other forces acting on them either were balanced or could be ignored.

The law of conservation of momentum is particularly useful in the kind of situation where a system of objects is sufficiently isolated

$$V_{Before} = 3 \text{ m/s}$$

FIGURE 8-11 *An ice skater carrying a clown has a certain amount of momentum before throwing him.*

from external forces to allow us to worry only about internal forces acting between objects in the system. We looked at one such situation when we considered the wreck between the car and the truck—when two things come together and stick. That example developed the concept of momentum. We will now consider the opposite kind of situation to see how the conservation of momentum can be applied. We will look at two things that are together to begin with and consider their relative velocities after an internal force throws them apart.

Suppose, for example, that you go to the Ice Follies and notice one of the performers throwing a midget clown across the ice. This supposedly humorous event comes to your attention because all the lights in the house are off except for one spotlight that follows the full-sized skater as he is about to throw the little clown. As an artifice to build suspense, the spotlight follows the big guy and you are left to wonder what happens to the little clown.

Let us further suppose that you had already noticed from the program that the little clown has a mass of only 25 kg. You quickly estimate the mass of the big skater as about 75 kg and his velocity before and after he throws the clown. You observe him going along at about 3 m/s toward your right when he had the clown in his arms. After throwing the clown, you observe him going backward, toward your left, at about 1 m/s.

The law of conservation of momentum tells you that the momentum before the skater throws the clown is equal to the momentum after. The momentum before, $(mv)_{before}$, is the total mass times the velocity before, v_{before}:

$$(mv)_{before} = (m_{skater} + m_{clown})v_{before}$$

The momentum after the skater throws the clown, $(mv)_{after}$, is the sum of the momentums of the skater, $(mv)_{skater}$, and that of the clown, $(mv)_{clown}$:

$$(mv)_{after} = (mv)_{skater} + (mv)_{clown}$$

The law of conservation of momentum states that the momentum before is equal to the momentum after:

$$(mv)_{before} = (mv)_{after}$$

Inserting the expressions for the momentums before and after,

$$(m_{skater} + m_{clown})v_{before} = (m_{skater}\, v_{skater}) + (m_{clown}\, v_{clown})$$

where m_{clown} and m_{skater} are the masses of the little clown and the big skater, v_{before} is their velocity before the skater throws the clown, and v_{clown} and v_{skater} are the velocities of the clown and skater after the clown is thrown. We can solve this expression for the velocity of the clown by rearranging terms.

$$m_{clown}\, v_{clown} = (m_{skater} + m_{clown})v_{before} - m_{skater}\, v_{skater}$$

$$= m_{skater}\, v_{before} + m_{clown}\, v_{before} - m_{skater}\, v_{skater}$$

Dividing both sides of the equation by the mass of the clown,

$$v_{clown} = \frac{m_{clown}\, v_{before} + m_{skater}\, v_{before} - m_{skater}\, v_{skater}}{m_{clown}}$$

$$v_{after} = 1 \text{ m/s} \qquad v_{clown} = ?$$

FIGURE 8-12 *The total momentum after the clown is thrown is equal to their momentum before.*

This is the general solution for the final velocity of the little clown, but it can be simplified with a bit of factoring. Notice that the first term in the numerator contains the mass of the clown, just as it appears in the denominator, while the other two terms contain the mass of the big skater. We can therefore separate the fraction and factor out the mass of the skater from the last two terms.

$$v_{clown} = \frac{m_{clown} \, v_{before}}{m_{clown}} + \frac{m_{skater} \, (v_{before} - v_{skater})}{m_{clown}}$$

We can therefore divide the mass of the clown into itself in the first fraction to show that the final velocity of the clown is equal to the initial velocity, v_{before}, plus a term that is the ratio of the masses times the change in velocity of the skater.

$$v_{clown} = v_{before} + \frac{m_{skater}}{m_{clown}} \, (v_{before} - v_{skater})$$

To see how this works in our numerical example, we put in values for mass and velocity. The masses of the big skater and little clown are, as you remember, 75 kg and 25 kg. If we say that the initial velocity of both skater and clown is $+3$ m/s, however, we must call the final velocity of the skater -1 m/s. Assigning a positive value to the initial velocity assumes that the positive direction is from left to right. The final velocity of the skater must then be negative, since it is in the opposite direction.

$$v_{clown} = 3 \text{ m/s} + \frac{75 \text{ kg}}{25 \text{ kg}} \, [(3 \text{ m/s}) - (-1 \text{ m/s})]$$

The double minus in the last term gives a physically reasonable result. The skater's change in velocity is truly 4 m/s. Not only did he stop going forward at 3 m/s but he kept on accelerating in the same direction to start going 1 m/s backwards.

$$v_{clown} = 3 \text{ m/s} + (3)(4 \text{ m/s})$$

The skater's change in velocity is multiplied by three before adding it to the clown's initial velocity, because the skater has three times as much mass.

$$v_{clown} = 3 \text{ m/s} + 12 \text{ m/s}$$
$$= 15 \text{ m/s}$$

We should pause for a moment to see what is important about this solution. The numerical result might give us a nice solid, satisfied feeling that we know the answer, but it is not nearly as valuable as the general solution in terms of understanding what is going on. Just looking at the algebraic result, it is clear that the final velocity of the clown would be increased by either increasing the mass of the big skater or decreasing the mass of the clown. By just subtracting the initial velocity v_{before} from both sides of the general solution, we can see that the change in velocity of the clown is directly proportional to the change in velocity of the big guy.

$$v_{clown} - v_{before} = \frac{m_{skater}}{m_{clown}} (v_{before} - v_{skater})$$

It is also directly proportional to the mass of the big skater, which is in the numerator on the right-hand side, and inversely proportional to the mass of the little clown, which is in the denominator.

The importance of this algebraic solution to the problem does not mean that nothing is to be learned from putting in the numbers. The trick of using a minus value for one of the velocities, for example, might well escape you unless you were to try it for the purpose of seeing whether or not the answer turns out to be reasonable. What is more, this trick is not a trivial thing, as its proper use depends on understanding the vector nature of velocity. The point is, however, that the algebraic solution has a great deal to say by itself and deserves your attention before you move on to putting in the numbers.

You should not, on the other hand, look upon the general solution of a problem like this as being so important that you try to memorize it as an equation you need to know. You should instead be able to derive it. There are many occasions when the law of conservation of momentum is useful. The preceding solution comes directly from that law as it is applied to a problem at hand. If we were to see another performance at the Ice Follies in which the spotlight followed the clown, a slightly different application of the same law would yield the recoil velocity of the skater. In yet another performance, if both the final velocities were observed, we could apply the law of conservation of momentum to calculate the mass of the skater to see if the estimated value was correct.

If you were not lucky enough to attend more than one performance of the Ice Follies in one season, consider another example. Suppose you were sentenced to be shot for a political crime. You might well wish to estimate the recoil of the gun with which you are to be shot. The momentum before the gun is fired is zero. The law of conservation of momentum tells us that the total momentum of the bullet and recoiling gun is also zero.

$$0 = m_{bullet} \, v_{bullet} + m_{gun} \, v_{gun}$$

where m_{bullet} is the mass of the bullet, m_{gun} is the mass of the gun,

FIGURE 8-13 *The velocity of recoil is proportional to the ratio of the masses.*

v_{bullet} is the velocity of the bullet, and v_{gun} is the unknown recoil velocity of the gun. Solving for v_{gun}, we get

$$m_{\text{gun}} \, v_{\text{gun}} = -m_{\text{bullet}} \, v_{\text{bullet}}$$

$$v_{\text{gun}} = - \left(\frac{m_{\text{bullet}}}{m_{\text{gun}}} \right) v_{\text{bullet}}$$

Immediately upon realizing that the recoil velocity of the gun increases with the mass of the bullet, you would know the proper response to give when your antagonist offers you your last wish. You might claim that your crime demands more severe punishment than being struck by a tiny bullet. Insist that the mass of the bullet match the magnitude of your crime, that the mass of the bullet be much more massive than the gun, and that your antagonist should be the one to pull the trigger.

CHECK QUESTION

A cannon ball having a mass of 5×10^3 kg is fired from a cannon having a mass of 3×10^3 kg. The velocity of the cannon ball is 6 m/s. What is the velocity of the cannon?

Answer: -10 m/s

The problems in this chapter have to do with impulse and momentum. The **momentum** of an object is defined as the product of its mass times its velocity.

$$\text{Momentum} = (mv)$$

In terms of momentum, Newton's second law of motion states that force is equal to the rate of change of momentum with respect to time.

$$F = ma = \frac{\Delta(mv)}{\Delta t}$$

This may be restated as **impulse** equals change of momentum, where impulse is defined as the average force times the change of time.

$$F\,\Delta t = \Delta(mv)$$
$$(\text{Impulse}) = \text{Change of momentum}$$

Momentum is a conserved quantity. The total momentum of a system can only change if there is an outside force applied to it. The law of conservation of momentum for systems isolated from outside forces states that the momentum before an interaction is equal to the momentum after.

$$(mv)_{\text{before}} = (mv)_{\text{after}}$$

The law of conservation of momentum agrees with Newton's second and third laws of motion. Newton's third law states that for every force of action there is an equal and opposite force of reaction somewhere else in the system. Assuming this is true for every instant of time, this third law is equivalent to saying that for every action impulse there is a reaction impulse. Thus, for every change of momentum there is an equal and opposite change of momentum of some other body in the system such that the total amount of momentum in the system is constant.

8-1 Two monkeys, named Ed and Marshal, are hanging from the opposite ends of a rope that goes up and over a pulley. The monkeys balance each other, because they both have the same weight. They are both stationary when Ed decides to scamper up the rope.

Given: Ed and Marshal each have a mass of 35 kg. When Ed decides to scamper up the rope, he gives himself an upward acceleration of 5.2 m/s².

Find: (a) How much does each monkey weigh?
(b) How great is the force of action necessary to make Ed accelerate up the rope?
(c) How great is the force of reaction and what does it act on?

Your Solution

Discussion

This problem illustrates Newton's third law for two situations—one in which the forces all balance to produce a static situation and the other in which unbalanced forces cause acceleration.

Both situations involve equal but opposite forces of action and reaction. These equal and opposite forces of action and reaction called for in the third law are not the same as the forces that balance each other in equilibrium problems. The rules for equilibrium refer to forces that act on the same body; the third law refers to forces that act on different bodies.

Although the problem does not specifically call for it, you might address yourself to what happens to Marshal. Ignoring the mass of the rope and friction in the pulley, can Ed ever get away from Marshal?

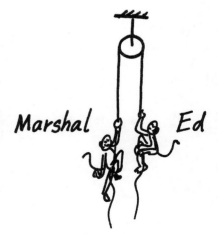

Marshal   Ed

Sample Solution 8-1

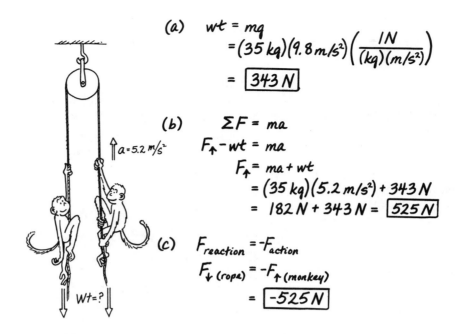

(a) $wt = mg$

$= (35\,kg)(9.8\,m/s^2)\left(\dfrac{1N}{(kg)(m/s^2)}\right)$

$= \boxed{343\,N}$

(b) $\Sigma F = ma$

$F_\uparrow - wt = ma$

$F_\uparrow = ma + wt$

$= (35\,kg)(5.2\,m/s^2) + 343\,N$

$= 182\,N + 343\,N = \boxed{525\,N}$

(c) $F_{reaction} = -F_{action}$

$F_{\downarrow\,(rope)} = -F_{\uparrow\,(monkey)}$

$= \boxed{-525\,N}$

Discussion

Part (a) is solved by the now familiar relationship between mass and weight. If we think of the interaction between one monkey, say Ed, and the rope, we could call this the force of action that the rope must exert on Ed to balance his weight and hold him stationary. The force of reaction acting on the rope has the effect of balancing Marshal's weight. If the force of action acting on Ed is increased, as in part (b), to give him an upward acceleration, the equal but opposite force of reaction acting on the rope also increases, as shown in part (c). The pulley changes the direction of this downward force acting on the rope, represented by a negative sign in part (c), into an upward force acting on Marshal. This increased upward force acting on Marshal gives him the same upward acceleration as experienced by Ed.

8-2 A rifle exerts a constant force of action on a bullet while it is in the barrel.

Given: An 11.7-g bullet is accelerated at 3.8×10^5 m/s² through a 0.51-m rifle barrel.

Find: (a) How great is the force of action of the gun on the bullet?
(b) For how long does the bullet experience this force of action?
(c) For how long does the gun experience an equal but opposite force of reaction?

Your Solution

Discussion

This problem points out that Newton's third law holds true at every instant of time. Not only do the action and reaction forces equal each other in magnitude, they are equal for the same period of time. The impulse produced by one is therefore equal and opposite to the impulse produced by the other.

Sample Solution 8-2

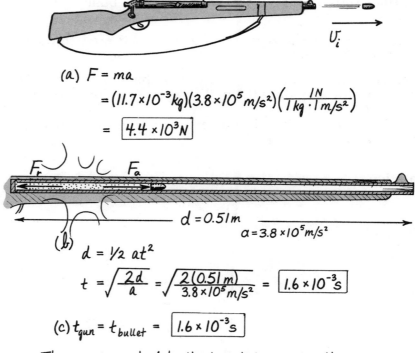

(a) $F = ma$

$$= (11.7 \times 10^{-3} kg)(3.8 \times 10^{5} m/s^{2})\left(\frac{1N}{1 kg \cdot 1 m/s^{2}}\right)$$

$$= \boxed{4.4 \times 10^{3} N}$$

$d = 0.51 m$

$a = 3.8 \times 10^{5} m/s^{2}$

(b) $d = \frac{1}{2} at^{2}$

$$t = \sqrt{\frac{2d}{a}} = \sqrt{\frac{2(0.51 m)}{3.8 \times 10^{5} m/s^{2}}} = \boxed{1.6 \times 10^{-3} s}$$

(c) $t_{gun} = t_{bullet} = \boxed{1.6 \times 10^{-3} s}$

The gun is pushed by the bullet for exactly the same amount of time that the bullet is pushed by the gun.

Discussion

In part (a) the unbalanced force applied to the bullet causes a certain acceleration, which can be calculated from Newton's second law. Assuming that this acceleration is constant and that the bullet starts from rest, the time it remains in the gun barrel can be calculated from the formula for distance traveled under constant acceleration. The gun exerts a force on the bullet only while it is in the gun barrel. The bullet, on the other hand, exerts a reaction force on the gun for exactly the same period of time. They cease to push against each other at the moment that the bullet departs from the muzzle.

8-3   An astronaut working in outer space tosses away a flashlight when he is finished using it. A few moments later, he tosses a screwdriver in the same direction.

Given: The astronaut exerts the same force of 20 N for the same period of time, 0.35 s, on both the 0.73-kg flashlight and the 0.28-kg screwdriver.

Find: (a) What is the momentum of the flashlight?
(b) What is the momentum of the screwdriver?
(c) How much faster is the screwdriver thrown than the flashlight?

Your Solution

Discussion

This problem illustrates Newton's second law expressed in terms of impulse and momentum. Impulse is defined as the product of force and the time over which it is applied. The second law of motion can either be expressed as saying $F = ma$ or as saying that impulse equals change of momentum.

Sample Solution 8-3

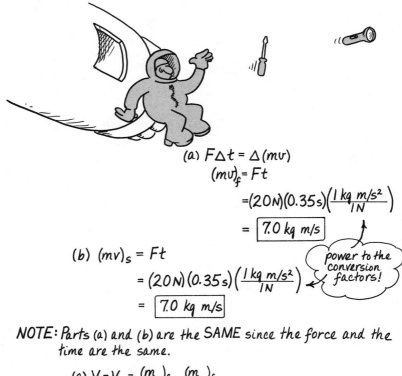

(a) $F\Delta t = \Delta(mv)$

$(mv)_f = Ft$

$= (20N)(0.35s)\left(\dfrac{1\,kg\,m/s^2}{1N}\right)$

$= \boxed{7.0\,kg\,m/s}$

(b) $(mv)_s = Ft$

$= (20N)(0.35s)\left(\dfrac{1\,kg\,m/s^2}{1N}\right)$

$= \boxed{7.0\,kg\,m/s}$

power to the conversion factors!

NOTE: Parts (a) and (b) are the SAME since the force and the time are the same.

(c) $V_s - V_f = \dfrac{(m\;)_s}{m_s} - \dfrac{(m\;)_f}{m_f}$

$= \dfrac{(7.0\,kg\,m/s)}{(0.28\,kg)} - \dfrac{(7.0\,kg\,m/s)}{(0.73\,kg)} = 25\,m/s - 9.6\,m/s$

$= \boxed{15\,m/s}$

Discussion

Parts (a) and (b) are found by solving Newton's second law for momentum. The delta notation can be ignored in this case, because the change of momentum is the momentum itself, since the screwdriver and flashlight are assumed to start from rest. The same is true of the time, since we can start measuring time at the instant that force is applied to each object. The result is the same for parts (a) and (b), because the same force is applied for the same period of time. Even though both objects have the same momentum, however, they have different velocities, since they have a different mass.

8-4 A tennis ball, traveling north one fine day, encounters a tennis racket and reverses its journey and begins to go south.

Given: The racket exerts a force of 24 N on the 55-g tennis ball to change its velocity from 15 m/s north to 25 m/s south.

Find: (a) By how much does the momentum of the ball change?
(b) Over what period of time does this change of momentum take place?

Your Solution

Discussion

This problem illustrates the vector nature of velocity and momentum. Note that the change of velocity, and hence momentum, is greater when the ball reverses its direction than if it had simply come to a stop.

Sample Solution 8-4

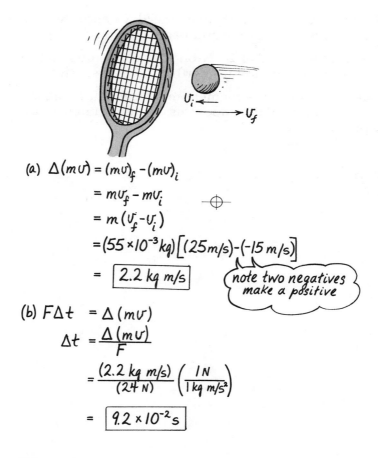

(a) $\Delta(mv) = (mv)_f - (mv)_i$

$\qquad = mv_f - mv_i$

$\qquad = m(v_f - v_i)$

$\qquad = (55 \times 10^{-3} kg)\left[(25 m/s)-(-15 m/s)\right]$

$\qquad = \boxed{2.2\ kg\ m/s}$ *note two negatives make a positive*

(b) $F\Delta t = \Delta(mv)$

$\quad \Delta t = \dfrac{\Delta(mv)}{F}$

$\qquad = \dfrac{(2.2\ kg\ m/s)}{(24\ N)}\left(\dfrac{1\ N}{1\ kg\ m/s^2}\right)$

$\qquad = \boxed{9.2 \times 10^{-2}\ s}$

Discussion

The negative value for the initial velocity (-15 m/s) comes from the vector nature of velocity. If one direction is assumed to be positive, the other direction must be negative. We can assume that the positive direction is that of the final velocity in this case, since that is the direction that the momentum is changing toward; we could as well assume the initial velocity to be positive, however, and the result would be the same but with a negative value because of the convention we would have adopted.

Part (b) is found by solving Newton's second law for the change in time associated with the impulse that must produce the change of momentum given by part (a).

8-5 A lumberjack, standing on the end of a stationary log floating in calm water, decides to take a little walk. The log reacts by floating in the opposite direction.

Given: The 100-kg lumberjack walks east at 1.5 m/s relative to the water. The log has a mass of 500 kg.

Find: (a) What is the momentum of the lumberjack?
(b) What is the momentum of the log?
(c) How fast does the log move in the opposite direction?

Your Solution

Discussion

This problem illustrates the law of conservation of momentum. The log and the lumberjack form an isolated system in that the force applied by the log to the lumberjack is equal and opposite to the reaction force applied by the lumberjack to the log, assuming that outside forces can be ignored in the forward and backward directions.

Sample Solution 8-5

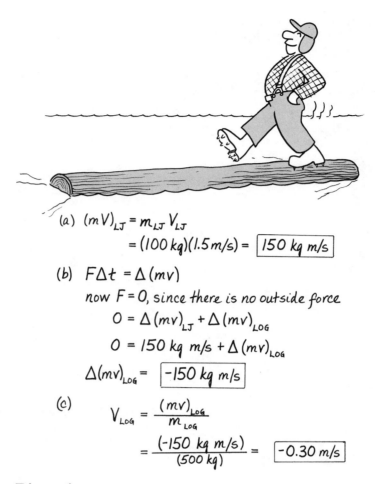

(a) $(mV)_{LJ} = m_{LJ} V_{LJ}$

$$= (100 \, kg)(1.5 \, m/s) = \boxed{150 \, kg \, m/s}$$

(b) $F\Delta t = \Delta(mv)$

now $F = 0$, since there is no outside force

$$0 = \Delta(mv)_{LJ} + \Delta(mv)_{LOG}$$

$$0 = 150 \, kg \, m/s + \Delta(mv)_{LOG}$$

$$\Delta(mv)_{LOG} = \boxed{-150 \, kg \, m/s}$$

(c)

$$V_{LOG} = \frac{(mv)_{LOG}}{m_{LOG}}$$

$$= \frac{(-150 \, kg \, m/s)}{(500 \, kg)} = \boxed{-0.30 \, m/s}$$

Discussion

Part (a) is solved from the definition of momentum. The momentum of the lumberjack is just the product of his mass and velocity.

Part (b) is solved by a special application of Newton's second law, where zero outside force is applied to the system as a whole. Any change in momentum of the lumberjack must therefore be accompanied by an equal and opposite change in the momentum of the log if the total change in the momentum of the system is to be zero.

The velocity of the log can be found from its momentum, as in part (c), by dividing momentum by mass, since momentum is defined as the product of mass and velocity.

8-6 During a snowball fight, a small girl is retreating across a nearly frictionless frozen pond toward her snow fort. She catches a snowball in the middle of her back. The snowball sticks.

Given: The child has a mass of 14.7 kg and an initial velocity of 3.0 m/s. The snowball has a mass of 0.3 kg and an initial velocity of 20 m/s.

Find: (a) What is the child's momentum before catching the snowball?

(b) What is the snowball's momentum before finding the child?

(c) What is the final velocity of the child and snowball combined?

Your Solution

Discussion

This problem is an example of the law of conservation of momentum applied to an inelastic collision. The little girl and the snowball may be assumed to be an isolated system whose total momentum after collision is equal to the total momentum of the parts before collision.

Sample Solution 8-6

(a) $(mv)_c = m_c v_c$

$\qquad = (14.7 \text{ kg})(3.0 \text{ m/s}) = \boxed{44 \text{ kg m/s}}$

(b) $(mv)_{sb} = m_{sb} v_{sb}$

$\qquad = (0.3 \text{ kg})(20 \text{ m/s}) = \boxed{6. \text{ kg m/s}}$

(c) $\quad \Sigma v = \dfrac{\Sigma (mv)}{\Sigma m}$

$\qquad = \dfrac{(mv)_c + (mv)_{sb}}{m_c + m_{sb}}$

$\qquad = \dfrac{(44 \text{ kg m/s}) + (6 \text{ kg m/s})}{(14.7 \text{ kg}) + (0.3 \text{ kg})} = \dfrac{50 \text{ kg m/s}}{15 \text{ kg}}$

$\qquad = \boxed{3.3 \text{ m/s}}$

Discussion

The momentums of the child and the snowball are found in parts (a) and (b) by the definition of momentum. Momentum is the product of mass and velocity.

The velocity in part (c) is found by dividing the total momentum after collision by the total mass after collision. The vector sum of the momentums is the same as the numerical sum of the momentums found in parts (a) and (b), since both momentums are in the same direction. The problem states that the little girl and the snowball were going in the same direction at the time of impact. The sum of the masses is always the numerical sum of the parts, because mass is a scalar quantity. The result indicates that the little girl is helped along by the snowball, her speed increasing by about 10%.

ESSENTIAL PROBLEMS

8-7 A horse is pulling a wagon equipped with nearly frictionless wheel bearings and excellent brakes. The horse exerts a constant force of action on the wagon.

Given: The force exerted by the horse is 2,200 N. The wagon has a mass of 500 kg.

Find: (a) How great is the reaction force with which the wagon pulls on the horse when the brakes are set and the wagon won't budge?

(b) When the brakes are released, how fast will the wagon accelerate?

(c) How great is the reaction force when the brakes are released?

8-8 An archer sets an arrow to his bow string, pulls it back to his ear, and then lets it fly toward the target.

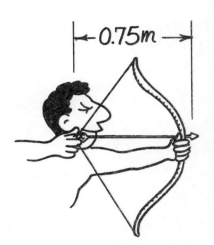

Given: The average force of action exerted by the bow on the 0.12-kg arrow is 120 N. The force acts over a distance of 0.75 m.

Find: (a) What is the average acceleration of the arrow while the force of action is applied?

(b) For how long does the force of action last?

(c) For how long does the force of reaction last?

8-9 A fisherman has been casting a heavy spoon lure all morning without luck. He decides to switch to a light fly lure, casting it with the same force of action applied for the same amount of time.

Given: Both the spoon and the fly experience a force of 5.3 N for 0.73 s. The spoon has a mass of 150 g, and the fly has a mass of only 25 g.

Find: (a) What is the momentum of the spoon lure?

(b) How fast is the spoon lure moving after being cast?

(c) What is the momentum of the fly lure?

(d) How fast is the fly lure moving after being cast?

8-10 A karate expert is breaking a board. He concentrates momentum transfer into a short period of time to optimize the force of a blow.

Given: The karate expert can put an effective mass of 15 kg moving at 22 m/s behind the blow. The momentum transfer to the board takes place in 3.3×10^{-2} s. (Assume all the momentum is needed to break the board.)

Find: (a) How much momentum does the blow transfer?

(b) What is the average force of the blow?

8-11   A woman in a bathing suit dives out of a rowboat into the water. The woman and the boat are initially stationary in calm water.

Given: The woman has a mass of 65 kg and the boat has a mass of 30 kg. The woman gives herself a horizontal velocity relative to the water of 0.85 m/s.

Find: (a) What is the woman's momentum?
   (b) What is the momentum of the boat?
   (c) How fast does the boat move in the opposite direction?

8-12   A mountain lion is pouncing on a hunter. The hunter shoots repeatedly at the mountain lion with a large caliber pistol in an attempt to stop it dead in the air.

Given: The mountain lion has a mass of 30 kg and a velocity of 10 m/s. Each bullet has a mass of 20 g and a velocity of 400 m/s.

Find: (a) What is the initial momentum of the mountain lion?
   (b) What is the momentum of each bullet?
   (c) How many bullets would it take to stop the beast dead in the air?
   (d) What would be the mountain lion's velocity if it was hit by just one bullet?

MORE INTERESTING PROBLEMS

8-13   In one of the nuclear reactions that produce solar energy, a neutron smashes into a deuteron and sticks to form a triton. A deuteron is just about twice as heavy as a neutron, so a triton is about three times as heavy.

Given: The deuteron is stationary and has a mass of 3.34×10^{-27} kg. The neutron is moving along at 2.5×10^6 m/s and has a mass of 1.67×10^{-27} kg.

Find: (a) What is the momentum of the neutron before the collision?
   (b) What is the velocity of the triton after the collision?

8-14   A cannon shell bursts at the top of its parabolic path due to a defect in manufacture and splits into two equal parts. The force of the explosion happens to be such that half the shell casing stops in mid-air and then falls straight downward after having given all of its momentum to the other half.

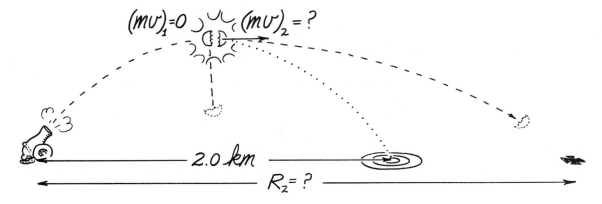

Given: The original cannon shell had a mass of 10.0 kg and a velocity of 300 m/s all in the horizontal direction just before the explosion. Had it not burst, it would have had a range of 2.0 km.

Find: (a) What was the momentum of the shell casing just before it prematurely exploded?

(b) What was the momentum of the second half of the shell casing right after the explosion?

(c) How far does the second half of the shell casing land from the cannon?

8-15 A sailboat loses headway on a calm day. The crew sets up a large electric fan on the deck. They naturally direct the fan toward the bow, or front of the boat, and erect a sail called a spinnaker to catch the wind.

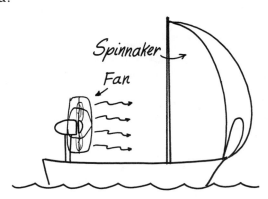

Given: The sailboat has a mass of 2,000 kg. The force of the wind against the spinnaker is 300 N.

Find: (a) What is the force of the wind against the fan (assuming that all of the wind is caught by the spinnaker)?

(b) What would be the acceleration of the boat with the spinnaker sail in place?

(c) What would be the acceleration of the boat if the fan were pointed sternward, toward the back of the boat, with no sail in place?

8-16 A little girl is watering the weeds. As she points the hose at the weedbed, she notices that the force of action of the hose against the water is accompanied by a reaction force of the water against the nozzle, which she can feel in her hand.

Given: The water stream carries water at the rate of 0.30 kg/s. The nozzle velocity of the water is 10 m/s.

Find: (a) How much momentum is transferred to the water in 1.0 s?
   (b) What is the reaction impulse necessary to transfer this momentum?
   (c) What is the reaction force of the water against the nozzle?

ANSWERS

8-7   (a) $-2,200$ N; (b) 4.4 m/s^2; (c) $-2,200$ N.
8-8   (a) 1,000 m/s^2; (b) 3.9×10^{-2} s; (c) 3.9×10^{-2} s.
8-9   (a) 3.87 kg · m/s $\cong$ 3.9 kg · m/s; (b) 26 m/s; (c) 3.87 kg · m/s $\cong$ 3.9 kg · m/s; (d) 155 m/s.
8-10 (a) 330 kg · m/s; (b) 1.0×10^4 N.
8-11 (a) 55 kg · m/s; (b) -55 kg · m/s; (c) -1.8 m/s.
8-12 (a) 300 kg · m/s; (b) 8 kg · m/s; (c) 38 bullets; (d) 9.7 m/s.
8-13 (a) 4.2×10^{-21} kg · m/s; (b) 8.3×10^5 m/s.
8-14 (a) 3.00×10^3 kg · m/s; (b) 3.00×10^3 kg · m/s; (c) 3.0 km.
8-15 (a) -300 N; (b) 0 m/s; (c) 0.15 m/s^2.
8-16 (a) 3.0 kg · m/s; (b) -3.0 kg · m/s; (c) -3.0 N.

Energy

9

In Chapter 8 we dealt with a quantity which, while somewhat esoteric, is quite real in the sense that it cannot be destroyed or created. We saw that momentum is absolutely conserved in every interaction—at least all of those that people have carefully examined. One might still engage in a philosophical discussion on the true existence of this quantity. Certainly momentum is not as tangible as mass, nor is it as directly measurable as velocity. One might argue that momentum is merely the product of mass and velocity, that it cannot exist without either member of that product, and that it therefore has no independent existence of its own. Still, there does seem to be this thing that is passed from body to body, divided into parts, added together, and constantly manipulated, with the total amount remaining perfectly constant for all time. From a problem-solving point of view, there are certain advantages to thinking of such a conserved quantity as having an existence in its own right. Whether or not that existence is real or imagined may be contemplated when you have an idle moment with nothing more pressing to do. Until that luxury comes your way, you would do well to treat momentum as a *thing*. Momentum lives, as it were.

Another quantity just as abstract as momentum, and yet as real in the sense that it is absolutely conserved, is **energy.** As in the case with momentum, you might well argue that energy is more of a condition than a thing—a condition that remains absolutely constant in an isolated system even as other things within the system change. Most people, however, prefer to think of this esoteric quantity as a nearly material substance. We buy the stuff from the electric company. We worry about whether or not there is going to be enough stored in fossil fuel to go around if we keep using it as fast as we do. Few people ever stop to wonder if the stuff really exists at all. The very fact that it is conserved, that you cannot create it out of nothing, that you have to get it from somewhere if you want some, makes it a commodity that is bought and sold like a material substance.

This idea of energy is new. It is so sophisticated that it has only been with us for about a century and a half. Neither Galileo nor Newton knew about this stuff. Of course, there were some general ideas kicking around concerning such things as vitality and work, but no one seems to have thought of tying these various types of related things together into the single quantity we call energy until about two centuries ago. Even then, it took half a century for the idea to catch on. As we will see, the key to this level of understanding was the fact that heat is just another form of this same kind of thing. It was not until the 1840s that James Joule was able to prove that the same amount of heat is always produced by the same amount of work no matter how the work is done. This provided the missing link. Even the word *energy* was not in common use before that. The word was, in fact, coined to give a name to all the different kinds of things that describe the vitality of inert objects.

The word *energy* was derived from the Greek *energos,* meaning "active." This was a particularly appropriate name, since it is derived in turn from the older Greek term *ergon,* which stands for "work."

▶ Work

Energy is often defined as the ability to do work. That might seem like a fine definition until you discover that *work* is frequently defined as the transfer of energy. Such is the circular nature of definitions one runs across when dealing with such truly basic quantities as time, distance, and force. Like these other basic quantities, energy is so fundamental that it cannot be defined in terms of anything more basic. There just isn't anything more basic. Still, there are relationships between basic quantities that allow us to get a handle on one by using the others. We can get a handle on energy by looking at how it is transferred from one form to the other, or from one object to the other.

The only way that energy can be transferred is by a force acting through a distance. This **transfer of energy** is called **work,** and its relationship to force and distance agrees with most people's common understanding of the word.

If a horse is pulling on a cart, for example, most people would agree that the horse is doing work on the cart. Most people would also agree that the amount of work is proportional to the force with which the horse must pull:

$$\text{Work} \propto F$$

$$WORK = F \cdot d = (55\,lb)(10\,ft) = 550\,ft\text{-}lb$$

FIGURE 9-1 *Work is proportional to both force and distance.*

This seems reasonable, because one horse pulling at, say, 55 lb, would do a certain amount of work, while two horses pulling together with twice the force, say, 110 lb, would do twice as much work.

It also seems reasonable to most people that the amount of work done by the horse is directly proportional to the distance over which the force is applied.

$$\text{Work} \propto d$$

If the horse pulled the cart 10 ft, for example, most people would agree that he had done half as much work as if he had pulled it 20 ft. Here again, the reasoning proceeds from the consideration of the two horses which could divide the work evenly between them. We could let the horses run relays, having one horse pull for 10 ft, and then allowing the other horse to take over to pull the other 10 ft.

Therefore, the work done (energy transferred) is directly proportional to both the force and the distance. Anything that is proportional to two separate things is also proportional to their product. We usually define work as the product of force and distance.

$$\boxed{\text{Work} \equiv F \cdot d}$$

The proportionality constant, by this definition, is just equal to unity (1), so that we don't even need to bother writing it down. If the horse in our example pulled with a force of 55 lb over a distance of 10 ft, we just say that it did 550 foot-pounds (ft-lb) of work.

CHECK QUESTION

Two horses wish to help each other pull a cart. They try teaming up in two different ways. In the first way, they take turns; in the second, they pull together. (a) How much work is done if one horse pulls the cart with a force of 55 lb for a distance of 5 ft and then the second horse pulls, also with a force of 55 lb, for a second distance of 5 ft? (b) How much work is done if they pull together with a combined force of 110 lb for a distance of 5 ft?

Answer: (a) 550 ft-lb, (b) 550 ft-lb

The **foot-pound** is a unit of energy in the English system of units; the corresponding unit of energy in the metric system is called the **joule** (J), defined as that amount of work done by 1 N of force when acting through a distance of 1 m.

$$\text{Work} \equiv F \cdot d$$
$$1 \text{ ft-lb} \equiv (1 \text{ lb})(1 \text{ ft})$$
$$1 \text{ J} \equiv (1 \text{ N})(1 \text{ m})$$

There are some peculiarities inherent in the definition of work. The first is that the force must act over some distance or else exactly zero work is done. It may not surprise you that you do no work on a wall when you merely lean against it, but it might be surprising to learn that the same thing is true even if you push forcefully against the wall. Sweat may stand out on your forehead, but unless the wall moves some, you are doing zero work on the wall. The wall will be in the same place where it was before you pushed. You will not have changed its condition at all. Any work you might have done, therefore, will have been only on yourself. Energy must be transferred when work is done, at least as the word *work* is used in physics. No force, however great, will transfer energy from one object to another unless it acts through some distance.

Another peculiarity that must be included in our definition of *work* is the requirement that the force and the distance must be in the same direction. It would seem reasonable to assume that a horse would be concerned only in how hard he must pull in the forward direction to make a cart move. The same force might be needed to pull a heavily loaded cart having good wheel bearings or a lightly loaded cart having lousy wheel bearings. The amount of work the poor horse must do would be the same in either case, although the weight, or force perpendicular to the ground, would be much greater in one case. The only part of the force that does any work is therefore the component that is **in the direction of the displacement.**

A woman pushing a lawnmower applies a force that pushes the mower forward and also against the ground. The part of the force that is forward does the work, while the part of the force that is perpendicular to the ground only helps to hold the wheels against the ground. Although this downward component serves a useful function in helping to keep the lawnmower's wheels from slipping, it still does no work. Although this might seem strange at first, the same resultant force applied by the woman could be supplied as well by the combined efforts of a large stone to hold the lawnmower down and a small beast of burden to supply the forward force. Most people would agree that the small beast does all the work in this situation. The stone would not seem to do any work as long as the lawnmower is pulled on level ground. The same is true with the woman pushing downward at an angle. The downward component does no work; all the work is done by the forward component.

This applies also to work done on an object when you carry it across the room. You might think that you do a lot of work when you carry a very heavy object horizontally from one place to another. That is not true. You do work on an object when you lift it, because the force and the displacement are in the same direction. But you do no work in holding an object stationary, however tired you might become, and you do very little more work if you move it sidewards as you hold it up. The energy you use in holding an object at a stationary height is like the isometric exercises you might do in pushing

$F \cdot 0 = $ no work

FIGURE 9-2 *Force times zero distance equals no work done.*

$F \cdot 0 = $ still no work

FIGURE 9-3 *Even a great force times zero distance produces no work. At least no work is done on the wall, as work is energy transfer and the wall remains in the same condition as it was before any force was applied.*

FIGURE 9-4 *The resultant force applied by a woman pushing a lawnmower is made up of two components. Only the forward component, that component in the direction of the distance, does any work.*

against a stationary wall that refuses to budge. No energy is transferred to the object. No more energy is transferred if you happen to be moving sidewards, since the force and the displacement are perpendicular. We might understand this result by supposing that you

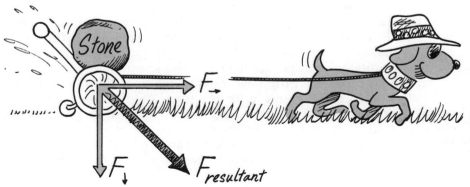

FIGURE 9-5 *The forward component could be provided by a worker, while the downward component could be provided by the weight of an inert object, and the same resultant force would act on the lawnmower as was applied by the woman.*

are wearing roller skates while holding the heavy object. If someone were to give you a sideward shove, you would do no more work on the object while coasting along than you would if you had just stood there holding your burden stationary.

Of course, a person who is walking with a heavy burden does a little more work than someone who is coasting on roller skates. A person taking a step bobs up and down a little bit, since he is a little closer to the ground when his legs are at an angle as he steps out than when his legs are vertical as his feet come together. This work is not permanently transferred to the object, however, because the work you do on an object in bobbing upward is returned to you as you bob downward in completing the step. Unfortunately, this energy is wasted. Your body has no way of storing this returned energy; you must do a fresh amount of work with each step you take. The best you can hope to do is carry the object smoothly and reduce the energy wasted in bobbing up and down.

FIGURE 9-6 *Little or no work is done in carrying something horizontally.*

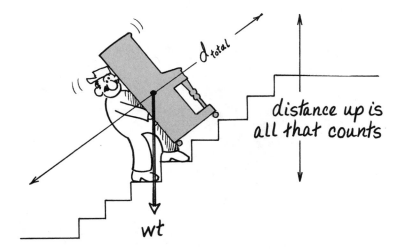

FIGURE 9-7 *The only distance that counts for doing work in carrying something upstairs is the component that is in the upward direction.*

The situation is a bit different if we carry a piano upstairs. Then there is a displacement in the direction of the force. The force necessary to lift the piano is upward, equal in magnitude to the piano's weight, and part of the total displacement is upward. It is only the upward component that counts for doing work, however; the sideward component still does no work. The same amount of work is done on the piano in carrying it to a certain height no matter how steep the stairs are or whether they are straight or spiraled. In any case, the work done is just the weight of the piano times the height, which is the upward component of the displacement. You would notice the additional energy given the piano in raising it up the stairs if you should chance to drop it from the greater height. If the piano were to make but one crashing bounce, the loudness of the sound would be equally great no matter what the steepness of the stairs.

In the example with the lawnmower, we split the force up into components. In the example with the piano, we split the displacement up into components. We used that component of the force in the direction of the distance in the one case and that component of the distance in the direction of the force in the other. We should note at this point that these two approaches give exactly the same result.

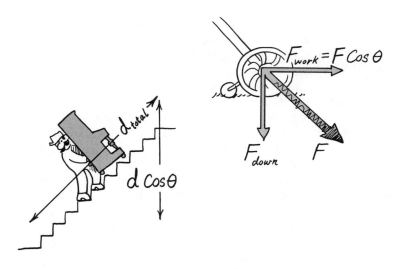

FIGURE 9-8 *The force must be in the same direction as the distance or the distance in the same direction as the force. Either case leads to the same result.*

They are algebraically equivalent. They are just two different ways of looking at the same thing.

$$\text{Work} = (F \cos \theta)d \qquad \text{(Lawnmower case)}$$
$$\text{Work} = F(d \cos \theta) \qquad \text{(Piano case)}$$
$$\text{Work} = F \cdot d \cdot \cos \theta \qquad \text{(Either case)}$$

▶ Power

Returning to the horse cart example, notice that the amount of work done by the horse is independent of time. The horse does 550 ft-lb of work when he pulls with 55 lb over 10 ft, whether in 1 s or in 10 s. It might seem that the horse is working harder if he covers the distance in 1 s, but no more work gets done—at least not in the same sense that we have defined work.

We need to account for this idea of working harder when work is done faster. Because we live in a time-oriented society, we are frequently more interested in how fast work gets done than in the total amount that eventually is done. The rate of doing work is so important to us that we give it a separate name, **power.** Power is defined as the ratio of work done to the time necessary to do it.

$$\text{Power} \equiv \frac{\text{work}}{\text{time}}$$

Power is analogous to velocity. Both concepts are rates. Whereas one is the rate of traveling distance, the other is the rate of doing work. The shorter the time, in both cases, the greater the rate. The horse would only be creeping along, for example, if he took 10 s to travel the 10 ft we were talking about. A velocity of only 1 ft/s would be slow for a horse, even pulling with a force of 55 lb. He wouldn't seem to be working hard at all at that rate. It turns out that an average horse can pull with a 55-lb force over a distance of 10 ft in only 1 s of time. Working at that rate, the average horse has a power output of 550 ft-lb/s.

$$\text{Power} \equiv \frac{\text{work}}{\text{time}} = \frac{(55 \text{ lb})(10 \text{ ft})}{(1.0 \text{ s})} = 550 \text{ ft-lb/s}$$

We know that most horses can actually produce 500 ft-lb/s of power for hours at a time, day after day, because James Watt actually measured the rate at which horses work, back in the reign of King George III. Watt had invented improved steam engines and was trying to sell them to English coal miners. At that time, horses were used to pump water in draining the mines. The mine operators were interested in how many horses a particular steam engine would replace. Actual measurements on poor hard-working horses showed that an average horse could do 550 ft-lb of work each second. We therefore call this rate of doing work a horsepower.

$$1 \text{ hp} \equiv 550 \text{ ft-lb/s}$$

James Watt's study of power as well as his technical contributions to the industrial revolution have led us to name the metric unit

FIGURE 9-9 *A 100-W lamp uses 100 J of energy each second.*

of power after him. As you recall, the metric unit of work is named after Sir James Joule. One joule of energy is the amount of work done when a force of 1 N acts over a distance of 1 m.[1]

$$1 \text{ J} \equiv (1 \text{ N})(1 \text{ m})$$

A watt is then defined as 1 J/s.

$$1 \text{ W} \equiv 1 \text{ J/s} = 1 \text{ N} \cdot \text{m/s}$$

People rarely hear of a joule of energy except in a physics course, whereas the watt is a household word. A 100-W light bulb uses 100 J of energy per second. Our familiarity with Watt's name is further evidence that ours is a time-oriented society.

CHECK QUESTIONS

1. A man lifts a 110-lb barbell to a height of 5.0 ft in 1.0 s. How much power does he produce for this brief period of time?

    Answer: 550 ft-lb/s = 1.0 hp

2. A person pushes with a force of 300 N over a distance of 0.50 m in 1.0 s. (a) How much work is done? (b) If one horsepower is equal to 746 W, what is the power produced in horsepower?

    Answer: (a) 150 J, (b) 0.20 hp

▶ Forms of Energy

We started this chapter by considering the work done by a force acting through a distance. The definition of *work*, as the word is used in physics, was shown to be reasonable by an example in which a horse was pulling a cart at a constant rate. Other examples showed that the force must act through a distance greater than zero and that the force and the distance must be in the same direction in order to do any work. Also, the amount of work done is independent of time, so that we need to define power to handle situations in which time is a factor.

In all our examples, energy has been *transferred* from one object to another. We have thus defined work, but what is energy? Work changes the energy content of an object, but how can you tell? The fact is that there is no single measure of the energy content of an object, because energy can take on different forms. The energy added to the horse cart shows up as heat in the wheel bearings. Part of the energy added to the lawnmower goes into cutting the grass. Energy added to a piano can show up in its height as it is carried upstairs, or in its velocity as it falls back down. A moment later, the piano's energy might be liberated in sound if it should come to a crashing halt as it meets the ground.

Heat and sound are two forms of energy that will be considered in Chapters 10 and 11. For the remaining part of this chapter, we will discuss forms of energy associated with an object's motion or position.

[1] A newton is, as you may recall, that amount of force required to accelerate a 1-kg mass at 1 m/s².

► Kinetic Energy

Moving objects have energy by virtue of their motion. This energy of motion is called **kinetic energy** after the Greek word for "moving." The examples we have considered so far were selected to ignore the change in kinetic energy. The objects were moving at a constant velocity or were lifted from a position at rest to another position at rest. The forces were all balanced, so that all the work done would either go into frictional heating or gravitational potential energy. But what of the energy necessary to produce the motion in the first place? Kinetic energy is produced by an *unbalanced force* that accelerates the object from rest to its final velocity. The amount of kinetic energy an object has depends only on how much mass it has and on its velocity.

The relationship between kinetic energy, mass, and velocity can be found by considering the work done in accelerating the object from rest to its final velocity. Assume that the kinetic energy of an object comes entirely from a constant accelerating force acting over a distance.

$$E_k = \text{work done in accelerating the body}$$
$$= F_{\text{acceleration}} \cdot d$$

The amount of acceleration produced by the force F is related to the mass of the object by Newton's second law:

$$F_{\text{acceleration}} = ma$$

And the amount of distance, d, over which this force is applied can be found from the average velocity of the object:

$$d = v_{\text{ave}} t$$

In terms of acceleration and average velocity, the kinetic energy of the object can be stated thus:

$$E_k = (ma)(v_{\text{ave}} t)$$

Assuming that the object started from rest and accelerated uniformly up to the final velocity, acceleration is simply final velocity over time and average velocity is simply one-half the final velocity.

$$a = \frac{v_f}{t}$$

$$v_{\text{ave}} = \frac{v_f}{2}$$

Putting these relationships into the expression for kinetic energy,

$$E_k = m \left(\frac{v_f}{t}\right) \left(\frac{v_f}{2}\right) t$$

$$\boxed{E_k = \tfrac{1}{2} mv^2}$$

Thus, the kinetic energy of an object is one-half the mass times the square of the velocity. This relationship is worthy of being remembered. The one-half comes from using average velocity to find the distance over which the force acts, and the velocity is squared because it figures into both the distance and the acceleration produced.

FIGURE 9-10 *The kinetic energy of a moving body is ½mv².*

The squared velocity in the kinetic energy expression has great importance in transportation energy uses. As the velocity doubles, the kinetic energy becomes four times as great. In an automobile this energy is lost when it is changed into heat by the brakes. Sixty-five miles per hour is only 18% greater than 55 miles per hour, but $(65)^2 = 4225$ is 40% greater than $(55)^2 = 3025$. Thus an 18% increase of speed is accompanied by a 40% increase in kinetic energy. That is one reason that slightly lower speed limits are more economical than you might think in terms of energy.

Kinetic energy considerations also give rapid transit trains an advantage over private automobiles in terms of energy economy. Each time an automobile stops, its brakes convert the kinetic energy to heat. The kinetic energy of a rapid transit train, on the other hand, can be partly recovered when the train stops. The brakes on commuter trains are actually electrical generators that feed power back into the system for use by other trains.

CHECK QUESTION

An elephant having a mass of 100 slugs is traveling along at 6.0 ft/s. What is the kinetic energy of the elephant?

Answer: 1,800 ft-lb

▶ Potential Energy

An object can have energy by virtue of its position as well as by virtue of its velocity. A massive object elevated some distance above the ground clearly has some energy associated with its height. The amount of energy can be figured from the work that was necessary to get it up there. Work is, as always, force times distance.

$$\text{Work} \equiv F \cdot d$$

The force necessary to lift an object is its weight, and the distance is its height.

$$\text{Work} = \text{wt} \cdot h$$

The energy imparted to an object by lifting it to a height h is called its **gravitational potential energy.** As long as the height is small compared to the radius of the earth, we can think of the weight of the object as constant, given by the product of mass and the acceleration due to gravity.

$$E_p = mgh$$

CHECK QUESTION

An elephant having a mass of 100 slugs is 6.0 ft above the ground. What is the potential energy of the elephant?

Answer: 19,200 ft-lb

This kind of energy is called potential because it can be converted to kinetic energy if the object should fall. Consider the example of a group of students transfixed on the edge of their seats by a physics lecture. If one of those students should happen to fall, there would be a little less energy stored in their collective positions.

Assume for a moment that you are the student. You fall asleep while listening to a physics lecture and we wish to know how hard you will hit the floor. We already know how to calculate the velocity of an object in free fall over a certain distance. You may recall the formula we derived by combining the definition of acceleration with the definition of average velocity:

$$v_f{}^2 = v_i{}^2 + 2ad$$

Solving this formula for final velocity, assuming that the initial velocity is zero, we can get the answer from our understanding of motion under constant acceleration:

$$v_f = \sqrt{2gh}$$

FIGURE 9-11   *The gravitational potential energy of an object is its weight times its height.*

We can get the same result, however, from the idea of conservation of energy. We can assume that your kinetic energy as you near the ground will be equal to the potential energy you had when you were perched on your seat.

$$E_k = E_p$$

Kinetic energy can be expressed in terms of velocity, and potential energy in terms of height.

$$\tfrac{1}{2}mv^2 = mgh$$

Solving this expression for velocity, we get the same result as we did from the assumption of constant acceleration.

$$v = \sqrt{2gh}$$

The nice thing about the energy calculation is that it doesn't assume constant acceleration. Let us invent a situation in which this difference becomes important. We had you falling from your seat in your sleep before; let us construct a dream in which you slide limply out of your seat and on to the floor. Since you are dreaming, it would be easy for us to eliminate friction entirely and imagine what would happen if all of your potential energy were to change into kinetic energy as you slip from your seat. The conservation of energy would send you out the door and down the hall with the same velocity that you would have had if you had fallen from your seat straight downward with the acceleration due to gravity. Your kinetic energy at the floor level would be the same whether you slide limply from your seat with non-constant acceleration or whether you got there by the most direct route.

The conservation of energy rule is always useful in situations where the acceleration is not constant. A common example in which frictional losses can be ignored for most purposes is the pendulum. The energy of the pendulum bob is traded back and forth from potential to kinetic and back to potential again as the pendulum swings back and forth. The acceleration is not constant, but the velocity of the pendulum bob can nonetheless be easily calculated from the conservation of energy.

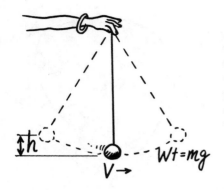

FIGURE 9-12 *The energy of the pendulum changes back and forth between potential and kinetic.*

The energy of a pendulum bob is all potential as the bob comes to a stop at the end of each swing, and it is all kinetic as the bob swings through the middle at the lowest point of its swing. The velocity of the bob as it swings through this lowest point can be calculated from the height of the bob at the ends of its path, measured vertically from its lowest level. Ignoring friction, we can assume that the kinetic energy at the bottom of the swing is equal to the potential energy at the top.

$$E_k = E_p$$
$$\tfrac{1}{2}mv^2 = mgh$$

where h is the vertical height measured from the lowest point in the swing. This relationship can be solved for velocity by dividing both sides by $\tfrac{1}{2}m$.

$$v^2 = \frac{mgh}{\tfrac{1}{2}m} = 2gh$$

As in the previous example, an interesting thing happens. The mass divides out. Taking the square root of both sides of the equation.

$$v = \sqrt{2gh}$$

We see that the velocity at the bottom of the swing is proportional to the square root of the height of the swing and is *independent* of the mass of the pendulum. Most people find this to be an unexpected result. Your little brother, for example, probably wouldn't believe that light and heavy pendulums swing with the same velocity when raised to the same height unless you actually showed him. In fact, maybe you don't even believe it yourself. If not, try it and see. The mass in the kinetic energy expression $\tfrac{1}{2}mv^2$ really does balance out the effect of the mass in the potential energy expression mgh.

▶ Elastic Potential Energy (Optional)

Another form of potential energy is the kind that is stored in deforming a spring. This is a very common form of potential energy, because, as we pointed out in our discussion of Hooke's law, nearly any solid object acts at least a little like a spring. The amount of force, F, that is needed to deform a solid object, by stretching, compressing, bending, or twisting, is nearly proportional to the distance x through which the object is deformed, as long as the force and deformation remain within the so-called elastic limits of the object.

$$F = Kx \qquad \text{(Hooke's law)}$$

The energy stored in deforming a spring is called the **elastic potential energy** of the spring.

Let us consider the energy stored in the spring of a BB gun as an example of elastic potential energy. The force that must be applied to the cocking lever starts out small but increases in accordance with Hooke's law as the lever is pulled to its full extent. We can figure out the amount of work done, and hence the energy stored in the spring, if we look at a graph of force and distance, as shown in Figure 9-14. Such a graph may be looked upon as a diagram in a special mathematical space, where one dimension is represented by force while

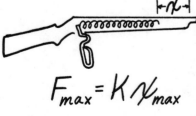

FIGURE 9-13 *Part of the elastic potential energy stored in the spring of a BB gun is given to the BB in the form of kinetic energy.*

the other is represented by distance. Work, which is the product of force and distance, would be represented in this space by area. All the work situations considered until now have been ones where the force was constant. The energy transferred by a constant force acting over a distance would be represented by a rectangle in this force-distance space. The work done on a spring, however, would be represented by a *triangular* area in this mathematical space. The work needed to cock the gun would be force times distance, as always, but the force actually needed to cock the gun is always smaller than the maximum value F_{max} that is reached when the gun is fully cocked. In accordance with Hooke's law, the actual force increases along a sloped line that forms the hypotenuse of a triangle in force-distance space. The area of this triangle, and hence the work actually done, is one-half the area of the rectangle formed by the maximum force acting over the same distance. It is therefore clear that the work done by the constantly increasing force is half as much as the work that would have been done by the maximum force if it had acted through the full distance.[2] From the definition of work,

$$\text{Work} \equiv F \cdot d$$

we find that the potential energy stored in a spring is

$$E_p = \left(\frac{F_{max}}{2}\right) x_{max}$$
$$= \tfrac{1}{2}(Kx)x$$
$$= \tfrac{1}{2}Kx^2$$

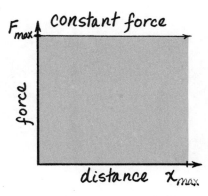

Shaded rectangular area is work done by $F_{max} \cdot x_{max}$.

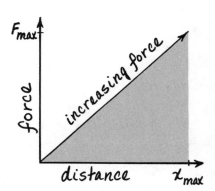

Triangular area is work done by $F \cdot x_{max}$

FIGURE 9-14 *Elastic potential energy is given by $\tfrac{1}{2}Kx^2$.*

How to Calculate the Velocity of a BB (Optional)

We have already seen how to use the conservation of energy to figure out the maximum velocity of a pendulum bob from its potential energy at the end of each swing. We can now see how to modify our conservation of energy argument to calculate the velocity of a BB pellet from the elastic potential energy of the spring in the BB gun.

We must bear in mind that not all the energy stored in the spring goes into the kinetic energy of the pellet. A good deal of the energy goes into heat and sound as the spring is released. Some of the energy goes into the recoil of the BB gun. Even so, we can get some idea of the velocity of the BB as it leaves the gun if we estimate the gun's efficiency. Assume that only about 30% of the elastic potential energy stored in the spring finds its way into the BB pellet in the form of kinetic energy. This may be a crude estimate, but it will show us how to proceed.

$$E_k = 30\% \ E_p$$

[2] The situation is similar to the case of motion undergoing constant acceleration. In that case, the distance traveled was the average velocity times the time. The average velocity turned out to be half the maximum velocity when the initial velocity was zero. In the same way, we could use average force in the work definition,

$$\text{Work} = F_{ave} \cdot d$$

where the average force is initial plus the final force divided by two. Since the initial force is zero, the average force is half the maximum force:

$$F_{ave} = \tfrac{1}{2}F_{max}$$

We now express the kinetic energy in terms of the mass, m, and velocity, v, of the BB pellet and the elastic potential energy in terms of the spring constant K and maximum stretch x_{max} of the spring:

$$\tfrac{1}{2}mv^2 = (0.30)(\tfrac{1}{2}Kx_{max}{}^2)$$

Solving for the velocity of the BB, we get an algebraic result into which we could insert numerical data for a particular BB gun:

$$v^2 = \frac{(0.30)(\tfrac{1}{2}Kx_{max}{}^2)}{\tfrac{1}{2}m}$$

$$v = \sqrt{\frac{(0.30)K}{m}} \cdot x_{max}$$

Of course, the estimate of 30% efficiency could be improved by making actual measurements on one particular model of BB gun, but the point is that the conservation of energy can still be used even when a known portion of the potential energy is lost because it changes into other forms of energy by such mechanisms as friction.

CHECK QUESTION

A BB gun with an efficiency of 30% has a spring with a spring constant of 533 N/m and shoots BB pellets with a mass of 1.0 g. When fully cocked, the spring is pulled back a distance of 20 cm. This gun can be fired from a half-cocked position, however, in which the spring is only pulled back a distance of 10 cm. (a) What is the velocity of the BB pellet as it leaves the gun after having been fired from the fully cocked position? (b) What is the velocity of the BB pellet when fired from the half-cocked position?

Answer: 80 m/s, 40 m/s

▶ Conservation of Energy and Momentum

People sometimes get the idea that energy and momentum are really different aspects of the same thing. Ask someone what happens to the momentum of a pendulum bob as the pendulum comes to the end of its swing. Our society is sufficiently sophisticated that most people seem to know that we should be able to account for the momentum that the pendulum bob has at the bottom of its swing but doesn't have when it comes to a stop at the end of its swing. The most frequent wrong answer is that the momentum changes into potential energy.

The fact is that energy and momentum are separate things, which, though closely related, are separately conserved. Momentum turns out to be different from energy in the same sense that space is different from time. It is therefore not possible for the momentum of an ordinary pendulum bob to turn itself into some form of energy. While the kinetic energy of the pendulum changes into potential energy, something else must happen to the momentum. The proper answer to the question is that the momentum is transmitted through the string to the support and actually pulls the world along a little faster. This peculiar fact can be demonstrated by mounting a pendulum on a skate board. If started from rest, the skate board support will oscillate back and forth in the opposite direction from the pendulum bob, keeping the total momentum of the system constant.

One fine demonstration of the separate nature of energy and momentum is to allow one pendulum bob to bang into a whole row of stationary bobs. The apparatus for doing this classic physics demonstration was popularly available in toy stores and gift shops a few years ago. Perhaps you remember seeing a string of five metal balls, all separately hung in a straight line from a wooden frame. When one ball is lifted and allowed to swing down and bang into the others, its momentum is completely and dramatically absorbed. It just sits there, while one ball at the other end of the string bounces out with the same velocity, or at least nearly the same velocity, in the same direction.

This is clearly a demonstration of the conservation of momentum, and many store clerks were quick to point out the educational advantage of the apparatus. This apparatus (which we might call a series impact pendulum) could be made to do other tricks as well. When two balls are lifted and allowed to collide with the others, two balls come out the other side. When three balls come in, three balls go out. Always the same number of balls come in as bounce out the other side. Furthermore, the number of balls that come out the other side does not seem to depend on the velocity of the incoming balls. It never seems to happen, for example, that two balls come in with a small velocity and one leaves with twice the velocity on the other side.

It used to be a great deal of fun to try to persuade the store clerk that this apparatus really demonstrates the "law of conservation of balls." Clearly, there is more involved than the conservation of momentum. One ball could have the same momentum as two incoming balls if it had twice the velocity, but this just never seems to happen. Somehow balls at one end seem to know how many balls come in at the other end. There must be some difference between the momentum contained in one fast-moving ball and the same amount of momentum contained in two slower-moving balls.

The difference turns out to be the energy. The same momentum is involved if one ball bounces out with twice the velocity as if two balls come out with the same velocity as the incoming balls:

$$(2m)v = m(2v)$$

However, the kinetic energy of one ball bouncing out would be twice as great as the energy of the two incoming balls ($\neq$ means "is not equal to"):

$$\tfrac{1}{2}m(2v)^2 \neq \tfrac{1}{2}(2m)v^2$$

$$\frac{m \cdot 4v^2}{2} \neq \frac{2mv^2}{2}$$

$$2mv^2 \neq mv^2$$

Thus, two balls in and one ball out twice as fast would conserve momentum but not energy. The ball coming out would have twice the energy as the balls going in. There would be a net energy gain.

In a similar fashion, one ball coming in and two balls bouncing out with half the velocity would conserve momentum but not energy. The momentum of the two balls going out would be the same as the momentum of the incoming ball, but the energy would be

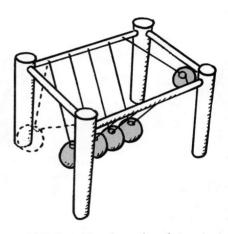

FIGURE 9-15 *A series impact pendulum demonstrates the conservation of both energy and momentum.*

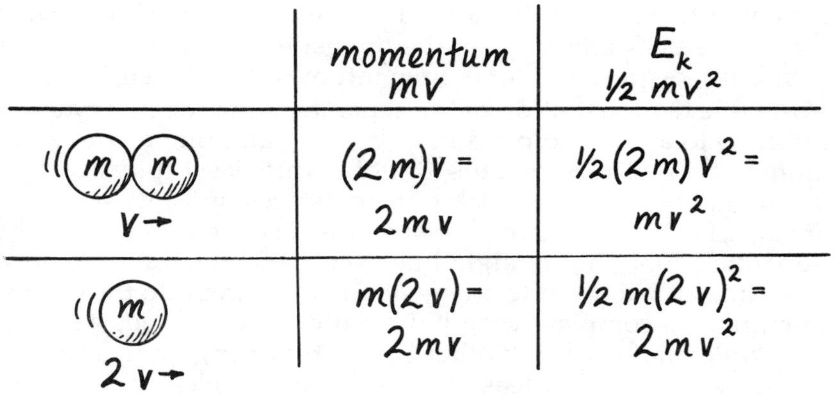

	momentum mv	E_k $\frac{1}{2}mv^2$
((m)(m)) $v \rightarrow$	$(2m)v =$ $2mv$	$\frac{1}{2}(2m)v^2 =$ mv^2
((m)) $2v \rightarrow$	$m(2v) =$ $2mv$	$\frac{1}{2}m(2v)^2 =$ $2mv^2$

FIGURE 9-16 *One ball with twice the velocity of two balls would have the same momentum but twice the kinetic energy.*

only half as great. The balls are made of a hard material, such as steel, which is springy and doesn't absorb energy by permanently changing shape. Most of the kinetic energy of the ball as it comes in is therefore changed back into kinetic energy of whatever comes out of the collision since little energy is converted into other forms. A little kinetic energy is lost to heat and sound, but nothing like half of it. Thus, you always observe the same number of balls coming out as going in.

Actually, you can sneak in a small change that will fix it so that two balls will come out when one ball goes in. If you put chewing gum between the last two balls, they will obviously stick together and they will both come out. You get the same result, however, even if you put the gum someplace down the line. The effect of the chewing gum is to cushion the impulse and absorb some of the energy. Momentum is conserved, but not the kinetic form of energy. Some of the kinetic energy is used in "chewing" the gum. While the total amount of energy is conserved, some of it is changed into the form of heat by the collision process.

The word *elastic* is used to describe how much of the energy retains its form as kinetic energy after a collision. A **perfectly elastic collision** would be an idealized collision in which there would be just as much kinetic energy after the collision process as before. A perfectly elastic ball, for example, would be one that would bounce from a surface with the same kinetic energy after the bounce as before. Such a ball dropped on the floor would bounce up to the same height that you dropped it from. Obviously, the steel balls in the toy-store-variety series impact pendulum are not perfectly elastic, but the proper operation of the device depends on the collision between balls being *nearly* elastic.

An **inelastic collision** is one in which some of the kinetic energy changes into heat, sound, or some other form of energy. Momentum is conserved in an inelastic collision, but not kinetic energy. One example of a perfectly inelastic collision would be our example in Chapter 8, where a little car collided with a truck and the wreckage fused together. As we saw in that example, not all the kinetic energy is lost in a completely inelastic collision. The wreckage may be moving after the collision. Much of the energy, however, does change form in an inelastic collision. The chewing gum between two balls of the momentum demonstration changes the largely elastic collision process into one that is largely inelastic.

The problems in this chapter are related to work, power, and mechanical energy. **Work,** or energy transfer, is defined as the product of force times the distance over which the force acts:

$$\text{Work} \equiv F \cdot d$$

The force and the distance must be in the same direction, because no work is done by any component of the force perpendicular to the distance. Thus, either the force or the distance must be resolved into components and only the parallel component be used to calculate work.

The unit of mechanical energy in the British system of units is simply the foot-pound, but a special name is given to the unit of energy in the metric system: a newton-meter is called a **joule.**

$$1 \text{ ft-lb} \equiv (1 \text{ lb})(1 \text{ ft})$$
$$1 \text{ J} \equiv (1 \text{ N})(1 \text{ m})$$

The rate at which work is done is called **power.**

$$\text{Power} \equiv \frac{\text{work}}{\text{time}}$$

The unit of power in the British system is the **horsepower,** defined as 550 ft-lb/s. The unit of power in the metric system is the **watt,** defined as 1 J/s.

$$1 \text{ hp} \equiv 550 \text{ ft-lb/s} = 746 \text{ W}$$
$$1 \text{ W} \equiv 1 \text{ J/s}$$

The two common forms of mechanical energy are **kinetic** and **potential.** The kinetic energy of an object of mass m and velocity v is given by

$$E_k = \tfrac{1}{2}mv^2$$

The potential energy of an object is the energy stored in a restoring force, such as in a gravitational field or a spring. The potential energy of an object at a height h, near the surface of the earth (where its weight $wt = mg$ is constant) is given by the definition of work, where the force is the weight and the distance is the height:

$$E_p = mgh$$

The potential energy stored in a spring, which obeys Hooke's law, $F = Kx$, with a spring constant K, turns out to be

$$E_p = \tfrac{1}{2}Kx^2$$

Energy, like momentum, is a conserved quantity. The relationship between velocity and the position of an object can be found from the conservation of energy in a whole class of problems where the loss of mechanical energy to other forms, such as heat, is either negligible or known.

$$E_{\text{before}} = E_{\text{after}}$$

In a completely elastic collision, for example, the frictional losses of energy are negligible; the kinetic energy before the collision equals the kinetic energy after. In a completely inelastic collision, on the other hand, much of the kinetic energy is changed into heat as the two objects stick together. In either case, however, momentum is conserved.

9-1   A small boy is helping his father change the tire on the family automobile. The car is not high enough for the fully inflated tire. The conservation of energy allows the boy's puny little force applied to the jack to lift the massive auto.

Given: The boy exerts a force of 80 N to move the jack handle a distance of 30 cm. Seventy per cent of the energy put into the jack comes out as useful work in lifting the automobile, which weighs 12,000 N, the rest going into heat.

Find: (a) How much work does the boy do in one stroke of the jack?
(b) How much work goes into lifting the automobile?
(c) How high does one stroke of the jack lift the car?

Your Solution

Discussion

This problem illustrates the concept of work in a situation where a percentage of the energy is lost because of friction. From the percentage of the energy that does useful work, the distance over which the output force acts can be determined.

Sample Solution 9-1

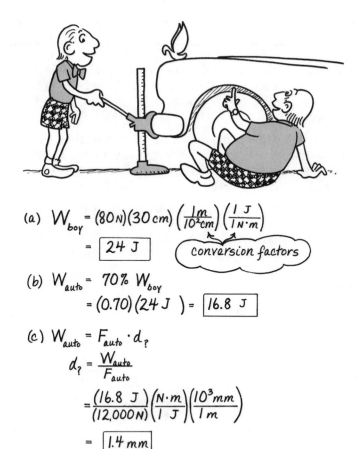

(a) $W_{boy} = (80\text{N})(30\text{ cm})\left(\frac{1\,m}{10^2 cm}\right)\left(\frac{1\,J}{1\,N\cdot m}\right)$

$= \boxed{24\text{ J}}$ *conversion factors*

(b) $W_{auto} = 70\%\ W_{boy}$

$= (0.70)(24\text{ J}) = \boxed{16.8\text{ J}}$

(c) $W_{auto} = F_{auto}\cdot d_?$

$d_? = \dfrac{W_{auto}}{F_{auto}}$

$= \dfrac{(16.8\text{ J})}{(12,000\text{N})}\left(\dfrac{N\cdot m}{1\ J}\right)\left(\dfrac{10^3 mm}{1\ m}\right)$

$= \boxed{1.4\text{ mm}}$

Discussion

The work done by the boy can be found by taking the product of his force times the distance over which he acts. The units are changed into joules by the fact that a joule is defined as that amount of work done by a newton of force acting over a meter of distance.

The work coming out of the jack into the automobile is a simple percentage of the work done by the boy on the jack, as found in part (a).

The distance the automobile moves can be found by solving the definition of work, using the result of part (b) as the work and the weight of the automobile as the force.

9-2 A carpenter's apprentice is dragging a wooden toolbox across a stone floor at constant velocity. (Most of the work he does goes into heating the floor and the toolbox, although some of it goes into sound, which is radiated in all directions.)

Given: The apprentice pulls with a force of 200 N on a rope that makes an angle of 30° with the horizontal floor. He pulls the toolbox a distance of 10 m along the floor.

Find: (a) How much of the force is in the direction the box actually moves?

(b) How much work does the carpenter's apprentice do?

Your Solution

Discussion

This problem is an illustration of work done by a force that is not entirely in the direction of the distance over which the force acts. Part of the force does work, therefore, while the other does not.

Sample Solution 9-2

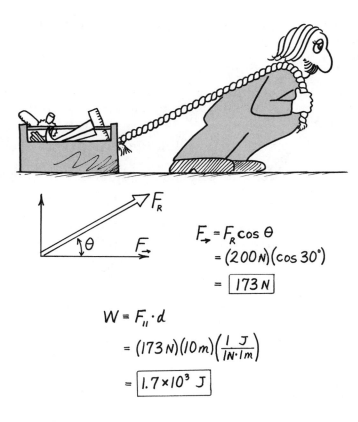

$$F_{\rightarrow} = F_R \cos \theta$$
$$= (200\,N)(\cos 30°)$$
$$= \boxed{173\,N}$$

$$W = F_{\parallel} \cdot d$$
$$= (173\,N)(10\,m)\left(\frac{1\ J}{1\,N \cdot 1\,m}\right)$$
$$= \boxed{1.7 \times 10^3\ J}$$

Discussion

The force acting on the toolbox is resolved into horizontal and vertical components, using the sine and cosine of the angle. Only that component of the force in the direction of the distance is of interest in this problem, however, since that is the only component that does work. The other component tends to reduce drag by helping to lift up on the toolbox, but it does not transfer energy, as there is no displacement in that direction.

9-3 A hoist is lifting concrete in the construction of a high-rise building. You wonder how much energy goes into lifting each bucket of concrete.

Given: The hoist lifts each bucket of concrete a vertical distance of 10 m before swinging it over a horizontal distance of 45 m. Each bucket of concrete has a mass of 2.0 metric tons (2,000 kg).

Find: (a) What is the weight of a bucket of concrete?
(b) How much work goes into lifting each bucket of concrete?
(c) How much work goes into swinging the concrete sidewards.

Your Solution

Discussion

This is a work problem involving a transfer of energy into potential energy. The force involved is the weight of the bucket. Both vertical and horizontal displacements are separately considered.

Sample Solution 9-3

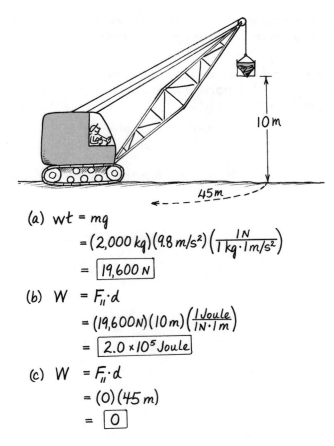

(a) $wt = mg$

$$= (2,000\,kg)(9.8\,m/s^2)\left(\frac{1N}{1\,kg\cdot 1m/s^2}\right)$$

$$= \boxed{19,600\,N}$$

(b) $W = F_{\parallel}\cdot d$

$$= (19,600N)(10\,m)\left(\frac{1\,Joule}{1N\cdot 1m}\right)$$

$$= \boxed{2.0\times 10^5\,Joule}$$

(c) $W = F_{\parallel}\cdot d$

$$= (0)(45\,m)$$

$$= \boxed{0}$$

Discussion

The weight in part (a) is the product of mass and the acceleration due to gravity. This relationship between weight and mass was considered as a proportionality in Chapter 1, and again later, as a special case of Newton's second law.

The work done is the product of force times distance. A parallel sign is used as a subscript on the force to emphasize the fact that the force and distance must be in the same direction. When the vertical distance is considered, the vertical force, which is the weight found in part (a), must be used. When the horizontal distance is considered, as in part (c), there is no force to be used, since the weight is all in the vertical direction. The hoist transfers no energy to the concrete bucket when swinging it sidewards. The only work that is done is in lifting it.

9-4   A tractor is pulling a plow at a uniform constant velocity; all the energy developed by the tractor is used in frictional losses and in lifting the soil.

Given: The tractor is delivering power at the rate of 36 hp to pull the plow at a velocity of 8.9 ft/s. (1 hp ≡ 550 ft-lb/s.)

Find: (a) How much work is done each second, in foot-pounds?
(b) With how much force does the tractor pull?

Your Solution

Discussion

This is a power problem in which work is being done at a constant rate. All the energy delivered by the tractor goes into frictional losses and in lifting the soil, so that no change in kinetic energy need be considered.

Sample Solution 9-4

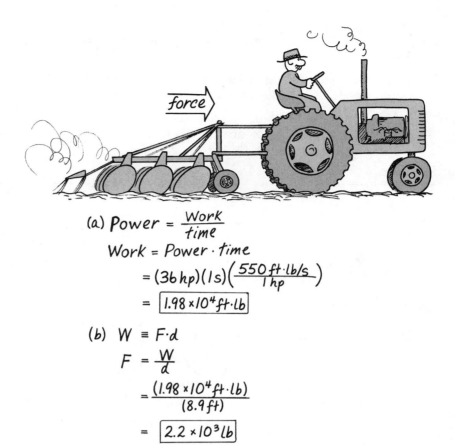

(a) $Power = \dfrac{Work}{time}$

$Work = Power \cdot time$

$= (36\,hp)(1\,s)\left(\dfrac{550\,ft\cdot lb/s}{1\,hp}\right)$

$= \boxed{1.98 \times 10^4\,ft \cdot lb}$

(b) $W \equiv F \cdot d$

$F = \dfrac{W}{d}$

$= \dfrac{(1.98 \times 10^4\,ft \cdot lb)}{(8.9\,ft)}$

$= \boxed{2.2 \times 10^3\,lb}$

Discussion

Power is defined as the rate at which work is done. The amount of work done, or change of work energy, in a unit of time can therefore be found as the product of power and the time increment involved. The units can be changed by using the definition of 1 hp, namely 550 ft-lb/sec.

The force exerted on the plow can be found by solving the definition of work, using the energy and distance for the same period of time. An alternative approach would be to express power in terms of force and velocity and solve for force.

$$Power = \frac{work}{time} = \frac{F \cdot d}{t} = F\,\frac{d}{t} = Fv$$

The result is a little more than 2 tons of force either way.

9-5   A large water tank is filled by pumping water into it from a reservoir below, using electric motor-driven pumps.

Given: 3.00×10^6 kg of water are lifted an average distance of 80 m in 12 min. The tank holds 1.5×10^7 kg of water.

Find: (a) How much work is done in the first 12 min?
(b) How much power is delivered by the pumps?
(c) How long will it take to fill the tank?

Your Solution

Discussion

This is a power problem involving a continuous flow rate. A certain amount of energy is converted into the potential energy of the water in the tank in a certain amount of time. The amount of time it takes to fill the tank can be found from the power, or rate at which work is being done.

Sample Solution 9-5

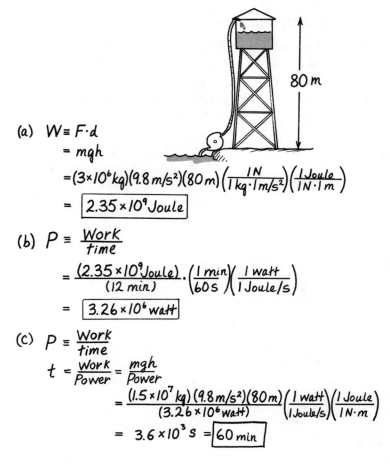

(a) $W \equiv F \cdot d$

$= mgh$

$= (3 \times 10^6 \, kg)(9.8 \, m/s^2)(80 \, m)\left(\dfrac{1 \, N}{1 \, kg \cdot 1 \, m/s^2}\right)\left(\dfrac{1 \, Joule}{1 \, N \cdot 1 \, m}\right)$

$= \boxed{2.35 \times 10^9 \, Joule}$

(b) $P \equiv \dfrac{Work}{time}$

$= \dfrac{(2.35 \times 10^9 \, Joule)}{(12 \, min)} \cdot \left(\dfrac{1 \, min}{60 \, s}\right)\left(\dfrac{1 \, watt}{1 \, Joule/s}\right)$

$= \boxed{3.26 \times 10^6 \, watt}$

(c) $P \equiv \dfrac{Work}{time}$

$t = \dfrac{Work}{Power} = \dfrac{mgh}{Power}$

$= \dfrac{(1.5 \times 10^7 \, kg)(9.8 \, m/s^2)(80 \, m)}{(3.26 \times 10^6 \, watt)}\left(\dfrac{1 \, watt}{1 \, Joule/s}\right)\left(\dfrac{1 \, Joule}{1 \, N \cdot m}\right)$

$= 3.6 \times 10^3 \, s = \boxed{60 \, min}$

Discussion

The work done in a given amount of time is just the potential energy of that amount of water that is lifted in that amount of time. Potential energy is given by mgh, since the force involved in the definition of work is then weight, mg, and the distance is the height, h.

The power found in part (b) is the work done in part (a) divided by the given amount of time necessary to do that much work.

The time necessary to fill the tank of water can be found from this power by solving the definition of power, assuming that the work being done is the potential energy of a full tank of water.

9-6   A good part of the power developed by an automobile engine is lost in the transmission. A typical sports car might have an engine that produces 100 hp, for example, but the power available at the rear wheels to accelerate the automobile turns out to be only a small part of that.

Given: This sports car has a weight of 2,100 lb and can accelerate to a velocity of 88 ft/s in a measured time of 11 s.

Find: (a) What is the mass of the sports car?
　　　　(b) What is the kinetic energy of the car at the end of this acceleration?
　　　　(c) How much average power is delivered to the rear wheels during the acceleration (neglecting friction)?

Your Solution

Discussion

This is a power problem in which work is being converted into kinetic energy at a constant rate. It is particularly important to distinguish between mass and weight when the kinetic energy of something is being considered.

Sample Solution 9-6

(a) $wt = mg$

$$m = \frac{wt}{g} = \frac{2,100\ lb}{32\ ft/s^2} = \boxed{65.6\ Slug}$$

(b) $E_k = \frac{1}{2}mv^2$

$$= \frac{1}{2}(65.6\ Slug)(88\ ft/s)^2 \left(\frac{\left(\frac{1\ lb}{1\ ft/s^2}\right)}{1\ Slug}\right)$$

$$= \boxed{2.54 \times 10^5\ ft\text{-}lb}$$

(c) $Power \equiv \dfrac{Work}{time}$

$$= \frac{(2.54 \times 10^5\ ft\text{-}lb)}{(11\ s)} \left(\frac{1\ hp}{550\ ft\text{-}lb/s}\right)$$

$$= \boxed{42\ hp}$$

Discussion

The mass of the sports car is found by solving the relationship between mass and weight. The unit of mass in the English system of units is the slug, defined as that amount of mass accelerated at the rate of 1 ft/s² by a force of 1 lb.

$$F = ma$$
$$1\ lb = (1\ slug)(1\ ft/s^2)$$

The kinetic energy of the car at a given velocity can be found from this mass by using the formula for kinetic energy, which is derived in the text.

All of the kinetic energy found in part (b) comes from the power delivered to the rear wheels. We can find this power since we know the work done, or energy transfer, from part (b) and we are given the time. We are, of course, neglecting frictional losses that take place while the automobile is accelerating. The result of our power calculation can be converted to more familiar units by using the definition of a horsepower (1 hp = 550 ft-lb/s) as a conversion factor.

Since the answer to part (c) is only 42 hp, which is only 42% of the rated horsepower of the engine, over half the power is lost in the sports car transmission in this example.

9-7 A pile driver has a hammer that is lifted and dropped a considerable distance above the top of the pile. Most of the kinetic energy of the hammer is converted upon impact into work used in forcing the pile into the ground.

Given: The pile hammer has a mass of 380 kg and is lifted 7 m above the top of the pile before being dropped. The impact drives the pile 20 cm into the ground.

Find: (a) What is the potential energy of the pile hammer before being dropped?
(b) What is the velocity of the hammer as it strikes the pile?
(c) With what average force does the hammer force the pile into the ground?

Your Solution

Discussion

This problem illustrates the use of the work concept to find the force of impact from the energy delivered by the blow. The pile driver is simpler than some hammer problems in that this energy all comes from the potential energy of the hammerhead before it is dropped. As in all hammer problems, however, the energy of the hammerhead is all in the form of kinetic energy just before impact.

Sample Solution 9-7

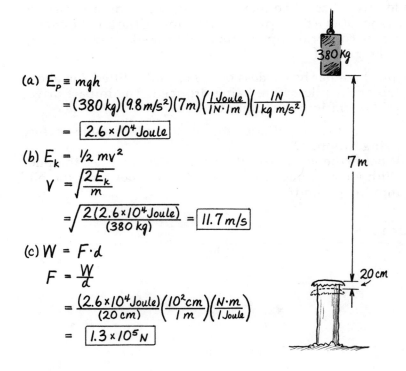

(a) $E_p \equiv mgh$

$$= (380\,kg)(9.8\,m/s^2)(7m)\left(\frac{1\,Joule}{1N\cdot 1m}\right)\left(\frac{1N}{1\,kg\,m/s^2}\right)$$

$$= \boxed{2.6 \times 10^4\,Joule}$$

(b) $E_k = \frac{1}{2} mv^2$

$$v = \sqrt{\frac{2E_k}{m}}$$

$$= \sqrt{\frac{2(2.6\times 10^4\,Joule)}{(380\,kg)}} = \boxed{11.7\,m/s}$$

(c) $W = F \cdot d$

$$F = \frac{W}{d}$$

$$= \frac{(2.6\times 10^4\,Joule)}{(20\,cm)}\left(\frac{10^2\,cm}{1\,m}\right)\left(\frac{N\cdot m}{1\,Joule}\right)$$

$$= \boxed{1.3 \times 10^5\,N}$$

Discussion

Potential energy is weight, *mg*, times height, *h*. This energy all shows up as kinetic energy at the time of impact, and the velocity of the hammerhead can be found from it, as shown in part (b). An alternative approach would be to find the velocity from one of the equations for motion under constant acceleration. You may remember from Chapter 5 that

$$v_f^2 = v_i^2 + 2ad$$

It is instructive to try both approaches to show that they give the same result.

The units in part (b) work out to meters per second if you express joules in terms of kilograms, meters, and seconds:

$$\sqrt{\frac{J}{kg}} = \sqrt{\frac{N\cdot m}{kg}} = \sqrt{\frac{(kg\cdot m/s^2)\cdot m}{kg}} = \sqrt{\frac{m^2}{s^2}} = \frac{m}{s}$$

9-8 The brakes start to slip on a car parked on a hill without its wheels having been curbed. The automobile gains momentum as it travels down the hill in spite of the fact that the brakes absorb part of its energy.

Given: The automobile has a mass of 2,500 kg, and its brakes absorb 40% of its energy. The street is 160 m long and is inclined at an angle of 18° to the horizontal. (Thus only 60% of the car's potential energy changes into kinetic energy.)

Find: (a) What is the automobile's potential energy at the top of the street?
   (b) What is the automobile's kinetic energy at the end of the street?
   (c) How fast is the automobile going at the end of the street?

Your Solution

Discussion

This problem illustrates conservation of energy in a situation where potential energy changes into kinetic energy with a known frictional loss. That portion of the potential energy that changes into kinetic energy determines the car's velocity.

Sample Solution 9-8

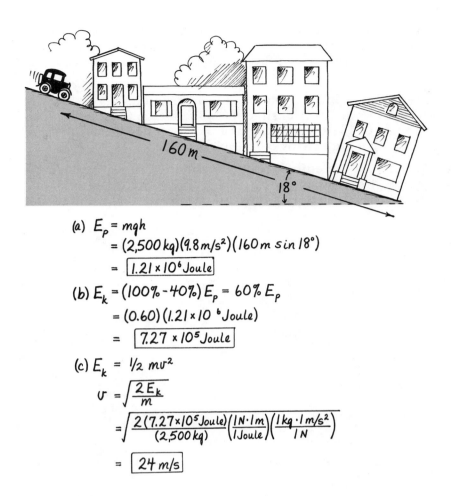

(a) $E_p = mgh$

$= (2,500\,kg)(9.8\,m/s^2)(160\,m\,\sin 18°)$

$= \boxed{1.21 \times 10^6\,Joule}$

(b) $E_k = (100\% - 40\%)\,E_p = 60\%\,E_p$

$= (0.60)(1.21 \times 10^6\,Joule)$

$= \boxed{7.27 \times 10^5\,Joule}$

(c) $E_k = \frac{1}{2}\,mv^2$

$v = \sqrt{\dfrac{2E_k}{m}}$

$= \sqrt{\dfrac{2(7.27 \times 10^5\,Joule)}{(2,500\,kg)}\left(\dfrac{1\,N \cdot 1\,m}{1\,Joule}\right)\left(\dfrac{1\,kg \cdot 1\,m/s^2}{1\,N}\right)}$

$= \boxed{24\,m/s}$

Discussion

The potential energy of the automobile can be expressed in terms of the length of the roadway by taking the height as the hypotenuse of a right triangle times the sine of the angle opposite the height. The units of kilograms times meters per second squared times meters work out to joules, as follows:

$$(kg)(m/s^2)(m) = (N)(m) = J$$

The kinetic energy is found in part (b) as a simple percentage of the potential energy found in part (a), as stated in the problem.

Velocity can be found by solving the kinetic energy expression, using the result of part (b).

9-9 A small boy is flipping a piece of cake across the lunchroom at school. He uses a fork as a catapult by holding the handle on the table and pressing on the back of it with his thumb. Only part of the energy delivered to the fork goes into accelerating the cake, however, as the rest goes into moving the fork.

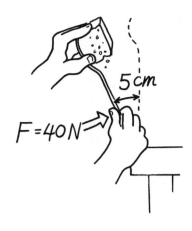

Given: The tines of the fork move through a distance of 15 cm as the boy applies a force of 40 N with his thumb through a distance of 5 cm. Only 60% of the energy is transmitted to the cake.

Find: (a) How much energy does the boy impart to the fork and cake together?

(b) How much energy is imparted to the cake?

(c) What force acts on the cake?

9-10 Paul Hew is cutting the grass again. The force he applies to the lawnmower handle performs two functions. It helps to hold the wheels down on the ground so they don't slip, and it pushes the lawnmower forward at a constant velocity.

Given: Paul is pushing with a force of 250 N at an angle of 38° to the (horizontal) ground. The lawnmower weighs 100 N. Paul pushes a distance of 8 m before stopping to wipe his brow.

Find: (a) What is the horizontal component of the force applied by Paul?

(b) How much work does Paul do before wiping his brow.

9-11 A circus acrobat climbs onto her brother's shoulders while he, in turn, is standing on her father's shoulders. They practice the feat while her father is standing on the ground, although the whole thing is to be done on the back of a galloping horse in an actual performance.

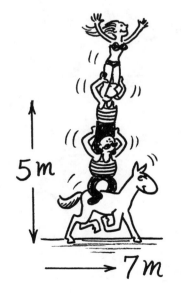

Given: The female acrobat has a mass of 50 kg and must climb a height of 3.0 m while the horse gallops (sidewards) a distance of 7.0 m around the ring.

Find: (a) How much does the female acrobat weigh?

(b) How much work must she do in practice while her father is standing stationary on the ground?

(c) How much work must she do in an actual performance while the horse is carrying her (and everyone else) sidewards?

9-12 Much of the time and expense in drilling an oil well goes into changing the drill bit as it wears out. The drilling pipe must be pulled up and disassembled one section as a time in order to pull the drill bit out of the hole and replace it.

Given: A pipe 0.65 miles long weighs 2,400 lb. It takes the pulling works 7.5 s to lift this pipe 20 ft, the height of one section, in order to disassemble it. (One horsepower equals 550 ft-lb/s.)

Find: (a) How much work is done in lifting the pipe the length of one section?
(b) What is the power delivered by the pulling works, in horsepower?

9-13   An escalator in a department store is driven by a large electric motor whose power consumption changes with the load to keep the escalator moving at a constant velocity, up to a certain maximum power capacity.

Given: An escalator working at 0.22 of its maximum power capacity lifts 18 people per minute a vertical height of 6.8 m. These people have an average mass of 73 kg.

Find: (a) How much work is done in 1 min?
(b) What is the power delivered by the escalator, in watts?
(c) What mass of people can the escalator deliver in a minute, working at its maximum power capacity?

9-14   Annette Racer is measuring her acceleration on her ten-speed bicycle.

Given: Annette produces an average power output of 0.10 hp for 20 s (1 hp = 550 ft-lb/s). Almost all the energy goes into the kinetic energy of her and her bicycle, which have a combined weight of 110 lb.

Find: (a) How many foot-pounds of kinetic energy does she produce?
(b) What is the combined mass, in slugs, of Annette and her bicycle?
(c) To what velocity does she accelerate?

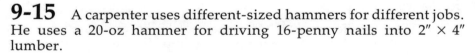

9-15   A carpenter uses different-sized hammers for different jobs. He uses a 20-oz hammer for driving 16-penny nails into 2″ × 4″ lumber.

Given: A carpenter exerts a horizontal force of 35 N to accelerate a 0.57-kg hammer toward a nail. When this force is exerted over a distance of 50 cm, the hammer is found to drive the nail a horizontal distance of 2.0 cm.

Find: (a) How much work does the carpenter do in accelerating the hammer?
(b) If all this work goes into the kinetic energy of the hammer, how fast is the hammer traveling when it hits the nail on the head?
(c) If all this kinetic energy goes into work in driving the nail, with what average force does the hammer push on the head of the nail (assume the carpenter hits the nail)?

9-16 A little person is going down a slide at the playground. She discovers that she has some control over her velocity by pressing her feet outward against the side rails of the slide so as to change the rate at which her energy is converted into the heat of friction.

Given: The little person has a mass of 20 kg and starts at the top of a 15-m slide, which is inclined at an angle of 50° to the ground. Only 25% of her potential energy is converted into kinetic energy.

Find: (a) What is the little person's potential energy when she is at the top of the slide?

   (b) How much kinetic energy does she have at the bottom of the slide?

   (c) What is her velocity at the bottom of the slide?

MORE INTERESTING PROBLEMS

9-17 A sports car is entered in a hill-climbing race. In a long straight section of the race the car gets up to a constant velocity, which it can just maintain in fourth gear.

Given: The car has a mass of 900 kg and is climbing a 12° incline (to the horizontal) at a velocity of 30 m/s. (One horsepower ≡ 550 ft-lb/s = 746 W.)

Find: (a) How much work is done in 1 s?
   (b) What is the power delivered to the rear wheels, in watts?
   (c) What is the power delivered to the rear wheels, in horsepower?

9-18 A woodsman is using an ax to split a log. He swings the ax from above his head and maintains a constant downward force, which adds to the original potential energy in giving the ax head kinetic energy with which to split the log.

Given: The ax head has a mass of 4.5 kg, and the woodsman swings it from a height of 2.4 m through an arc having a length of 3.1 m with a relatively constant force applied to it of 80 N.

Find: (a) How much potential energy does the ax have at the top of its swing?
   (b) How much work does the woodsman do on the ax head?
   (c) With what velocity does the ax hit the log on the ground? (*Hint:* Its kinetic energy comes from its potential energy added to the work done.)

9-19 An elevator is lifting two people at a constant velocity from the first to the twentieth floor of their apartment building.

Given: The elevator has a mass of 1000 kg, while each of the 2 people have a mass of 120 kg. The elevator lifts them 90 m in 30 s (each floor is 4.5 m tall).

Find: (a) How much energy goes into lifting the elevator when empty?
(b) How much energy goes into lifting the two people?
(c) What is the power required to lift the elevator with the two on board?

9-20 Water in a fountain is squirted up into the air by a motor-driven pump.

Given: The pump draws electrical power at 200 W and is 85% efficient at imparting this energy to the water. It pumps water at the rate of 400 kg/min.

Find: (a) How much energy is used by the pump in 1.0 min?
(b) How much kinetic energy is given to 400 kg of water?
(c) What is the maximum height to which the water will squirt?

9-21 A rubber ball when dropped from a certain height will lose some of its energy on impact with a hard surface and will only bounce up to a certain percentage of its original height.

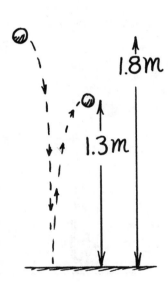

Given: A ball having a mass of 0.45 kg is dropped from a height of 1.8 m above a concrete sidewalk. It bounces back up to a height of 1.3 m.

Find: (a) What was the ball's initial potential energy?
(b) What percentage of this energy was changed into heat and sound by the impact?
(c) With what velocity did the ball bounce back up from the concrete?

9-22 A stone is thrown from the edge of a cliff into the ocean below. It is thrown somewhat upward, but mostly outward.

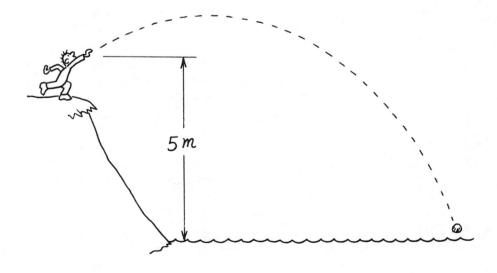

Given: The stone has a mass of 2.0 kg and is thrown with a speed of 10 m/s from an initial height of 5.0 m above the ocean.

Find: (a) What is the kinetic energy of the stone when it is thrown?
(b) What is the potential energy of the stone relative to the ocean when it is first thrown.
(c) What is the kinetic energy of the stone just as it hits the water?
(d) What is the speed with which the stone hits the water?

ANSWERS *(Some answers may be reported to one more than the proper number of significant figures justified by the data so as to avoid round-off errors.)*

9-9   (a) 2.0 J; (b) 1.2 J; (c) 8 N.
9-10 (a) 197 N; (b) 1.6×10^3 J.
9-11 (a) 490 N; (b) 1.5×10^3 J; (c) 1.5×10^3 J (same as part (b)).
9-12 (a) 4.8×10^4 ft-lb; (b) 12 hp.
9-13 (a) 8.76×10^4 J; (b) 1.46×10^3 W; (c) 6.0×10^3 kg.
9-14 (a) 1.1×10^3 ft-lb; (b) 3.44 slug; (c) 25 ft/s.
9-15 (a) 17.5 J; (b) 7.8 m/s; (c) 8.7×10^2 N.
9-16 (a) 2.25×10^3 J; (b) 5.6×10^2 J; (c) 7.5 m/s.
9-17 (a) 5.5×10^4 J; (b) 5.5×10^4 W; (c) 74 hp.
9-18 (a) 106 J; (b) 248 J; (c) 13 m/s.
9-19 (a) 8.8×10^5 J; (b) 2.1×10^5 J; (c) 3.6×10^4 W = 49 hp.
9-20 (a) 1.2×10^4 J; (b) 1.02×10^4 J; (c) 2.6 m.
9-21 (a) 7.94 J; (b) 28%; (c) 5.0 m/s.
9-22 (a) 100 J; (b) 98 J; (c) 198 J; (d) 14 m/s.

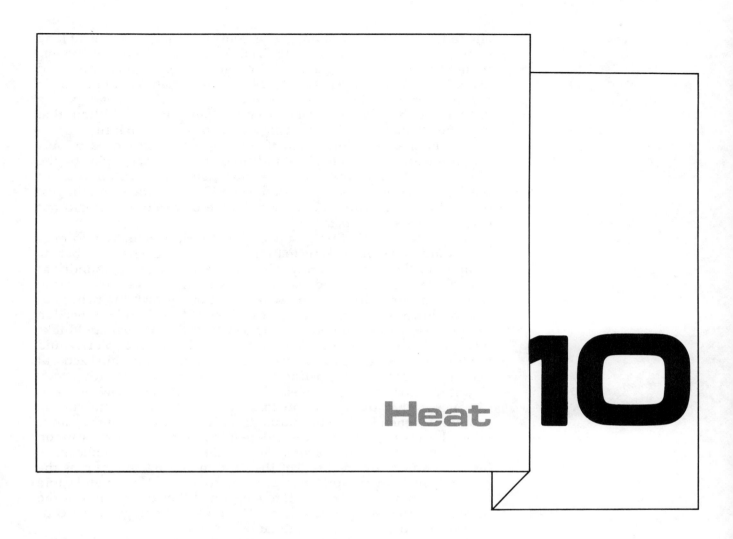

Heat **10**

▶ Heat as a Form of Energy

Heat is so well established in modern thinking as a form of energy that it is hard to imagine what people thought of it before it was thus understood. As mentioned in Chapter 9, not even the idea of energy as a conserved quantity was established until the 1840s, when Sir James Joule did some experiments that conclusively demonstrated that heat was the missing piece in the energy picture. Before that, most people thought of heat as a kind of fluid that permeated all real objects, flowing from hot objects to cold ones in the same way that water seeks its own level. They even had a name for this fluid. They called it caloric.

As long as the caloric fluid theory held, no one could explain what happened to the energy of a system after long periods of time. It was easy to see that the potential energy of a system could be converted into kinetic energy and, under the right circumstances, back again. The pendulum is a classic example. It swings back and forth, the potential energy of the pendulum bob changing into kinetic energy as it swings through the middle and then reappearing again as potential energy as it reaches the end of its swing on the opposite side. But it also seemed obvious that the mechanical energy, potential and kinetic, of this and other systems always seemed to die away after a while. The pendulum always comes to a stop if you wait long enough. Before heat was recognized as a form of energy, people thought that energy had a finite lifetime. It seemed that you could make it live longer by reducing friction, but it always appeared to fade away eventually. Even as late as a century and a half ago, people did not see the relationship between heat and energy. Without that key, they could not think of energy as a conserved quantity.

Under the old caloric fluid theory, heat was supposed to surround the atoms of both gases and solids to hold them apart by the mutual repulsion of their caloric atmospheres. Count Rumford put the first chink into this theory in the late 1700s, but he spent the rest of his life trying without success to prove that caloric heat did not exist as a material substance.

Count Rumford's first suspicions that heat is actually a form of energy are said to have come to him while he was in charge of boring cannon for the Bavarian army. Cannon were cast in the foundry as large metal cylinders, and the barrels were then bored by turning them on a horse-driven lathe with a stationary drill bit being advanced down the casting. Large quantities of heat were produced in the process, which limited the rate at which the process could take place without overheating the cutting tool. The old caloric fluid theory saw this heat as a material substance that was squeezed out from the surface of the casting by the pressure of the cutting tool. Count Rumford carefully collected the metal cuttings, however, and discovered that there was no loss of weight due to the cutting process. He then tried using a dull drill bit. He found that the rate at which heat was produced was independent of how long the cannon had been drilled. If heat were a material substance, it would have been eventually drained out of the cannon casting. As long as the horses kept at their work, however, he found that more and more heat was produced. He therefore concluded that the motion of the horses was the source of the heat. The idea that energy was a conserved quantity, however, still eluded him.

It was not until about 50 years later that anyone thought of the idea of conservation of energy. When this idea did come, it came to several people independently at about the same time. A number of people then set out to prove that energy is conserved. The most successful was James Joule, who did a whole series of experiments in the early 1850s. His experiments showed that the same amount of heat is always produced by a certain amount of frictional work, no

FIGURE 10-1 *Count Rumford's experiments with boring cannon led him to believe that the motion of the horses was the source of the large quantities of heat produced.*

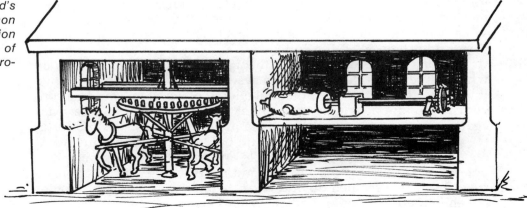

matter how the work is done. That is, he established the mechanical equivalence of heat. We can express this mechanical equivalence of heat as a proportionality between the work done and the heat, Q, produced.

$$\text{Work} \propto Q$$

Count Rumford noticed this proportionality earlier, but James Joule showed that the proportionality constant was always the same for all different kinds of work. (Note that J, symbolizing the constant, is different from J, the abbreviation for a joule of energy.)

$$\boxed{\text{Work} = JQ}$$

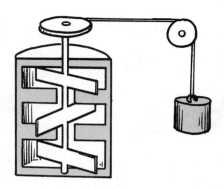

FIGURE 10-2 *Joule measured the mechanical equivalent of heat by using falling weights to turn paddles in a container of water.*

Joule's main experiment was to measure this proportionality constant, J, very carefully using falling weights to turn a set of paddles in a well-insulated container of water. The idea was to measure the amount of heat produced by stirring the water with the known amount of potential energy stored in the weights at the start of the experiment. He then showed that the same proportionality constant results from all the other friction experiments he could think of. He tried stirring mercury instead of water. He tried rubbing iron rings together. He even tried rubbing iron rings together in a mercury bath. He went so far as to do experiments using electrical energy. He found that he always got the same result, as long as the work was frictional in the sense that it is "all used up" and not stored in some form of kinetic or potential energy.

We might say that Joule's experiments unlocked the secret of conservation of energy, a secret we take for granted these days. Count Rumford held the key to that secret, and you might wonder why he didn't try that key in the lock, so to speak. The answer must be that he didn't even recognize that the lock was there. The concept of energy as something that actually exists, but not as a material substance, is quite new to the human race.

▶ The Measure of Heat

We have established units for measuring work in both the British and the metric systems of units, but before we can give the value of the mechanical equivalent of heat, J, we must establish units for measuring heat. You might argue that heat is now well-accepted as a form of energy and that we don't really need to invent another set of units. Perhaps we could just go ahead and speak of heat in terms of foot-pounds or joules, but vestiges of the old fluid theory of heat remain even after all this time. It is still common to find heat measured in BTUs and calories rather than in foot-pounds or joules. What is even stranger, the word *calorie* can refer either to a certain amount of heat or to a thousand times as much heat, depending on whom you are talking to.

The idea of measuring heat in both systems of units was based on the change of temperature of a certain amount of a certain substance. Both systems used water as the substance, but the British used a pound of water whereas the metric system started out using a gram of water. Both systems are based on 1° of temperature change, but the British use a degree on the Fahrenheit scale while the metric system uses a degree on the Celsius scale.

Not surprisingly, the British refer to the amount of heat necessary to raise the temperature of 1 lb of water 1 °F as the British Thermal Unit, or the BTU. Actually some refinements have been made over the range of the temperature changes, because the heat capacity of water isn't exactly constant, but these small differences can be ignored for all practical purposes.

CHECK QUESTION

A 30-gallon hot water tank holds 240 lb of water. How much heat is required to change the 60 °F cold water that comes into your house into the nice, steamy hot 180 °F stuff you expect to see at the hot water valve in your shower?

Answer: 28,800 BTU

The first metric unit of heat was the small **calorie.** It was defined as the amount of heat necessary to raise the temperature of 1 g of water 1 °C. This original definition has been modified somewhat by popular usage. The small calorie represents such a small quantity of heat that people found the **kilocalorie** much more useful for most purposes. The kilocalorie may either be thought of as 1,000 small calories or as the amount of heat necessary to raise 1 kg of water 1 °C. The kilocalorie became so common in usage that people just shortened its name to Calorie with a capital C to distinguish it from the small calorie. It is this big Calorie that is used to measure the energy content of various quantities of food. People used this Calorie so much for food energy values that they probably forgot why they capitalize the C and over the years just decided that Calorie should not be capitalized. Thus, calorie stands for both the amount of heat necessary to heat 1 g of water 1 °C and for a thousand times that much heat, depending on who is using the word. The surprising thing about this state of affairs is that it seldom results in any permanent confusion.

FIGURE 10-3 (a) A big Calorie is the amount of heat necessary to raise the temperature of 1 kg of water 1 °C. (b) A BTU is the amount of heat necessary to raise the temperature of 1 lb of water 1 °F.

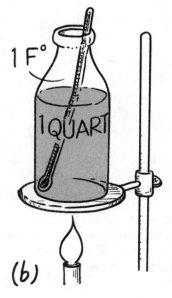

CHECK QUESTION

A peanut contains about 3 Cal of heat energy. If it is burned under a flask containing 0.5 liter of water and about half the heat goes into warming the water, how much will the temperature of the water increase?

Answer: 3 °C

▶ The Mechanical Equivalent of Heat

One reason why it took the old fluid theory of heat so long to die, and why it wasn't immediately evident that heat is a form of energy, is that a small quantity of heat, say a BTU or a Calorie, represents a great big bunch of energy. The value of the mechanical equivalent of heat, J, turns out to be a rather large number. That is to say, the proportionality constant between frictional work and heat energy, Q, is such that you would have to do a huge amount of work before you begin to notice the heat you generate.

$$\text{Work} = JQ$$

Such huge quantities of work were first available in industrial processes like the cannon boring done by Count Rumford, but it took the careful paddle-wheel experiments of Joule to put a number to J. In terms of the two systems of units we are now using, the mechanical equivalent of heat turns out to have the following values:

$$J = 4{,}185 \text{ J [joules]/Cal}$$
$$J = 777.9 \text{ ft-lb/BTU}$$

Let us use an example to illustrate the difference in size in the amount of energy represented by a normal quantity of heat as compared to a normal quantity of mechanical work. Suppose that you are served a lukewarm cup of coffee in a restaurant and that it occurs to you, after you fail to attract the waiter's attention by normal means, that you might be able to raise the temperature of the coffee by stirring it with a spoon. The question is whether you might not wind up by attracting the waiter's attention anyway if you were to try to stir the coffee violently enough to produce a significant temperature change, say, 4 °C. That would amount to 1 Cal of heat in a quarter-liter cup of coffee.

Assume that you stir with a force of about 1 N. (That would be the amount of force you would need to hold a quarter-pound stick of butter on the end of your spoon.) If you stir 5 cm in one direction and then 5 cm back again, you would stir a total distance of 10 cm—0.10 m—with each stroke. Ten such strokes would produce a joule of energy, but it would take almost 42,000 such strokes to produce a Calorie of heat energy. Therefore, the process of evaporation would cool your coffee more quickly than you could reasonably warm it by stirring.

We are fortunate in some ways that so much mechanical energy can turn into such a little bit of heat. In driving an automobile down the freeway, we have a tremendous amount of kinetic energy. All that energy must be dissipated if we should need to stop. Our brakes can handle all this energy only by converting it into the more concentrated form of energy known as heat.

FIGURE 10-4 *A little bit of heat represents a great deal of work.*

▶ Perpetual Motion Machines of the First Kind

Philosophically, probably the most important thing about the mechanical equivalent of heat is that the proportionality works both ways. Not only can work be changed into a certain amount of heat but heat can also be changed into a certain amount of work in exactly the same proportion. In this case, however, we would be talking about a change of heat, ΔQ, since work is the change of, or flow of, energy. The energy you can get out of any device

$$\text{Work} = J \, \Delta Q$$

in the form of work is just the mechanical equivalent of the amount of heat that went into the device as long as the internal energy of the device doesn't change. That is, the work put out is just the proportionality constant J times the difference between the heat, Q_{in}, which flows into the device from a region of high temperature and the heat, Q_{out}, which flows out into a region of lower temperature.

$$\text{Work} = J(Q_{in} - Q_{out})$$

If you have a box that you get work out of, no matter what is in the box, you must always put the same amount of heat energy into the box or else the mechanism in the box will run down. This statement seems reasonable to most people today, since it is really nothing more than the idea of the conservation of energy. You can't get more energy out of a box than you put in without the internal energy of the box changing. This is called the **first law of thermodynamics.** Nobody has figured out a way around it.

Many very clever people have spent many hours trying to get more energy out of a box than they put in. Some of these boxes contain such ingenious mechanisms that it is difficult to see why they wouldn't run on forever without running down. Such a device would be very useful in that it would provide a limitless supply of work. Many inventors have applied for patents on perpetual motion machines of this kind, but the United States Patent Office has never issued one. No one has ever been able to come up with a working model.

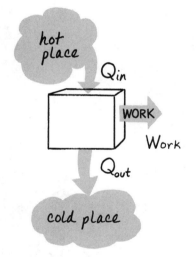

FIGURE 10-5 *The work that comes out of a device on a continuous basis is just the mechanical equivalent of the difference between the heat energy in and the heat energy out.*

▶ Temperature

We have been using the concept of temperature as though everyone knew what it meant. Heat units have been defined in terms of a temperature difference, as would be read on a thermometer. A heat engine has been described in terms of the difference in the heat energy coming out of a hot place and going into a cold place. The fact is that everyone is familiar with the concept of temperature. We all have daily experience with the operation of a thermometer, and everyone knows that heat flows from hot places to cold places. Most people could even tell you that temperature is some kind of measure of the concentration of heat energy. In view of the high level of general understanding most people have of this concept, it seems surprising that a formal definition of temperature is hard to come by. But very few people worry much about what temperature is; they are content to understand what it does. An operational understanding does in fact solve more problems than a theoretical definition. We will therefore go over the practical aspects of temperature before starting to initiate you into the small circle of people who understand temperature on a theoretical level.

One of the obvious indications of temperature is the volume of mercury or alcohol in a glass thermometer that expands or contracts to fill the capillary tube to a certain point. By marking the tube at a point where the mercury or alcohol stops rising, we indicate a certain temperature. A temperature scale could be established by marking the tube at two reproducible temperatures, say, the freezing point and the boiling point of water under standard atmospheric pressure, and then putting a certain number of equally spaced marks in between. We might select that number to be 100, for example. Since the Latin prefix for one hundred is *cent*, we might well call our scale the centigrade scale. Actually, the first person to do this was a Swedish astronomer by the name of Anders Celsius. We therefore call this scale the Celsius scale.

The choices of 0° and 100° as representing the freezing and boiling points of water are arbitrary. You might argue that any other two numbers, such as 32 and 212, would do just as well. Subtracting 32 from 212 would give 180° between these two temperatures:

$$212 - 32 = 180$$

You could then think up a name for 180, say, call it the "fahren," and it would then seem natural to call such a temperature scale the fahrengrade scale. If the first person to think of doing such a thing had happened to have the name Gabriel Daniel Heit, you might well reason that it would be even better to call this scale the fahrenheit temperature scale.

In truth, the first person to think of such a temperature scale was a German physicist named Gabriel Daniel Fahrenheit, and he did not really select arbitrary numbers for the freezing and boiling points of water. He just happened to select different temperatures for the 0 and 100 point marks on his scale. Perhaps Fahrenheit liked making hand-cranked ice cream or something, because the zero point he chose was the lowest temperature he could reach by mixing salt with ordinary ice. The salt forces the ice to melt at lower and lower temperatures, up to a point. That point, called the "brine point," is the temperature at which salt water freezes. The 100° mark, on the other hand, was selected to be near his own body temperature. Maybe he had a slight fever that day or something, but his temperature scale was well established and widely used before Fahrenheit heard of the fixed points used by Celsius. To his credit, Fahrenheit agreed that the temperatures of freezing and boiling of pure water were easier fixed points to use than the ones he had originally adopted, and he redefined his temperature scale in terms of them. The result is that average human body temperature is almost, but not quite, 100° on the Fahrenheit scale.

FIGURE 10-6 *A temperature scale might be based on the freezing point and boiling point of water.*

Once in a while, a person wants to convert from one temperature scale to the other. This usually happens when you have the wrong kind of thermometer. Someday people will go out and break all the Fahrenheit thermometers in the land so that we won't have this problem anymore. Before this happens, however, you may want to know how to do this conversion.

The conversion must take into account the differences in the degree size as well as the difference in the zero points on the two scales. The difference in size is easy. An ordinary conversion factor like we have been using all along will handle a *difference* in temperature very nicely. A temperature difference of 180° on the Fahrenheit scale, from 32 °F for freezing to 212 °F for boiling of water, is equal to a temperature difference of 100° on the Celsius scale. Any other temperature difference is in the same ratio. A temperature difference of 9° on the Fahrenheit scale is equal to a temperature difference of 5° degrees on the Celsius scale.

$$\Delta T = 9 \text{ °F} \left(\frac{100 \text{ °C}}{180 \text{ °F}} \right) = 5 \text{ °C}$$

There are nearly twice as many Fahrenheit degree marks between the freezing and boiling points of water as there are on the Celsius scale. For rough work, you can therefore just multiply or divide by 2 and add or subtract a degree or two to compensate for the fact that 180 divided by 100 is not exactly equal to 2.

The difference between the zero points on the two scales is also easy to compensate for. You just add 32° to a Fahrenheit temperature difference relative to freezing to get the reading on a Fahrenheit thermometer. Going the other way, you just subtract 32° from a Fahrenheit reading to find out how many degrees it is relative to freezing and then convert this temperature difference to Celsius by an ordinary conversion factor.

CHECK QUESTION

Photographic developing times are often given for 68 °F. (a) How many Fahrenheit degrees is this above freezing? (b) What would this temperature read on a Celsius thermometer?

Answer: (a) $\Delta T = 36$ °F, (b) 20 °C

▶ Thermal Expansion

Temperature is easily measured by the changes in various properties of materials that happen as temperature changes. Different substances melt or boil at different temperatures. Other things undergo a change in color or electrical resistance as their temperatures change. However, one of the most familiar changes associated with an object's temperature is simply its size.

The thermal expansion of mercury or alcohol in a glass thermometer is one of the best-known measures of temperature. It happens that the rate of expansion of neither one of these substances is exactly constant with increasing temperature, but they can be calibrated against a gas thermometer. Gas thermometers were described in Chapter 3 when the pressure in fluids was discussed. It turns out that the best thermometer, in terms of a constant rate of expansion with increasing temperature, would be one that would use the expansion of an ideal gas under constant pressure. Of course, an ideal gas does not exist, but people have spent a great deal of time calibrating thermometers on what they figure an ideal gas would do if only it did exist.

It also turns out that ordinary solid objects generally expand with increasing temperature. We do not ordinarily think of the length of various objects around us as depending much on temperature, but a careful look reveals many situations where it must be taken into account. Cracks in sidewalks are really thermal expansion joints put there so that forces generated by changes in temperature will not break the concrete. Freeway overpasses must be provided with much larger expansion joints. You can entertain yourself as you ride down the freeway by studying the overpasses to see if you can find the expansion joints. Sometimes one end of a bridge will be mounted on rollers, and sometimes the whole support structure will be on rockers. You can frequently spot the expansion joint from the top by the metal plate in the roadway used to cover the joint. Perhaps you have noticed the metal plates with a zig-zag tooth pattern running across the roadway and wondered what they were for. Everyone has noticed the cracks in a railroad track that make the wheels go clickity-clack on both trains and streetcars, but few people realize that these, too, are put there deliberately as expansion joints.

We can figure out exactly how big an expansion joint must be to handle certain change in temperature if we know the coefficient of linear expansion of the object in question. Let us consider an experiment that you could do in your own home to figure out how much of a crack you would need to leave in, say, a railroad track. Go down to your basement, at least in your mind's eye, and find an old piece of galvanized steel water pipe and a garden hose. Attach one end of the pipe to the drain on your hot water heater with the hose. Nail this end of the pipe down onto a block of wood and rest the other end on another block some distance away so that it is free to slide. You might think that running hot water through the pipe would cause it

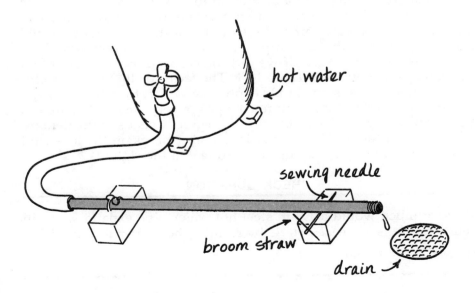

FIGURE 10-7 *The thermal expansion of steel could be measured at home in your basement.*

to expand such a tiny amount that you would never be able to measure it, but there is a way to see the pipe expand before your very eyes. Just put a sewing needle under the free end of pipe to act as a roller, and stick a broom straw through the eye of the needle to act as a pointer. Surprisingly, a piece of pipe about a meter long will expand enough to make the broom straw turn through something like 30 to 50 degrees, depending on the temperature of your hot water and the diameter of your sewing needle.

The distance that the pipe expands, Δl, is directly related to the angle through which the needle rolls. It would be reasonable to assume that this change in length of the pipe would be proportional to the temperature change, ΔT. The hotter the water flowing through the pipe, the more the pipe expands.

$$\Delta l \propto \Delta T$$

There is, however, something besides temperature that affects this change in length. You could verify that the change in length, Δl, is also proportional to the original length of the pipe, l. If you were, for example, to move the block with the sewing needle to the middle of the pipe, the angle through which the broom straw would twist would be only half as large as when if it were at the end. The pipe would actually expand the same as before, but the rolling needle would only record the amount by which the first half of the pipe expands. Thus, we have a second proportionality.

$$\Delta l \propto l$$

We can put these two proportionalities together in a single statement by just saying that Δl is proportional to the product of l and ΔT.

$$\Delta l \propto l \, \Delta T$$

The proportionality constant between Δl on one side of the equation and l times ΔT on the other is called the **coefficient of linear expansion,** and the symbol usually used for it is the Greek letter alpha, α.

$$\boxed{\Delta l = \alpha l \, \Delta T}$$

The value of the cofficient of linear expansion, α, is different for different materials. We might look around in your basement for a piece of copper tubing and a piece of aluminum electrical conduit to use in place of the steel water pipe. We would find that the copper would expand more than the steel pipe and that the aluminum would expand over twice as much. If we did the calculation, we would find the values for α given in Table 10-1. The factor of 10^{-5} in the coefficient of linear expansion simply expresses the fact that the change in length for ordinary temperature changes is very small compared to the original length. The unit $(°C)^{-1}$, or degrees Celsius in the denominator, is just the unit that α needs to make things on the left-hand side of the equality match up with things on the right.

TABLE 10–1

Material	α
Aluminum	$2.8 \times 10^{-5}(°C)^{-1}$
Copper	$1.7 \times 10^{-5}(°C)^{-1}$
Steel	$1.2 \times 10^{-5}(°C)^{-1}$

CHECK QUESTION

By how much would a 1.0-m steel hot water pipe expand when its temperature is increased 50 °C by running hot water through it?

Answer: $\Delta l = 0.6$ mm

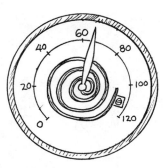

FIGURE 10-8 *Some thermometers use a coiled bimetallic strip to turn a pointer.*

Some thermometers use the difference in α of different metals to measure temperature. A bimetallic strip is composed of two metals welded together. The two metals, such as brass and iron, expand differently. The difference in the expansion causes the strip to bend as it is heated. If you look inside a thermometer and see what seems to be a coil spring, the chances are that this spring is made of a bimetallic strip wound up into a coil. Heat the coil by breathing on it, and see if it doesn't unwind itself a little.

Bimetallic strips are frequently used in thermostats to open and close electrical contacts. The thermostat that controls the furnace in your house probably works this way.

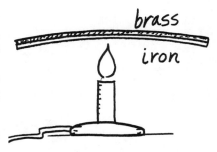

▶ Heat Capacity

The ideas of heat and temperature are so closely related that people sometimes forget there is a difference. At this point, we are still working with an operational definition of temperature. Everyone knows what temperature is—it's what you read on a thermometer. We have seen how it is associated with thermal expansion. We have even defined units of heat in terms of it. One Calorie was the amount of heat necessary to raise the temperature of 1 kg of water 1 °C. Almost everyone also knows, however, that we can get the same temperature in a little water by adding a little heat as we can in lots of water by adding lots of heat. So heat and temperature, while related, are two different things.

FIGURE 10-9 *A bimetallic strip bends, because the two metals expand differently.*

Let us go back to the lukewarm cup of coffee. Having abandoned our attempts to warm it by stirring, we proceed to warm it by holding it over the candle that the management has put on our table for just such emergencies. Right away we would notice that the amount of heat necessary to warm the coffee is directly proportional to the amount of coffee in the cup. If you were to drink half the coffee, you would need to warm only that which remains, and the temperature would go up much more quickly. That is, the change in heat, ΔQ, necessary to produce a certain change in temperature is directly proportional to the mass, m, of the coffee:

$$\Delta Q \propto m$$

Of course, the change in heat, ΔQ, is also proportional to the change in temperature, ΔT. Colder coffee would require more heat to warm to the same temperature:

$$\Delta Q \propto \Delta T$$

Since the heat, ΔQ, is proportional to both the mass of the coffee, m, and the change in temperature, ΔT, it is also proportional to their products:

$$\Delta Q \propto m \, \Delta T$$

The proportionality constant is called the **specific heat** of the material. People usually use c to stand for specific heat:

$$\boxed{\Delta Q = cm \, \Delta T}$$

The specific heat of coffee is very close to that of water, because that is what coffee mostly is, and you should already know the specific heat of water if you think about it. Our units of heat in both systems of units are defined in terms of the specific heat of water. One Calo-

FIGURE 10-10 *The amount of heat necessary to warm a cup of coffee depends on the amount of coffee in the cup and the temperature of the coffee to begin with, as well as on the kind of coffee you are drinking.*

TABLE 10–2

Substance	Specific Heat	
Alcohol	0.58	Cal/kg-°C
Aluminum	0.217	
Brass	0.094	
Copper	0.093	
Glass	0.199	
Ice	0.55	
Iron	0.113	
Lead	0.030	
Marble	0.21	
Mercury	0.033	
Water	1.00	

rie, for example, is defined as the amount of heat that raises the temperature of 1 kg of water 1 °C:

$$1 \text{ Cal} \equiv c_{H_2O}(1 \text{ kg})(1 \text{ °C})$$

Therefore, the specific heat of water is equal to 1 in both systems of units by the very definition of the Calorie and the BTU.

$$c_{H_2O} \equiv 1 \text{ Cal/kg-°C} = 1 \text{ BTU/lb-°F}$$

Actually, it would be safe to say that the specific heat of coffee is a little less than that of water, even without knowing the exact organic chemicals that make it into coffee. It happens that the specific heats of all common substances are smaller than that of water. Irish coffee, for example, would have a specific heat significantly smaller than that of regular coffee. The specific heat of alcohol is less than six-tenths that of water. Just for interest, the specific heats of several common substances are listed in Table 10-2.

You might notice from looking at the table that the specific heat of a substance seems to get smaller as the density increases. This rule of thumb results from the fact that the heat capacity of a substance represents the ways in which heat can be stored by the atoms in the substance vibrating in different motions. One atom can only store so much energy. If that atom happens to be very heavy, it can still only store so much energy. The heat capacity per atom of many substances is about the same, but the heat capacity per kilogram tends to decrease as the number of atoms or molecules represented by a kilogram decreases.

There are some exceptions to this rule of constant heat storage per molecule, and these exceptions occur when something about the molecule gives it additional ways of storing heat. Water, for example, has a high specific heat because of its physical structure. The funny shape of the water molecule gives it more ways to vibrate than is common for other substances. When the same water freezes into ice, however, its specific heat is drastically reduced because it loses some of its degrees of freedom, or ways to vibrate.

▶ Calorimetry Problems

Many situations in daily life involve combining things and allowing them to come to the same final temperature. Suppose that the waiter in our example finally notices you and offers to warm your coffee. He adds piping hot coffee to your half-filled cup of lukewarm coffee, and the mixture comes to a temperature somewhere in the middle depending on how much piping hot coffee was added. It would seem obvious that the final temperature would be exactly halfway between the temperatures of the lukewarm and the piping hot parts if the parts of the mixture are equal to begin with. Also, the final temperature would be closer to that of the larger part if they were not equal. Let us assume that the parts are not equal, and we will figure out how to predict the final temperature of the mixture using the law of conservation of energy.

All the energy to be conserved in this case is in the form of heat. If we assume that none of the heat gets away from us by ignoring any cooling of the overall mixture during the short time that we do our experiment, the principle of conservation of energy comes down to the heat gained by the lukewarm coffee being equal to the heat lost by the piping hot coffee that was added.

> Heat gained = Heat lost

This particular form of the energy conservation principle turns out to be the key to solving a whole class of physics problems involving thermal equilibrium. Careful measurements are frequently done in a well-insulated container called a **calorimeter,** and the whole science of measuring heat in this way has come to be known as **calorimetry.** (If you reach into your pocket and pull out a thermometer with which to stir your coffee, the waiter will recognize you as a calorimetrist.)

We can express the heat gained by the cold coffee, ΔQ_{gained}, in terms of its mass, m_{cold}, and change in temperature in going from the cold temperature, T_{cold}, to the final temperature, T_{final}:

$$\Delta Q_{gained} = cm_{cold}(T_{final} - T_{cold})$$

Similarly, we can express the heat lost, ΔQ_{lost}, by the hot coffee in terms of its mass, m_{hot}, and the change in temperature going from hot, T_{hot}, to the final temperature, T_{final}:

$$\Delta Q_{lost} = cm_{hot}(T_{hot} - T_{final})$$

Setting the heat gained equal to the heat lost, we can solve for the final temperature:

$$cm_{cold}(T_{final} - T_{cold}) = cm_{hot}(T_{hot} - T_{final})$$

First we need to multiply out the temperatures and then move all terms containing the final temperature, T_f, to one side of the equation:

$$m_{cold}T_{final} + m_{hot}T_{final} = m_{hot}T_{hot} + m_{cold}T_{cold}$$

Factoring out the final temperature and solving, we see that the final temperature turns out to be a sort of average of the initial and final temperatures:

$$T_{final} = \frac{m_{hot}T_{hot} + m_{cold}T_{cold}}{m_{hot} + m_{cold}}$$

In fact, it is easy to see that the final temperature will be halfway between the initial and final temperatures if $m_{hot} = m_{cold}$. If the mass of the hot coffee, m_{hot}, equals the mass of the cold coffee, m_{cold}, there is no difference between the masses and we can ignore their subscripts.

$$T_{final} = \frac{m_{hot}T_{hot} + m_{cold}T_{cold}}{m_{hot} + m_{cold}} = \frac{m(T_{hot} + T_{cold})}{2m} = \frac{T_{hot} + T_{cold}}{2}$$

This is exactly what you would expect with equal amounts of coffee. The temperature change of the two half cups of coffee are equal, and they come to an equilibrium temperature halfway in the middle.

If the two masses m_1 and m_2 are not equal, a so-called weighted average results. If two-thirds of a cup of hot coffee is added to one-third of a cup of cold coffee, the contribution of the hot coffee to the temperature change is twice as important as that of the cold. The final temperature would still be between the hot and the cold, but it would be two-thirds of the way toward the hot end of the scale.

$$T_{\text{final}} = \frac{(\tfrac{2}{3}m)T_{\text{hot}} + (\tfrac{1}{3}m)T_{\text{cold}}}{\tfrac{2}{3}m + \tfrac{1}{3}m} = \frac{2T_{\text{hot}} + T_{\text{cold}}}{3}$$

CHECK QUESTION

A good husband adds 3.0 kg of boiling hot water from a tea kettle to the 12.0 kg of 30 °C warm water in his wife's bath. What is the final temperature of the resulting 15.0 kg of water?

Answer: 44 °C

The problems so far discussed are of the type where both the hot and cold substances have the same specific heat. If they didn't have the same specific heat, say the waiter was feeling remorseful for having neglected you and decided to top your cup with steaming hot Irish coffee, you would need to repeat the above reasoning with subscripts on the specific heats. The specific heat c would not drop out of the equation in this case, since the c_{hot} on one side of the equation would have a different value from the c_{cold} on the other. The general solution would therefore be similar to that shown previously, except that it would contain the specific heats.

You could also fix up this reasoning to account for the change in temperature of the cup itself. All you need to do is add in another heat term, ΔQ_{cup}, on the heat gained side of the equation to take care of the heat used in warming the cup.

There are many other variations on this same theme. Sometimes you are trying to find the final temperature, and sometimes you are trying to use the final temperature to find the specific heat of an unknown substance or the heat generated in a given process. The same approach can be extended to account for the heat of fusion and vaporization as water or other substances freeze or evaporate. All these problems, however, are solved in the same general way, starting from the premise that the heat gained equals the heat lost.

▶ Heat Flow

There is another class of problems in which the focus of attention is on the *rate* at which heat is gained or lost. The rate at which heat flows from a region of high temperature to a region of low temperature is found to follow a rule known as **Newton's law of cooling.** Sir Isaac Newton noticed that the rate at which something cools is proportional to how hot it is relative to the ambient room temperature. We can use algebraic notation to express the rate of heat flow, $\Delta Q/\Delta t$, as the ratio of the change of heat, ΔQ, to the change in time,

Δt. According to Newton's law of cooling, this rate of heat flow is proportional to the difference in temperature, ΔT, between the object and its surroundings:

$$\frac{\Delta Q}{\Delta t} \propto \Delta T$$

Let us first illustrate this proportionality with an example. We can then expand it to include surface area and the thickness of the container, which also affect the rate of heat flow.

Let's stop blaming the poor waiter for the temperature of your coffee and assume, for a change, that he serves it steaming hot promptly as you start your meal. The only problem is that you like to put cream and sugar in your coffee and have it for dessert, and you want it to be as hot as possible when you will be ready to drink it. Let us use Newton's law of cooling to help us decide whether you should put the cream and sugar in as soon as the coffee is served or wait until you are ready to drink it.

The cream will certainly cool off the coffee a certain amount whenever it is added. Assume that the cream starts off at room temperature. It has been sitting out on the table all day in a little metal pitcher; its temperature is going to start off the same whether you add it to the coffee right away or wait until later. Of course, the temperature difference is greater when the coffee is first served, and the largest sudden change in temperature will happen if you add the cream then. The question is, however, which course of action will produce the highest final temperature after a short time has elapsed, say 15 min.

If the cream is kept out until just before you are ready to drink it, the coffee will be hotter all that time. That doesn't do you any good, however. According to Newton's law of cooling, the greater the coffee's temperature, the greater will be its rate of heat loss. We should be more interested in keeping down the rate of heat loss than in keeping up the average temperature of the coffee. Keeping the cream out may result in warmer coffee at the last moment before the cream is finally added, but more of the heat will have gotten away from you, and there will be less heat to share with the cream when it is finally added.

Adding the cream right away may cause a large sudden temperature drop, but this very fact reduces the rate at which heat is lost, according to Newton's law of cooling. We must distinguish between the cooling effect of the cream and the cooling effect of the air in the room. Heat given up by the coffee to the cream is not really lost, since you plan to drink them both eventually, but heat given up by the coffee to the room is gone. Adding the cream when the coffee is first served reduces the heat loss to the room and provides a slightly higher final temperature.

The proportionality between heat flow and temperature difference is the reason that conservationists recommend lowering the thermostat in your house. The rate at which heat escapes through your walls and roof is proportional to the difference between the inside and outside temperatures. Another way to save heat is to add insulation, which is a layer of material having a low thermal conductivity. The meaning of this term will become apparent if we consider the other factors, besides temperature, that affect heat flow.

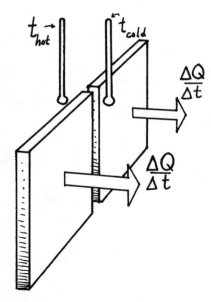

FIGURE 10-11 *Heat flow is directly proportional to area as well as to temperature difference.*

Consider a slab of insulating material such as you might buy at a lumber yard for use in insulating your house. It is easy to see that the rate of heat flow is directly proportional to the area of this slab of material. If you had two slabs of the same size side by side and maintained the same temperature difference across both of them, you would expect the same heat to flow through each of them. By doubling the area through which the heat can flow, you double the rate of heat flow. If you triple the area, you would expect triple the heat flow rate. Thus, we consider the rate of heat flow, $\Delta Q/\Delta t$, to be directly proportional to the area, A, through which the heat can flow:

$$\frac{\Delta Q}{\Delta t} \propto A$$

Instead of laying the two slabs of insulating material side by side, we could stack them one on top of the other. Instead of doubling the area, we would then be doubling the thickness. The result would be to cut the heat flow rate in half. This result seems reasonable if we stop to consider that laying the insulation on twice as thick has the effect of reducing the temperature drop across each slab of insulation. We would expect a thermometer stuck between the two blankets of insulation to read a value halfway between the hot temperature, T_{hot}, on one side of the stack and the cold temperature, T_{cold}, on the other. That is true because of the way in which we have these two slabs of insulation stacked. All the heat that passes through one slab must pass through the other. It therefore requires the same temperature difference across each slab to cause the same amount of heat to flow. If we stacked three slabs, one on top of another, we would expect the temperature across each one to be one-third of the difference between that on the hot side of the stack and that on the cold. Thus, the heat flow would only be one-third as great. We are therefore led to the conclusion that heat flow is inversely proportional to the thickness of the insulation. Expressed algebraically, an inverse proportion between the heat flow rate, $\Delta Q/\Delta t$, and the thickness, l, of the material is the same as a direct proportion between $\Delta Q/\Delta t$ and the reciprocal of l.

$$\frac{\Delta Q}{\Delta t} \propto \frac{1}{l}$$

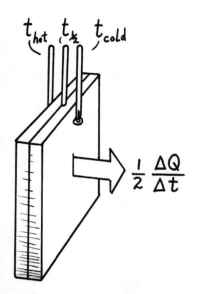

FIGURE 10-12 *Heat flow is inversely proportional to thickness.*

Combining this proportion with the ones for area, A, and temperature difference, ΔT, we get the proportionality

$$\frac{\Delta Q}{\Delta t} \propto \frac{A \, \Delta T}{l}$$

This can be expressed in equation form using a proportionality constant K, called the **thermal conductivity** of the material.

$$\boxed{\frac{\Delta Q}{\Delta t} = K \frac{A \, \Delta T}{l}}$$

This is the general expression by which the rate of heat flow through a solid may be calculated. The thermal conductivity of various materials may be found in reference handbooks. Just to give you an idea

TABLE 10–3 The Thermal Conductivities of Various Substances, in Metric and English Units.

Material	Thermal Conductivity Constant, K (Ordinary Temperature)	
	Cal/s-m-°C	BTU/hr-ft²-(°F/in.)
Copper	9.2×10^{-2}	2,640
Steel	1.1×10^{-2}	312
Glass	1.6×10^{-4}	5.5
Brick	1.7×10^{-4}	5.0
Wood, oak	0.4×10^{-4}	1.1
Wood, pine	0.29×10^{-4}	0.84
Rock wool	9.0×10^{-6}	0.27
Felt	8.6×10^{-6}	0.26
Kapok	8.0×10^{-6}	0.24

of the range of values, the thermal conductivities of a few common substances are listed in Table 10-3.

CHECK QUESTION

A certain glass window has an area of 1.0 m² and a thickness of 0.5 cm. Glass has a thermal conductivity $K = 2 \times 10^{-4}$ Cal/s-m-°C. How much heat goes through this glass in 1 hour (3,600 sec) if the temperature inside is 20 °C and the temperature outside is 10 °C?

Answer: 1,400 Cal

If you buy insulation for your house, a much more useful figure to consider is the **thermal resistance,** R, or R factor, of the insulating material. This depends not only on the thermal conductivity of the material but also upon its thickness. As you might expect, the R factor increases as the thickness of the material increases and as the thermal conductivity decreases. In fact, the R factor as it is used in the building trade is simply defined as the ratio of the thickness of the material to its thermal conductivity as defined in the preceding equation.

$$R \equiv \frac{l}{K}$$

In terms of the R factor, the expression for heat conduction may be stated as follows:

$$\frac{\Delta Q}{\Delta t} = \frac{A\,\Delta T}{R}$$

It is easy to see why people talk about this R factor when they are buying insulation to improve energy conservation in their homes. The greater the R factor, the smaller will be the heat flow through the area protected by this insulation.

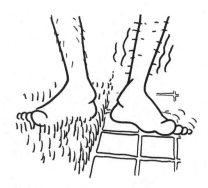

FIGURE 10-13 *A tile floor feels colder than a wool carpet at the same temperature.*

► Entropy

Heat flow is so closely associated with temperature that people sometimes forget they are not the same. A tile floor feels colder under foot than does a wool rug, because the rate of **heat flow** is greater for the tile, even though the rug and the tile are at the same actual temperature. Our sense of feel tells us that something is hot by the fact that heat flows from that object into our hand. Something is cold when heat flows from our hand into that object. The faster the heat flows, the colder it feels. We could even define temperature in terms of heat flow, but the definition would need some refinement to account for the rug and the tile feeling different even though they are really at the same temperature.

When we say that the rug and the tile are really at the same temperature, we mean that no heat will flow from one to the other. If we put a thermometer next to the rug and note its reading, then move it to the tile, the reading will not be changed by a flow of heat either into or out of the thermometer. Thus, we have come to look at temperature as a sort of thermal pressure that pushes heat from one place to another.

To understand what is really meant by temperature, it is necessary to understand why heat will not, of its own accord, flow from a cold place to a hot one. Why does heat always flow from hot to cold? The answer has to do with disorder in the universe, a topic one can have a lot of fun thinking about.

Some people like to think that there is a certain order in the universe and that this is the natural state of things. It is certainly true that there is order in the universe, but it seems that nature actually prefers disorder. Natural processes almost always seem to make disorder out of order. A new deck of cards, for example, comes out of its box in ordered suits. Shuffle it once and you have disorder. Shuffle it again and you have even more disorder. Think of the probability of shuffling it enough to spontaneously get order out of this disorder. This is theoretically possible. Shuffle a deck of cards enough times and it will eventually get back into ordered suits. The number of possible arrangements of the cards is $52! = (52)(51)(50) \ldots (3)(2)(1)$. This number turns out to be something very big, like 10^{68}. If you take into account that there are something like 50 different ways to have the cards arranged in ordered suits, the chances of your getting any one of these ordered arrangements in any one shuffle would be about 1 out of 10^{66}. The point is that while it is possible to get order out of disorder spontaneously, the likelihood of doing so is so slight that it is hard to imagine. The reason is simply that there is a vast number of possibilities with 52 cards.

Imagine the effect if we were to use more cards. Suppose we used a canasta deck having 106 cards, including jokers. The number of ways we could have ordered suits would increase significantly, but the number of possibilities would increase enormously. The number represented by $106! = (106)(105)(104) \ldots (3)(2)(1)$ is inconceivably large, even compared to $52!$, which, as we said, is already something like 10^{68}.

When speaking of a deck of cards, we are dealing with numbers much smaller than the number of particles in an ordinary object, say, a breath of air. Yet we can extend our arguments concerning the likelihood of order over disorder to a breath of air if we think of the molecules as cards. The value of these "cards" might represent the energy of the molecules. Blowing a breath of air into a small balloon could then be thought of as shuffling the cards, in the sense that you would be randomly distributing molecules having different energies into different parts of the balloon. Of course, the number of possible ways these molecules could be distributed is much greater than for the pack of cards. Instead of 52!, we are talking about $10^{22}!$, a big number indeed. Of course, there are a lot of "cards" that have the same value, which reduces the number of possible distributions using the same reasoning as for the canasta deck, but the number of possibilities is still nearly beyond belief.

You might think that it would be possible to accidently get more "hot" molecules, molecules having a high value of energy, on one side of the balloon and thereby leave the air on the other side of the balloon relatively cold. While it is true that such a possibility does exist, the number of random energy distributions is so vastly great in comparison that even a slight spontaneous departure from a state of disorder can be entirely disregarded.

It is therefore virtually certain that the energy of the molecules will start off being randomly distributed throughout the balloon, all parts of the balloon thus being at the same temperature. You might stop to consider what the chances would be that heat would spontaneously flow from one part of the balloon to the other. The molecules are, after all, in constant motion. They are constantly bouncing off each other, trading their kinetic energy around. Isn't it possible that the molecules on one side of the balloon would wind up at least temporarily with more than their average share of the heat? Here again, a spontaneous departure from disorder is possible but vastly improbable. The chances of the heat flowing spontaneously to one part of the balloon full of air would be something like the chances of a small object on the table beside you suddenly jumping up and taking flight on its own.

As an example of nature's preference for disorder over order, let us stop for a moment to consider the chances that an airmail stamp would jump up off the table and take flight because of thermal agitation. The airmail stamp is, after all, composed of molecules in constant motion. It is only the law of averages that says that half these molecules are moving downward at any one moment in time when the other half are moving upward. Certainly it would seem that there would be a finite probability that more than half the molecules would jump upward at the same moment and that the stamp would then leap up into the air. The fact is that such a probability does exist and can be figured out using a law of statistical mechanics known as Boltzmann's principle.[1] This same theory works very well for gas molecules and gives the right result for the density of air in the atmosphere of the earth. When applied to objects as massive as postage stamps, the theory also gives us results in keeping with our observations, although it pushes those results beyond what we might ever expect actually to see. This theory has it that the probability of finding a postage stamp at some height above the table decreases exponentially with the height, but the exponential curve is so sharp that the probability is nil for any measurable height. Even if we choose the smallest imaginable height that would be consistent with the idea of flight, one atomic diameter (about 10^{-10} m), the probabil-

[1] Kerson Huang, *Statistical Mechanics* (New York: Wiley, 1963), p. 91.

ity turns out to be smaller than the time it takes a molecule to complete one jump compared to the life of the universe. Nature so favors disorder that even the tiny degree of spontaneous order that would be necessary for all the molecules in an airmail stamp to get it together long enough to jump one atom high is not likely to have ever occurred.

The same argument that applies to motion of the molecules in the postage stamp can be extended to apply to the flow of energy of the molecules in a balloon filled with air, or to the energy of molecules, for that matter, in any object. One of the best ways to look at heat flow in a solid is to think of the heat energy as diffusing through the solid—like gas molecules moving about in a container. Heat energy becomes evenly distributed in the same way that gas in a balloon becomes evenly distributed. The even distribution is the state of maximum disorder, and maximum disorder is by far the most favored by probability. It is true that there are many ways for energy to be concentrated in only part of a system, just as there are many ways for all the red cards to be arranged in half a deck, but there are vastly more ways for that same energy to be evenly distributed. Just as you expect the cards to become mixed when you shuffle them, you expect heat to flow from a hot to cold place in a system for the same reasons of probability but vastly more compelling because of the huge number of "cards" involved.

Temperature difference is associated with a concentration of energy in a system. If one side of the balloon filled with gas is hotter than the other, the molecules on that side have a greater average energy than on the others. This is a state of order that is ripe to disintegrate into disorder if the system is left to itself. Heat spontaneously flows from a hot place to a cool place until it becomes evenly distributed. Then there is no hot place or cool place, and that is the state of maximum disorder of the system.

This idea of the disorder of a system in terms of the way energy is distributed in a system assumes prime importance in understanding how heat can be converted into mechanical energy. Heat can only be converted into work as long as the overall disorder of the system is increased. To get a handle on this idea of disorder, people have invented a mathematical function called the **entropy** of the system, usually represented by S.

The entropy of a system, being a measure of the disorder of the system, is a function of the energy of the system and how it is distributed. Its philosophical importance comes from the fact that it always increases in any closed system. If you want to lower the entropy of a system, you have to exert some sort of outside influence on the system. You have to sort the cards if you want to start out with some low degree of disorder. You have to get all the hot molecules on one side of the balloon, wind the spring of the clock, or put fuel in your car. You have to somehow concentrate the energy to start with, because all systems run on energy transfer. Once the system is isolated from outside influences, the energy can only flow from the places of concentration to become more evenly distributed in such a way as to increase the entropy of the system. The only way that the entropy of a part of a system can decrease is for the entropy of the rest of the system to increase by as much or more. In fact, exerting outside influence to initially lower the entropy of the system may be looked upon as making the system a part of a larger overall system in which the total entropy increases.

Our real concern with entropy is how it changes. For now, you need only worry about how the change in entropy, ΔS, is related to the flow of heat energy, ΔQ, and the temperature. (The actual entropy function is a logarithmic function of the probability that a system will exist in a particular state relative to all other possible states. But don't let that worry you.) As you might well expect, the change in entropy is directly proportional to the flow of heat from one part of the system to another:

$$\Delta S \propto \Delta Q$$

This means that, all else being equal, twice the entropy change is associated with twice the heat flow, and three times the entropy change will occur with three times the heat flow, and so on.

As it turns out, the "all else" that must be equal is the temperature. The more the heat is concentrated in a system, the greater is the degree of order in the system. But the entropy of the system is a measure of the *disorder* of the system, so that, as you might expect, a higher temperature is associated with a *lower* change in entropy. The change in entropy is inversely proportional to temperature:

$$\Delta S \propto \frac{\Delta Q}{T}$$

The temperature, T, involved in this proportionality must certainly be measured on an absolute temperature scale, which starts off at absolute zero ($-273\,°C$). The use of temperature scales that start off at the freezing point of water or brine would otherwise produce an artificial discontinuity at the zero point, since division by zero is undefined. At true absolute zero, on the other hand, the change of entropy, ΔS, should actually be undefined, according to the theory of thermodynamics.

A proper choice of units for entropy allows the proportionality constant to be selected as 1, so that we can just replace the proportionality sign with an equality sign:

$$\Delta S = \frac{\Delta Q}{T}$$

CHECK QUESTION

A heat engine accepts 900 Cal of heat from a heat source having a temperature of 450 K. Some of this heat energy is converted to work and some is released as waste heat dumped into a heat sink at a temperature of 300 K. (a) How much entropy decrease is associated with accepting heat from the heat source? (b) How much heat must be dumped into the heat sink if this entropy decrease is to be balanced by an equal increase of entropy due to heat flow at the lower temperature?

Answer: (a) 2.00 Cal/K, (b) 600 Cal

This relationship between entropy, heat, and temperature turns out to be more than an equality in the usual development of the theory of thermodynamics. In those more advanced courses, temperature is usually defined in terms of the ratio of heat flow, ΔQ, to entropy change, ΔS, in the limit where the latter is very small.

$$T \equiv \frac{\Delta Q}{\Delta S}$$

For our purposes, however, we can think of temperature as that which would be measured on an ideal gas thermometer. A good enough definition of the change of entropy of a system is therefore the ratio of the flow of heat, ΔQ, to the temperature, T. If the flow of heat is from a hot place, the entropy change represents a decrease, which must be compensated someplace else in the system by at least as great an entropy increase, represented by heat flow into a cold place.

▶ Perpetual Motion Machines of the Second Kind

When speaking of the principle of conservation of energy earlier in this chapter, we pointed out that no one has invented a machine that would put out more work energy than it takes in, in the form of heat. Such a machine is known as a perpetual motion machine of the first kind, and it would violate the principle of conservation of energy, which when expressed in terms of heat flow, is called the **first law of thermodynamics.**

There is nothing in this law, however, that would prevent *all* the energy taken in by a machine from being converted into work. Energy would be conserved; it would just be converted from heat into mechanical energy.

A machine of this second kind would be even more useful in some ways than a perpetual motion machine of the first kind. Not only would it provide a nearly limitless supply of work, but in absorbing this energy in the form of heat from its environment, it would make a nice refrigerator. As an air conditioner, it would absorb the heat from a building and use the energy to operate the fans that circulate the cooled air—and no outside power would be needed. As an ice cream machine, it would be great. Not only would it freeze the ice cream, but it would turn the crank as well.

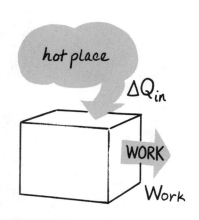

FIGURE 10-14 *A perpetual motion machine of the second kind would convert all the heat coming into it into work.*

Needless to say, people have tried to invent one of these perpetual motion machines of the second kind, and they have met with the same lack of success as they did with the first. Their failure comes from the fact that heat is disordered energy, whereas work is ordered energy. To get work out of heat, you need to make order out of disorder. Thus, their failure is not due so much to practical considerations as to theoretical ones. If you believe that disorder prevails over order simply because the number of disordered states in a system is vastly greater than the number of ordered states, you have to believe that it is quite impossible to build a perpetual motion machine of the second kind.

Just as the first law of thermodynamics prohibits the construction of a perpetual motion machine of the first kind, the impossibility of building a perpetual motion machine of the second kind is called the **second law of thermodynamics.** Another statement of this law is to

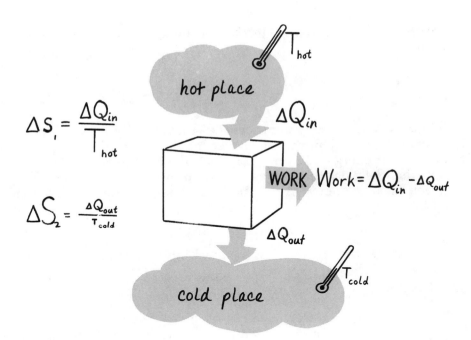

$$\Delta S_1 = \frac{\Delta Q_{in}}{T_{hot}}$$

$$\Delta S_2 = \frac{\Delta Q_{out}}{T_{cold}}$$

Work $= \Delta Q_{in} - \Delta Q_{out}$

FIGURE 10-15 *All heat engines that really work have to exhaust some heat to a cold place so that the total* entropy *of the system can increase.*

say that the **entropy of a closed system always increases to a maximum.**

You can stall. You can do things that will slow down the increase of entropy. Under ideal situations, it is theoretically possible to keep the entropy of a system constant. Nothing you do, however, will cause the total entropy of a closed system to decrease. You can get the entropy of part of a system to go down, but only by letting the entropy of another part of the system increase by as much or more to compensate. All heat engines do in fact work in this way. Changing heat into work represents entropy decrease in part of the system, but this is only possible by allowing an equal or greater entropy increase when some of the heat goes out, still in the form of heat, at a lower temperature. This lower temperature allows the entropy increase to more than compensate for the other decrease.

The entropy decrease, ΔS_{in}, associated with the heat flow into the machine from the hot place, ΔQ_{in}, is, as you recall, inversely proportional to the temperature of the hot place, T_{hot}:

$$\Delta S_{in} = \frac{\Delta Q_{in}}{T_{hot}}$$

The entropy increase associated with the heat dumped into the cold place, ΔS_{out}, can be as large or larger than the preceding entropy increase, even though the energy, ΔQ_{cold}, is smaller, because the temperature, T_{cold}, of the cold place is smaller in as much or greater proportion.

$$\Delta S_{out} = \frac{\Delta Q_{out}}{T_{cold}}$$

The heat that must be dumped into the cold place (called a heat sink) to satisfy the second law of thermodynamics represents an inherent inefficiency of the heat engine. By the efficiency, we mean the ratio of the work that we get out of the machine to the heat energy that we put in:

$$\text{Efficiency} \equiv \frac{\text{Work}}{\Delta Q_{\text{in}}}$$

The first law of thermodynamics tells us that the efficiency can never be greater than 1. You can never get more work out of a machine than the heat energy you put in. In fact, the conservation of energy tells us that the work that comes out is simply equal to the difference between the heat energy, ΔQ_{in}, that goes in and the heat, ΔQ_{out}, that must be dumped into the cold place.

$$\text{Work} = \Delta Q_{\text{in}} - \Delta Q_{\text{out}}$$

We can express the efficiency of the machine in terms of these heats:

$$\text{Efficiency} = \frac{\Delta Q_{\text{in}} - \Delta Q_{\text{out}}}{\Delta Q_{\text{in}}} = 1 - \frac{\Delta Q_{\text{out}}}{\Delta Q_{\text{in}}}$$

The second law of thermodynamics tells us that the heat, ΔQ_{out}, dumped into the cold place must always be something. How big it must be depends on the temperatures, but it must always be big enough to keep the entropy increase due to the heat dump as big as the entropy decrease due to converting energy from disordered heat into the ordered state of work. That is, ΔS_{out} must be equal to or larger than ΔS_{in}:

$$\Delta S_{\text{out}} \geqq \Delta S_{\text{in}}$$

Even if you had a perfectly ideal engine such that the entropy of the whole system remained constant, the best you could do would be to make these two changes in entropy equal. Let us assume the equality case for the moment and express these entropies in terms of heat and temperature:

$$\frac{\Delta Q_{\text{out}}}{T_{\text{cold}}} = \frac{\Delta Q_{\text{in}}}{T_{\text{hot}}}$$

From this we can see that the ratio of the heats in the ideal case is equal to the ratio of the absolute temperatures:

$$\frac{\Delta Q_{\text{out}}}{\Delta Q_{\text{in}}} = \frac{T_{\text{cold}}}{T_{\text{hot}}}$$

We can use this equivalence to get the maximum ideal efficiency of a heat engine in terms of the temperatures between which it is working:

$$\boxed{\text{Efficiency} = 1 - \frac{T_{\text{cold}}}{T_{\text{hot}}}}$$

The maximum efficiency of even a perfect heat engine is therefore limited by the temperature of the heat sink, T_{cold}, as well as the temperature of the heat source, T_{hot}. At any temperature of the heat sink above absolute zero, the efficiency must necessarily be less than 1.

The first law of thermodynamics may be interpreted as saying, "You just can't win." You cannot get any more energy out of something than you put in. The second law of thermodynamics tells us, "You can't even break even." You can't even get out as much as you put in.

The inherent need to waste part of their heat energy as well as the need for a cold place to dump it becomes a serious problem for large generating plants. Water-cooled steam electric generating systems can cause severe thermal pollution of local water resources. By 1980 the electric power industry will be using between 15% and 20% of the nation's flowing fresh water for cooling its power plants.

The earth itself is a heat engine operating on energy from the sun. Most people realize that the sun is our source of energy, but few people stop to think about the heat sink, which is just as essential in thermodynamics as the source. That heat sink is the cold night sky. It is essential to the thermodynamics of the situation that the dark side of the earth radiate its waste energy into the universe. We are fortunate that the universe does not radiate back. Some say that the reason it does not, why the average temperature of the night sky is but a few degrees Kelvin, is that the universe is expanding. From that point of view, the expansion of the universe is of great importance to us. Let's all hope it keeps going.

The problems in this chapter are related to heat as a form of energy. The relationship between heat and energy was only established once it was shown that mechanical energy, or work, is exactly proportional to heat, Q.

$$\text{Work} = JQ$$

The proportionality constant J, called the **mechanical equivalent of heat,** has been carefully measured for the two systems of units we deal with:

$$J = 4{,}185 \text{ J [joules]}/\text{Cal} = 777.9 \text{ ft-lb/BTU}$$

A Calorie is defined as the amount of heat necessary to raise the temperature of 1 kg of water 1 °C. A BTU is defined as the amount of heat necessary to raise 1 lb of water 1 °F. The Celsius and Fahrenheit temperature scales are now both defined in terms of the freezing point (0 °C = 32 °F) and boiling point (100 °C = 212 °F) of water.

Most solid objects increase in size as their temperatures increase. The linear expansion, Δl, of an object is proportional to its original length, l, as well as to the change in temperature, ΔT:

$$\Delta l = \alpha l \, \Delta T$$

The proportionality constant α is called the **coefficient of linear expansion** and depends on the composition of the material involved.

A change in the heat content, ΔQ, is also generally associated with a change in temperature, ΔT, but the proportionality also includes the mass, m, of the stuff being heated.

$$\Delta Q = cm \, \Delta T$$

The proportionality constant c is called the **specific heat** of the substance. The specific heat of water equals 1 in both systems of units because of the way that the units of heat are defined. Almost everything else has a smaller specific heat than water.

Heat flows from a hot object to a cold one until they both reach the same temperature. The final temperature can be predicted by saying that the heat gained equals the heat lost:

$$\Delta Q_{\text{gained}} = \Delta Q_{\text{lost}}$$

and expressing these heats in terms of the specific heats, masses, and changes in temperatures of the various objects involved. The rate at which heat flows by conduction from one object to another is proportional to the temperature difference, ΔT, and to the area, A, through which the heat can flow, but it is inversely proportional to the thickness, l, of the material conducting the heat:

$$\frac{\Delta Q}{\Delta t} = K \frac{A \, \Delta T}{l}$$

The proportionality constant K is called the **thermal conductivity** of the material. Its reciprocal times the thickness is sometimes known as the thermal resistance, or R factor, of a certain thickness of insulation.

A heat engine is a device that takes some of the heat flowing from a hot place and turns it into work, a more ordered form of energy. To balance order with at least an equal increase in disorder, the heat engine must dump some of its heat into a cold place. The maximum theoretical efficiency possible under the constraint that a closed system never becomes more orderly is determined by the temperature T_{cold}, of the cold place as well as by the temperature, T_{hot}, of the hot place:

$$\text{Efficiency} = 1 - \frac{T_{cold}}{T_{hot}}$$

These temperatures must be expressed on an absolute temperature scale, such as the Kelvin scale.

10-1 A Girl Scout is making fire by rubbing two sticks together. She gets a Brownie to stand on one of the sticks so that she doesn't have to exert any force perpendicular to the two surfaces but can concentrate her efforts on the frictional force.

Given: The Girl Scout exerts a frictional force of 120 N in pushing the stick a distance of 40 cm in one direction and then the same distance back again.

Find: (a) How much work does the Girl Scout do on each complete stroke going back and forth?

(b) How many Calories of heat energy are generated on each stroke?

(c) How many strokes would be necessary to generate 50 Cal?

Your Solution

Discussion

This problem illustrates the mechanical equivalent of heat. Mechanical energy is being converted into heat in given increments, and the number of increments in a given amount of heat is to be found.

Sample Solution 10-1

(a) $W = F \cdot d$

$\quad = (120N)(2 \times .40m)$

$\quad = \boxed{96 \text{ Joule}}$

(b) $W = JQ$

$\quad Q = \dfrac{W}{J}$

$\quad = \dfrac{96 \text{ Joule}}{(4,185 \text{ Joule/Cal})} = \boxed{2.3 \times 10^{-2} \text{Cal}}$

(c) $\text{Strokes} = \dfrac{Q_{total}}{Q_{per \, stroke}}$

$\quad = \dfrac{50 \text{ Cal}}{2.3 \times 10^{-2} \text{ Cal/Stroke}} = \boxed{2.2 \times 10^{3} \text{Strokes}}$

Discussion

The energy in a single stroke is found from the definition of work, where the frictional force is the effort exerted by the Girl Scout and the distance is the total displacement, forward and back, in a single stroke. This distance is two times 40 cm, or 2×0.40 m, since a complete stroke includes motion in both directions.

The work found in part (a) is expressed in terms of the heat generated, using the mechanical equivalent of heat solved in part (b).

Part (c) starts from the assumption that the number of strokes is the ratio of the total amount of energy necessary to start the fire divided by the amount of energy in one stroke.

10-2   Health experts agree that exercise is no substitute for a good diet. A man who hasn't heard this wisdom decides to have an extra piece of cake for lunch and to work it off by climbing the stairs to his office.

Given: The man weighs 1,100 N and he climbs a height of 75 m. The piece of cake has a food value of 95 Cal, but his body efficiency only makes 25% of this available as work. (The rest is given off as heat and is used to keep him alive.)

Find: (a) What is the heat equivalent of the work done by the man in climbing the stairs?
  (b) How high would the man need to climb to work off the cake?

Your Solution

Discussion

This problem illustrates the fact that there is a great amount of energy in an ordinary amount of heat. This large difference in scale is one of the main reasons that heat was only recently recognized as a form of energy. The efficiency gives that percentage of energy stored in the cake that the body can turn into work. The rest is all wasted in heat.

Sample Solution 10-2

(a) $W = F \cdot d$

$= (1,100\,N)(75\,m)\left(\dfrac{1\,Cal}{4,185\,Joule}\right)\left(\dfrac{1\,Joule}{1\,N\cdot m}\right)$

$= 19.7\,Cal \cong \boxed{20\,Cal}$

(b) $W = F \cdot d$

$d = \dfrac{W}{F}$

$= \dfrac{25\%\,Q_{cake}}{F}$

$= \dfrac{(0.25)(95\,Cal)}{(1,100\,N)}\left(\dfrac{4,185\,Joule}{1\,Cal}\right)\left(\dfrac{1\,N\cdot m}{1\,Joule}\right)$

$= \boxed{90\,m}$

Discussion

The work found in part (a) may be expressed in terms of heat, using the mechanical equivalent of heat as a conversion factor. An alternative approach would be to treat the mechanical equivalent of heat in a separate step as a proportionality, as in Problem 10-1.

The distance that you can go on a piece of cake is found in part (b) by solving the definition of work, using as the total energy available 25% of the heat energy in a piece of cake. This 25% efficiency is a bit higher than one would expect in an actual case.

The result in part (a) is rounded off to two significant figures, since distance is known to no better precision than that. The result of part (b) is good to only one significant figure, or maybe even less, because of the crude estimate of efficiency.

10-3 Many swimming pools are heated to make them more pleasant, but the cost of heating a pool may get unpleasantly high.

Given: A gasoline-fired water heater is used to warm a swimming pool to raise its temperature 8 °F. The pool contains 300,000 lb of water, and a gallon of gasoline produces about 80,000 BTU of heat energy.

Find: (a) How much heat is necessary to warm the pool if none is lost?

(b) How many gallons of gasoline does this represent?

Your Solution

Discussion

This problem illustrates the concept of heat capacity in a system containing only water. Water, as you remember, has a heat capacity of unity (1) in both systems of units because of the way in which the basic units of heat are defined in both systems.

Sample Solution 10-3

(a) $\Delta Q = cm\Delta T$

$\quad = \left(1\ \frac{BTU}{lb\ F°}\right)(3\times10^5 lb)(8\ F°)$

$\quad = \boxed{24\times10^5\ BTU}$

(b) $\Delta Q = (24\times10^5\ BTU)\left(\frac{1\ Gallon\ of\ Gas}{80{,}000\ BTU}\right)$

$\quad = \boxed{30\ Gallons\ of\ Gas}$

Discussion

The heat necessary to warm the pool is found from the proportionality between heat on the one hand and mass and the change of temperature on the other. The proportionality constant is unity because a BTU of heat is defined as that amount of heat necessary to raise one pound of water 1 °F in temperature.

One approach to the solution of part (b) is to treat it as a conversion factor problem. Since one gallon of gas represents 80,000 BTU of heat when it is burned, it can be said to be, in a restricted sense, "equal" to that much heat. An alternative approach would be to construct another proportional model using a proportionality constant called the "heat of combustion" of gas. This constant is in common use and may be found listed in reference tables for various substances.

10-4   A new refrigerator comes with directions on how to adjust the temperature in both the freezing and cooling compartments. The directions state the correct Fahrenheit temperatures, but you find that the thermometer you have is marked in degrees Celsius.

Given: The freezing compartment is supposed to get down to 0 °F, but you find that it is actually − 14 °C. You also find that the cooling compartment is at +2 °C.

Find: (a) What should the temperature of the freezing compartment be, in degrees Celsius?
(b) What is the actual temperature of both compartments on the Fahrenheit scale?

Your Solution

Discussion

This problem is an exercise in converting from one temperature scale to another. This is one conversion problem where the difference in zero point must be taken into account as well as the difference in size of the units.

Sample Solution 10-4

(a) $0°F = 32 F°$ below Freezing

$$\Delta T = 32 F° \left(\frac{100 C°}{180 F°} \right)$$

$$= 17.8 C° \text{ below Freezing}$$

$$T = -17.8°C \cong \boxed{-18°C}$$

(b) $-14 C° = 14 C°$ below Freezing

$$\Delta T = 14 C° \left(\frac{180 F°}{100 C°} \right) = 25.2 F° \text{ below Freezing}$$

$$T = T_{Freezing} - \Delta T$$

$$= (32°F) - (25.2°F) = 6.8°F \cong \boxed{7°F}$$

$2°C = 2 C°$ above Freezing

$$\Delta T = 2 C° \left(\frac{180 F°}{100 C°} \right) = 3.6 F° \text{ above Freezing}$$

$$T = T_{Freezing} + \Delta T$$

$$= (32°F) + (3.6 F°) = 35.6°F \cong \boxed{36°F}$$

Discussion

The temperatures are first expressed in terms of a temperature difference relative to the freezing point of water before being multiplied by the conversion factor (180 F°/100 C°), which changes the size of the degrees. This temperature difference is then related to the freezing point on the different scales.

Sometimes the notation F° is used to refer to a Fahrenheit degree of temperature difference to distinguish it from a °F, or degree Fahrenheit of temperature referred to zero point on that temperature scale.

10-5   The valves in a sports car engine can be adjusted to compensate for wear. The gap that should be left between the valve stem and the rocker arm (which actuates the valve by pushing on the valve stem at the proper time) is given by the manufacturer in anticipation of the thermal expansion of the valve.

Given: The proper setting of this gap is 0.025 cm when the engine is cold, assuming the temperature will increase by 22 °C. The valves are 15 cm long and have a coefficient of linear expansion of $1.1 \times 10^{-5}/°C$.

Find: (a) What is the change in length of the valve as the engine warms up?

(b) What would be the new gap after thermal expansion if the engine did not also expand as it got hot?

Your Solution

Discussion

This problem is an example of thermal expansion in an application that touches many people's daily lives. Thermal expansion is one reason that a driver should let her automobile engine warm up for a minute or so when starting it in the morning.

Sample Solution 10-5

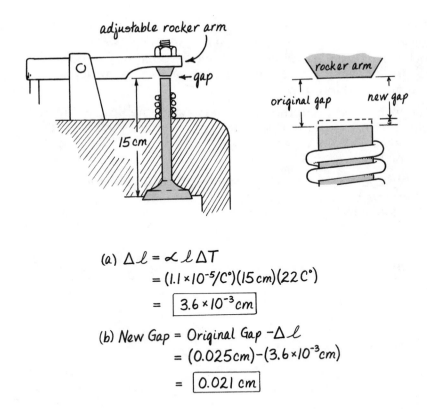

(a) $\Delta \ell = \alpha \ell \Delta T$

   $\qquad = (1.1 \times 10^{-5}/C°)(15\,cm)(22\,C°)$

   $\qquad = \boxed{3.6 \times 10^{-3}\,cm}$

(b) New Gap = Original Gap $- \Delta \ell$

   $\qquad = (0.025\,cm) - (3.6 \times 10^{-3}\,cm)$

   $\qquad = \boxed{0.021\ cm}$

Discussion

The change in length due to thermal expansion is given by the pro-portionality between it and the original length and the change in temperature.

The new gap will be smaller than the original gap by the amount that the valve expands, assuming that is the only thing in the engine to expand as it warms up. This calculation therefore represents an outside limit to the possible change in relative dimensions. In an actual case, the change in gap would be much smaller, since almost everything in the engine expands at nearly the same rate. Still, small differences in the rate of heating while the engine is warming up and differences in the coefficient of expansion of different kinds of metal used in the engine result in unusual wear if the engine is operated under heavy loads while it is warming up.

10-6 You get to wondering how much it costs to heat an auditorium in preparation for a performance. You figure the main things to heat are a volume of air about 50 m $\times$ 50 m $\times$ 30 m, the plaster in the walls, and the steel in the seats.

Given: The auditorium is heated from 12 °C to 20 °C. It contains 9.7×10^4 kg of air, 1.2×10^4 kg plaster, and 2,500 seats, each containing 12 kg steel. The specific heats of air, plaster, and steel are 0.24, 0.20, and 0.11 Cal/kg-°C. A Calorie of heat costs a little less than 10^{-3} cents.

Find: (a) How much heat is necessary to heat the air in the auditorium?

(b) How much heat does it take to warm the walls and the seats?

(c) How much does it cost to heat the auditorium?

Your Solution

Discussion

This problem demonstrates the idea of heat capacity in systems having several different kinds of material. People who rent auditoriums to other people frequently list the cost of heating among other things that seem to make it necessary to charge as much as they do. This problem gives you an opportunity to estimate this cost for yourself.

Sample Solution 10-6

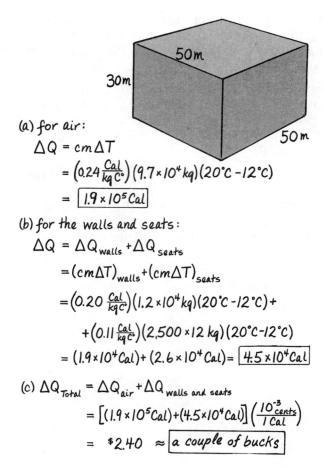

(a) for air:

$$\Delta Q = cm\Delta T$$
$$= \left(0.24 \tfrac{Cal}{kg\,C°}\right)(9.7\times10^4\,kg)(20°C - 12°C)$$
$$= \boxed{1.9\times10^5\,Cal}$$

(b) for the walls and seats:

$$\Delta Q = \Delta Q_{walls} + \Delta Q_{seats}$$
$$= (cm\Delta T)_{walls} + (cm\Delta T)_{seats}$$
$$= \left(0.20\tfrac{Cal}{kg\,C°}\right)(1.2\times10^4\,kg)(20°C - 12°C) +$$
$$+\left(0.11\tfrac{Cal}{kg\,C°}\right)(2,500\times12\,kg)(20°C - 12°C)$$
$$= (1.9\times10^4\,Cal) + (2.6\times10^4\,Cal) = \boxed{4.5\times10^4\,Cal}$$

(c) $\Delta Q_{Total} = \Delta Q_{air} + \Delta Q_{walls\ and\ seats}$

$$= \left[(1.9\times10^5\,Cal) + (4.5\times10^4\,Cal)\right]\left(\tfrac{10^{-3}cents}{1\,Cal}\right)$$
$$= \$2.40 \approx \boxed{a\ couple\ of\ bucks}$$

Discussion

Part (a) is a simple application of the idea of heat capacity as the proportionality constant between the heat necessary to warm some substance on the one hand and the mass of that substance and the change in temperature on the other. Part (b) is just an application of the same proportionality to a system containing two different materials.

You might be interested to note that this estimate indicates that it costs nearly five times as much to warm the air in the auditorium as it does to heat the walls and seats. The conversion factor given in the problem, that a Calorie of heat costs about a thousandth of a cent, is only an order-of-magnitude estimate, given the unstable energy economy these days. The result is therefore only good to an order of magnitude as well, but notice that it is still good enough for most purposes.

10-7 Jerry Housekin is sharpening a blade on his lawnmower by grinding it on an abrasive wheel. The heat generated makes it necessary to cool the blade from time to time by dipping it into a cup of water.

Given: The blade has a mass of 350 g and is made of steel having a specific heat of 0.11 Cal/kg-°C. Each time its temperature gets up to 190 °C, Jerry brings its temperature back down to 100 °C by dipping it into a cup containing 250 g of water. The water starts with a temperature of 20 °C, but gets hotter with each dip.

Find: (a) What would be the temperature of the blade and water if Jerry let them come to the same temperature on the first dip?

(b) How many dips would raise the temperature of the water to boiling?

Your Solution

Discussion

This is a calorimetry problem of the basic type, where a hot object and a cold object are put together and allowed to come to an equilibrium temperature. In the second part of the problem, you know the final temperature and you are to find the number of equal quantities of heat that will produce that result.

Sample Solution 10-7

(a) $\Delta Q_{gained} = \Delta Q_{lost}$

$$c_1 m_1 (T_f - T_1) = c_2 m_2 (T_2 - T_f)$$

$$c_1 m_1 T_f - c_1 m_1 T_1 = c_2 m_2 T_2 - c_2 m_2 T_f$$

$$c_1 m_1 T_f + c_2 m_2 T_f = c_2 m_2 T_2 + c_1 m_1 T_1$$

$$T_f = \frac{c_2 m_2 T_2 + c_1 m_1 T_1}{c_1 m_1 + c_2 m_2}$$

$$= \frac{(0.11 \frac{Cal}{kg\,C°})(350\ g)(190°C) + (1 \frac{Cal}{kg\,C°})(250\ g)(20°C)}{(0.11 \frac{Cal}{kg\,C°})(350\ g) + (1 \frac{Cal}{kg\,C°})(250\ g)}$$

$$= \boxed{43°C}$$

(b)

$$\Delta Q_{gained} = \Delta Q_{lost}$$

$$= n \Delta Q_{dip}$$

$$n = \frac{\Delta Q_{gained}}{\Delta Q_{dip}}$$

$$= \frac{c_1 m_1 (T_f - T_1)}{c_2 m_2 (T_2 - T_f)}$$

$$= \frac{(1 \frac{Cal}{kg\,C°})(250\ g)(100°C - 20°C)}{(0.11 \frac{Cal}{kg\,C°})(350\ g)(190°C - 100°C)}$$

$$= 5.77 \cong \boxed{6\ times}$$

Discussion

Calorimetry problems are mostly solved on the theory that the heat gained by one part of the system, say, the water, will be equal to the heat lost by the other part of the system, which is the lawnmower blade in this case. The unknown final temperature, T_f, is involved in both the heat gained and the heat lost in the first part of the problem. The heat gained on the left side of the equation can then be expressed in terms of this final temperature as well as the initial temperature, the specific heat, and the mass of the water, T_1, c_1, and m_1. The initial temperature, specific heat, and mass of the lawnmower blade, T_2, c_2, and m_2, are involved in the heat lost.

Part (b) is worked out on the assumption that the blade gives off the same amount of heat each time it is dipped in the water without regard to the temperature of the water, an approximation that is not quite true unless Jerry pulls the blade out of the water just as its temperature reaches 100 °C.

10-8   A good treatment for sinus congestion is to inhale the steam from a pan of hot water containing a few drops of eucalyptus oil. Boiling water direct from a tea kettle is too hot, however, and must be mixed with a bit of cool water.

Given: The tea kettle holds 3 qt (6.0 lb) of boiling hot water. You dilute it by pouring it into an aluminum pan, weighing 2.2 lb and having a specific heat of 0.22 BTU/lb-°F, which already contains 0.5 lb of cool water at 68 °F.

Find: (a) How much of the 3 qt of boiling water available should be put into the pan to get the ideal temperature of 190 °F?

   (b) What would be the temperature if you just poured in all 3 qt?

Your Solution

Discussion

This is a calorimetry problem in a situation where several things change temperature at the same time. Both the cool water and its container are warmed by the addition of hot water to the system. Algebra becomes a very useful tool in problems of this sort, because enough things are going on so that, while the reasoning is straight-forward, a pencil and paper are needed to keep track of everything.

Sample Solution 10-8

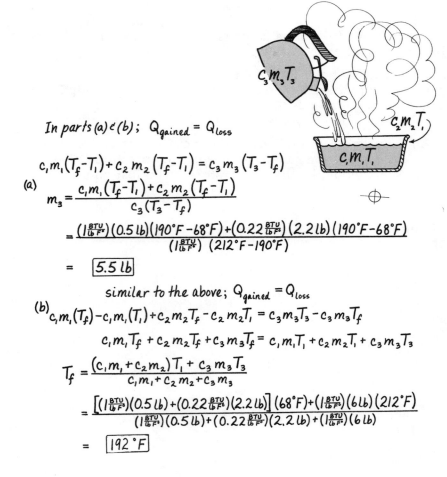

$$\text{In parts (a) \& (b); } Q_{gained} = Q_{loss}$$

$$c_1 m_1 (T_f - T_1) + c_2 m_2 (T_f - T_1) = c_3 m_3 (T_3 - T_f)$$

(a)
$$m_3 = \frac{c_1 m_1 (T_f - T_1) + c_2 m_2 (T_f - T_1)}{c_3 (T_3 - T_f)}$$

$$= \frac{(1\tfrac{BTU}{lb\,F^\circ})(0.5\,lb)(190^\circ F - 68^\circ F) + (0.22\tfrac{BTU}{lb\,F^\circ})(2.2\,lb)(190^\circ F - 68^\circ F)}{(1\tfrac{BTU}{lb\,F^\circ})(212^\circ F - 190^\circ F)}$$

$$= \boxed{5.5\,lb}$$

$$\text{similar to the above; } Q_{gained} = Q_{loss}$$

(b)
$$c_1 m_1 (T_f) - c_1 m_1 (T_1) + c_2 m_2 T_f - c_2 m_2 T_1 = c_3 m_3 T_3 - c_3 m_3 T_f$$

$$c_1 m_1 T_f + c_2 m_2 T_f + c_3 m_3 T_f = c_1 m_1 T_1 + c_2 m_2 T_1 + c_3 m_3 T_3$$

$$T_f = \frac{(c_1 m_1 + c_2 m_2) T_1 + c_3 m_3 T_3}{c_1 m_1 + c_2 m_2 + c_3 m_3}$$

$$= \frac{\left[(1\tfrac{BTU}{lb\,F^\circ})(0.5\,lb) + (0.22\tfrac{BTU}{lb\,F^\circ})(2.2\,lb)\right](68^\circ F) + (1\tfrac{BTU}{lb\,F^\circ})(6\,lb)(212^\circ F)}{(1\tfrac{BTU}{lb\,F^\circ})(0.5\,lb) + (0.22\tfrac{BTU}{lb\,F^\circ})(2.2\,lb) + (1\tfrac{BTU}{lb\,F^\circ})(6\,lb)}$$

$$= \boxed{192^\circ F}$$

Discussion

The first part of the problem is easier than the second in spite of the fact that algebraically they both start with the same two first steps. Both start from the assumption that heat gained equals heat lost, and this general rule of calorimetry is applied to both situations in about the same way. These heats are expressed in terms of the specific heats, masses, and initial temperatures, using the subscript 1 for the cold water, the subscript 2 for the container, and the subscript 3 for the hot water. In part (a), however, the only unknown is the mass of the hot water, m_3. The algebra is simpler, because the unknown is only on one side of the initial equation. In part (b), the unknown quantity is the final temperature, T_f, which occurs on both sides of the equation in every term. More algebra is necessary to solve for this unknown.

10-9

A good part of the heat lost from a house goes through the roof; this is frequently the easiest heat loss to reduce by adding insulation to an existing house. Walls are harder to insulate without making structural changes.

Given: Your gas bill shows that you used 7.9×10^6 BTU in heating your house one month (30 days) when the average outside temperature was 50 °F and the inside temperature was 68 °F. You assume that about half of this heat loss went through the roof while the other half went through the walls. You estimate the roof area as 1,500 ft² and the wall area as about 1200 ft² with a thickness of 8 in.

Find: (a) What is the thermal conductivity of your walls?
(b) What is the thermal resistance of your roof?

Your Solution

Discussion

This is a heat transfer problem that illustrates the concepts of thermal conductivity and thermal resistance. The rate of heat flow through a certain slab of material is assumed to be proportional to the area of the slab and the temperature difference between the faces and inversely proportional to the thickness of the slab. The proportionality constant is called the thermal conductivity constant, and the ratio of the thickness to this proportionality constant is called the thermal resistance of the material.

Sample Solution 10-9

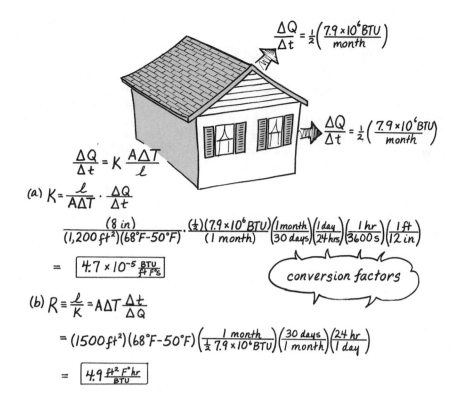

$$\frac{\Delta Q}{\Delta t} = \frac{1}{2}\left(\frac{7.9 \times 10^6 \, BTU}{month}\right)$$

$$\frac{\Delta Q}{\Delta t} = \frac{1}{2}\left(\frac{7.9 \times 10^6 \, BTU}{month}\right)$$

$$\frac{\Delta Q}{\Delta t} = K\frac{A\Delta T}{\ell}$$

(a) $K = \dfrac{\ell}{A\Delta T} \cdot \dfrac{\Delta Q}{\Delta t}$

$$\frac{(8 \, in)}{(1,200 \, ft^2)(68°F-50°F)} \cdot \frac{(\frac{1}{2})(7.9 \times 10^6 \, BTU)}{(1 \, month)}\left(\frac{1\,month}{30\,days}\right)\left(\frac{1\,day}{24\,hrs}\right)\left(\frac{1\,hr}{3600\,s}\right)\left(\frac{1\,ft}{12\,in}\right)$$

$$= \boxed{4.7 \times 10^{-5} \frac{BTU}{ft \, F°s}}$$

conversion factors

(b) $R \equiv \dfrac{\ell}{K} = A\Delta T \dfrac{\Delta t}{\Delta Q}$

$$= (1500 \, ft^2)(68°F-50°F)\left(\frac{1\,month}{\frac{1}{2}\,7.9 \times 10^6\,BTU}\right)\left(\frac{30\,days}{1\,month}\right)\left(\frac{24\,hr}{1\,day}\right)$$

$$= \boxed{4.9 \frac{ft^2 \, F°hr}{BTU}}$$

Discussion

The thermal conductivity is found in part (a) by solving the heat conduction proportionality for the proportionality constant. The heat flow is figured on only half the heat produced in a month, since only about half goes through the roof.

The thermal resistance is found in part (b) by solving the heat conduction proportionality for the ratio of thickness to thermal conductivity. The resulting expression is useful, because it gives a number that can be compared to the properties of insulating material. A layer of R-11 insulation would have more than twice the thermal resistance of the roof in this example. Adding that insulation would therefore decrease the heat loss through the roof to less than one-third the original value.

10-10 An inventor designs a heat engine that turns the waste heat from the cooling system of an automobile into useful work.

Given: The cooling system of an automobile operates at 85 °C whereas the temperature of the ambient atmosphere is 20 °C. The inventor claims that his machine will use these source and sink temperatures to provide an overall efficiency of 12%.

Find: (a) What is the maximum theoretical efficiency possible?
(b) What percentage of the maximum theoretical efficiency does the inventor claim?

Your Solution

Discussion

One comes across schemes for using or saving energy from time to time. Some are practical and some are not. The rules of thermodynamics provide a useful estimate of practicability by telling us what would be possible under perfectly ideal circumstances.

Sample Solution 10-10

(a) $Eff = 1 - \dfrac{T_{cold}}{T_{hot}}$

$ = 1 - \dfrac{(20°C + 273°K)}{(85°C + 273°K)}$

$ = 1 - \dfrac{(293°K)}{(358°K)}$

$ = \boxed{18\%}$

(b) % theoretical $= \dfrac{claimed}{theoretical} \times 100\%$

$ = \dfrac{12\%}{18\%} \times 100\%$

$ = \boxed{67\%}$

Discussion

The ideal thermodynamic efficiency is less than unity (1) by the ratio of the temperatures of the cold and hot places available for heat flow. A calculation as in part (a) shows that our inventor is claiming less efficiency than would be possible under ideal circumstances. In fact, part (b) shows that he is only claiming a bit less than 70% of what is possible. That is still very good, considering the extremes that one would need for perfectly ideal circumstances. You would be justified in asking to see a working model before investing any money in such a nearly ideal invention.

10-11 A horse is turning a cannon lathe in the process of cutting a bore into the cannon (J = 778 ft-lb/BTU).

Given: The horse walks 10 ft/s and pulls with a force of 55 lb.

Find: (a) How much work is done in 1 s?
(b) What is the heat generated in 1 s, expressed in BTUs?
(c) How long would it require the horse to produce 100 BTU?

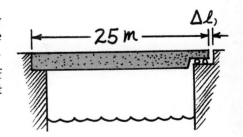

10-12 A coal miner lifts coal one shovel at a time into a rail car (J = 4,185 J/Cal).

Given: One day he loads 15 tonnes (1 metric tonne = 1,000 kg = 1.1 ton) by raising it to a height of 2.0 m.

Find: (a) How much work did he do?
(b) What is the heat equivalent in Calories of the work he did all day?
(c) How many Calories of food value must he eat a day, assuming a body efficiency of 7%?

10-13 John Cheap has a large 50-gallon hot water heater installed in his house so he can take a long shower in the morning, but he begins to feel guilty about the waste of energy and how much it must be costing him.

Given: The 50-gallon tank holds 400 lb of water and heats it from 65 °F to 140 °F when set on medium. The gas company sells one Therm of natural gas, enough to produce 100,000 BTU of heat, for 15 cents.

Find: (a) How much heat is necessary to bring one tank of water up to maximum temperature?
(b) How much does it cost to heat one tank of water?

10-14 The absolute temperature scale in the English system of units is the Rankine scale. It starts at absolute zero just like the Kelvin scale, but it has the same size degrees as the Fahrenheit scale.

Given: Absolute zero, which is 0 K, is just a fraction below −273 °C.

Find: (a) What is the freezing point of water on the Rankine scale?
(b) What is absolute zero on the Fahrenheit scale?

10-15 A bridge made of concrete has an expansion joint to allow for thermal expansion with daily fluctuations in temperature. One end of the bridge rests on rollers while the other is solidly attached.

Given: This particular concrete has a thermal expansion of 2×10^{-5}/°C. The bridge is 25 m long one morning, but that afternoon it is 12 °C warmer and therefore a bit longer.

Find: (a) How much longer is the bridge that afternoon?
(b) What is the new length of the bridge?

10-16 An electric steam iron is made mostly of steel, but you have to put a little water into its reservoir before plugging it in. While waiting for the iron to get hot, you wonder about how much heat goes into the iron and the water and the cost of all this.

Given: Your steam iron has a mass of 1.3 kg and a specific heat of 0.11 Cal/kg-°C. You add 100 g of water before turning it on. The iron heats from 20 °C to 100 °C on a low setting. One Calorie of heat costs 3.5×10^{-3} cents.

Find: (a) How much heat goes into the steel?
  (b) How much heat goes into the water?
  (c) What does it cost to heat the steam iron with the water in it to this low setting?

10-17 Ed Might finds his coffee too hot to drink and pours in some Irish whiskey to cool it off. He finds it is still not cool enough and adds more.

Given: Ed's cup originally contains 8 oz of coffee at 190 °F before he pours in a jigger (1.5 oz) of room temperature (68 °F) whiskey having a specific heat of 0.79 BTU/lb-°F. He then adds more whiskey to bring the final temperature down to 140 °F, which he finds to be a drinkable temperature. (1 lb = 16 oz)

Find: (a) What is the temperature of the coffee mixed with one jigger of whiskey?
  (b) How much whiskey is needed altogether to bring the mixture down to the temperature that Ed considers drinkable?

10-18 The Ahwahnee Indians used to boil water by putting it in a closely woven water-tight basket and adding hot stones. They had to stir to keep the stones from scorching the basket.

Given: An Indian puts 2.5 kg of rocks having a temperature of 210 °C and a specific heat of 0.20 Cal/kg-°C into a basket containing 6.0 kg of water at 20 °C. After stirring, more rocks are added and the process repeated until the water begins to boil.

Find: (a) What is the temperature after the first bunch of rocks are added?
  (b) What is the total mass of rocks necessary to boil the water?

10-19 A coffee mug is nice and warm to the touch when it is filled with piping hot coffee. You get to wondering about the rate of heat flow through the mug.

Given: The mug has an outside area of 250 cm² and is made of material having a thermal conductivity of 2×10^{-4} Cal/s-m-°C. The coffee inside the mug is at 86 °C, and the outside is at 50 °C. The mug if 0.5 cm thick. (Assume constant temperatures.)

Find: (a) What is the rate of heat flow through the mug?
  (b) At this rate, how much heat would flow through the mug in the 5.9 s it takes to process your thoughts?

10-20 Some people have developed experimental engines just to see how close they could get to maximum theoretical efficiency.

Given: One such engine operated between a heat source at 1,727 °C and a heat sink at 427 °C. It produced an actual efficiency of 40%.

Find: (a) What is the maximum theoretical efficiency of this engine?

(b) What percentage of the maximum theoretical efficiency did the engine actually produce?

MORE INTERESTING PROBLEMS

10-21 The limiting factor in the speed of metal working tools is the rate of heat production. If the cutting tool becomes too hot, it will lose its temper—and we all know what happens when you lose your temper. (J = 4,185 J/Cal)

Given: A steel cutting tool has a mass of 350 g and a specific heat of 0.11 Cal/kg-°C. It can be heated up to 218 °C before it loses its temper. It starts at room temperature (20 °C) and is acted upon by a frictional force of 50 N.

Find: (a) How many Calories of heat can the tool absorb before getting too hot?

(b) What is the energy equivalent of this much heat?

(c) Over what distance would the frictional force need to work to convert this much energy into heat?

10-22 The steel cables on the suspension section of the San Francisco Bay Bridge expand and contract with changes in temperature. One cold summer day, you get to wondering how much the cables would expand if it were to change into a hot summer day. You look up the coefficient of expansion of steel in a book.

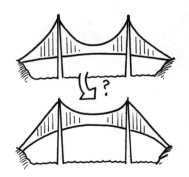

Given: It is 55 °F that day, and you are wishing it would get up to 85 °F. The book says the coefficient of linear expansion of steel is 1.1×10^{-5}/°C. Someone tells you the suspension cables are 3.0 km long.

Find: (a) What is the change of temperature you are thinking of, in degrees Celsius?

(b) How many centimeters would the cables expand if the day were to warm up as desired?

10-23 Lead shot for shotgun shells is made by pouring molten lead through a sieve above a pan of water. The trick is to get the lead droplets to harden in the air before they hit the water so that they will be round instead of tear-drop shape. For this reason, the sieve is usually placed some distance above the water at the top of a shot tower. (J = 4,185 J/Cal)

Given: 1.0 kg of lead shot is dropped from a height of 80 m above 5.0 kg of water at 20 °C. The lead is just barely solid at 325 °C and has a specific heat of 0.03 Cal/kg-°C as it enters the water.

Find: (a) How much is the water warmed by cooling off the shot?
(b) How much is the water warmed by the kinetic energy of the shot due to its fall from the top of the tower?

10-24 A good tea kettle with a copper bottom is outstanding both in absorbing heat from the flame and in conducting the heat through to the water inside.

Given: A tea kettle has a round copper bottom 20 cm in diameter and 0.12 cm thick. The flame keeps the outside surface of the bottom at 106 °C, while the inside surface is at 105 °C when the water is boiling. It requires 540 Cal to boil away 1 kg of water. (K = 9.2 × 10^{-2} Cal/s-m-°C)

Find: (a) What is the rate of heat flow through the bottom of the kettle?
(b) How much water boils away in 5 min?

10-25 A 1956 Lincoln is going down the freeway at the speed limit when the driver decides it is time to stop. He throws on the brakes and converts all the kinetic energy into heat. (J = 4,185 J/Cal)

Given: The car has a mass of 2.3 × 10^3 kg and a velocity of 25 m/s. The combined mass of all four brake drums is 20 kg, and they are made of steel with a specific heat of 0.11 Cal/kg-°C. Their initial temperature is 20 °C.

Find: (a) How much heat is generated in stopping the car?
(b) What is the final temperature of the brake drums, assuming they absorb all the heat generated.

10-26 The basal metabolism of a person is defined as the rate of heat production of that individual at rest some 12 to 15 hours after having eaten a meal, although it is frequently measured indirectly in terms of the rate at which oxygen is absorbed by the lungs and carbon dioxide given off.

Given: A classroom containing 75 people and 500 m³ of air becomes stuffy after a 50-min lecture. The basal metabolism of an average person is 65 Cal/h. Air has a density of 1.2 kg/m³ and a specific heat of 0.24 Cal/kg-°C.

Find: (a) How much heat is given off by all these people during the lecture, assuming they remain at rest.
(b) By how many degrees Celsius does the temperature of the room increase if the ventilation system is on the blink?

10-27   An ice-cube tray contains water at a temperature of 0 °C. This temperature does not change as heat is withdrawn through the aluminum tray, but the water changes into ice at the same temperature of 0 °C.

Given: It requires 80 Cal of heat to be drawn through the tray to freeze each kilogram of water. The tray bottom is 10 cm wide, 30 cm long, and 0.2 cm thick. Aluminum has a thermal conductivity of 4.9×10^{-2} Cal/s-m-°C. The tray contains 250 ml of water at 0 °C.

Find: (a) How much heat must pass through the tray before the water is frozen?
(b) What temperature difference must be maintained across the tray to cause this much heat to pass in 1 hour?

ANSWERS

10-11   (a) 550 ft-lb; (b) 0.707 BTU; (c) 141 s = 2.4 min.
10-12   (a) 2.94×10^5 J; (b) 70 Cal; (c) 1,000 Cal.
10-13   (a) 30,000 BTU; (b) 4.5¢.
10-14   (a) 491 °R; (b) −459 °F
10-15   (a) 6mm; (b) 25 m (The change is small.).
10-16   (a) 11.4 Cal; (b) 8.0 Cal; (c) 6.8×10^{-2} cents.
10-17   (a) 174 °F; (b) 7 oz.
10-18   (a) 34.6 °C; (b) 22 kg.
10-19   (a) 3.6×10^{-2} Cal/s; (b) 0.21 Cal.
10-20   (a) 65%; (b) 62%.
10-21   (a) 7.6 Cal; (b) 3.2×10^4 J; (c) 6.4×10^2 m.
10-22   (a) 17 °C; (b) 55 cm.
10-23   (a) 1.8 °C; (b) 3.7×10^{-2} °C.
10-24   (a) 2.4 Cal/s; (b) 1.3 kg.
10-25   (a) 172 Cal; (b) 98 °C.
10-26   (a) 4.1×10^3 Cal; (b) 28 °C.
10-27   (a) 20 Cal; (b) 7.6×10^{-3} °C.

Sound and Other Oscillations

11

Delights of the ear rank with delights of the eye in the collection of enjoyments generally referred to as culture. Just as artists have been supplemented and aided in modern times by the technical skill of the photographer, so have musicians seen their art modified and magnified by the electronics technician. Some of the very shortcomings of electronic distortion have even grown into positive contributions at the hands of the artist in that advanced and esoteric form of music called rock.

FIGURE 11-1 *A rock singer is aided by electronic distortion.*

The modern electronics enthusiast will spend thousands of dollars and weeks of time working on a high fidelity system whose chief function is to wiggle the eardrums back and forth in just the right way. We will start our study of this elegant and complex wiggling of the eardrums by considering the simpler motion associated with a pure tone. We will then study the tone itself in terms of the propagation of a wave through a medium. Finally, we will be in a position to see how waves can be added together to make any motion imaginable—or even no motion at all.

► Simple Harmonic Motion

The kind of motion associated with a pure tone is called **simple harmonic motion.** It is also sometimes called **sinusoidal motion,** because a graph of position against time for this kind of motion produces a sine curve. If you were to look at the position of your eardrums, for example, when you were listening to a single note produced by flute or a tuning fork, you would find that your eardrums move back and forth in very much the same way that the sine of an angle increases and decreases as the angle increases. An easy way to produce a graph of this kind of motion would be to attach a marking pen to a weight that is bobbing up and down on the end of a spring and to use this pen to mark the position of the weight on a card as the card is pulled past at a constant velocity.

This kind of motion turns out to be very common. Just about anything may be regarded as a mass on a spring. A leaf on a tree swings back and forth because of the springiness of the leaf stem. Every support, from building beams to spider webs, obeys Hooke's law if the displacements are small enough. Any object restored to its equilibrium position by a force proportional to displacement will undergo simple harmonic motion. If the frictional force is small, just about anything will bounce around undergoing simple harmonic motion—sometimes several simple harmonic motions at the same time.

A pendulum bob swinging back and forth on a string also undergoes simple harmonic motion. The pendulum bob is restored to its lowest position by the horizontal component of the force acting in the string. As long as the bob swings only a small distance compared to the length of the string, this horizontal component of force, as well

FIGURE 11-2 *Sinusoidal motion results when a mass oscillates on a spring.*

as the resulting acceleration, is nearly proportional to the displacement. The bob decelerates to a stop as it reaches the end of its swing, but the acceleration doesn't stop when the pendulum bob stops. The horizontal component of the force continues to accelerate the bob back in the other direction toward the equilibrium position. The acceleration only decreases as the bob gets closer to this midpoint, picking up velocity along the way. When the bob finally reaches the lowest point and has its maximum velocity, the acceleration momentarily goes through zero. The bob then coasts on through the equilibrium position and the acceleration goes negative, or, if you like, the deceleration begins to build up in the other direction, tending to bring the bob to a stop. The further it swings, the greater this retarding acceleration becomes until the bob finally does stop. Even so, the acceleration continues and the pendulum bob starts picking up velocity in the other direction back toward the equilibrium position. The acceleration diminishes to zero only as the bob returns to the bottom of its swing. Thus, the acceleration and the velocity go through zero at different times. The acceleration goes through zero at the middle of the swing, while the velocity goes through zero at the ends of the swing.

A mass bounces up and down on a spring with the same kind of motion just described for a pendulum. Instead of the acceleration being produced by one component of the force acting in the pendulum string, the acceleration is produced by the changing Hooke's law force produced as the spring stretches. In both cases, the acceleration is proportional to the displacement and opposite in direction. That relationship between displacement and acceleration turns out, in fact, to be a necessary and sufficient condition for simple harmonic motion.

If you want to describe the motion of a pendulum or a mass on a spring, you would need to specify how far it swings or bounces and how much time it takes. The usual way of describing distance in simple harmonic motion is to measure from the equilibrium position to one of the peaks. This is called the **amplitude** of the motion. Since the equilibrium position is halfway between the peaks in opposite directions, the amplitude is half the total peak-to-peak distance traveled by the pendulum or the mass.

The time necessary for one complete cycle of the simple harmonic motion is called the **period** of the motion. The letter T is usually used to signify this length of time. The only point of caution is to measure a complete cycle with the motion going in both directions. If the period is measured between peaks, make certain that the peaks are on the same side of the equilibrium position. If the period is measured between times that the motion passes through the equilibrium position, make certain that the direction of motion is the same at the end as it is at the start. Simple harmonic motion passes through the equilibrium position twice in one period, but it is going the other direction as it passes through at the halfway point.

An alternative to describing the amount of time for one cycle is to describe the number of cycles for a given period of time. The **frequency**, f, of a simple harmonic motion is the reciprocal of its period.

$$f \equiv \frac{1}{T}$$

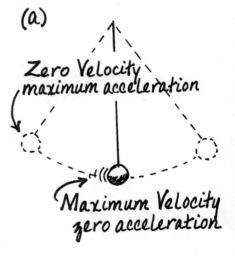

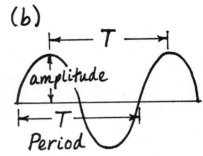

FIGURE 11-3 (a) A pendulum has maximum velocity when it has minimum acceleration, and vice versa. (b) The amplitude describes the size of the motion in space and period describes its size in time.

If the period of a pendulum is $\frac{1}{3}$ s, for example, its frequency is three times per second.

CHECK QUESTIONS

1. Tarzan is swinging from tree to tree at the end of a long vine attached to a tall tree located midway between his starting point and his destination. He swings from one tree to the same height in another tree 12 m away, and it takes him 4.0 s to make his journey. What are the period, T, and the amplitude, A, of the simple harmonic motion involved?

    Answer: $T = 8.0$ s
    $A = 6.0$ m
    (Tarzan only travels through half a cycle, but he goes through twice the amplitude since he travels from one peak to the peak on the opposite side.)

2. An AM rock radio station broadcasts at a frequency of 750 kHz (750,000 cycles per second). What is the period of the simple harmonic motion associated with this radio frequency?

    Answer: (1/750,000) s

Years ago, the widely accepted unit for frequency was the cycle per second (cps). It seemed to make a lot of sense to say that the frequency of a radio station was 750,000 cycles per second if, in fact, the carrier wave went through 750,000 complete cycles every second. In the 1960s, however, some electrical engineers noticed that we had named a lot of units after great men who had pioneered different fields, but there was one great man who seemed to have been left out. They thought that Heinrich Hertz was a very important man because he had discovered radio waves. So they decided to name a cycle per second after him. Henceforth, the unit Hz, pronounced hertz, will stand for a cycle per second.

$$1 \text{ Hz} \equiv \left(\frac{1 \text{ cycle}}{1 \text{ second}}\right)$$

That leaves the unit for period unaccounted for. We don't have a good name for a second per cycle. Henceforth the unit Wl, pronounced Wall, will stand for the second per cycle.

▶ Sinusoidal and Circular Motion

You may have wondered why there seems to be a connection between trigonometry and the behavior of a pendulum and that of a mass bouncing up and down on a spring. If you did more experiments with pendulums and bouncing masses, you would find even more similarities. You would find, for example, that the period of a pendulum is directly proportional to the square root of the length, l, of the string and inversely proportional to the square root of the acceleration due to gravity, g. The proportionality constant turns out to be 2π.

$$T_{\text{pendulum}} = 2\pi \sqrt{\frac{l}{g}}$$

By doing experiments with a mass bouncing up and down on a spring, you would find that the period of oscillation is directly proportional to the square root of the mass, m, and inversely proportional to the square root of the spring constant k of the spring. The proportionality constant turns out to be the same 2π.

$$T_{\text{mass on spring}} = 2\pi \sqrt{\frac{m}{k}}$$

We shall see that the similarity of these two relationships, as well as the fact that both motions trace out a sine curve, is more than a coincidence. The two equations result naturally from the relationship between simple harmonic and circular motion. The fact that there is a relationship can be easily seen by imagining a simple experiment with a pendulum.

Suppose you had a pendulum swinging back and forth from an overhead support, and you were to stop it and start it swinging instead from side to side. As you would expect, not having changed either the length of the pendulum or the acceleration due to gravity, the period of oscillation would be exactly the same in both directions. But then suppose you decide to verify this result by setting the pendulum swinging in both directions at once. Is that possible? Such an arrangement is called a **conical pendulum.** The pendulum bob will sweep out a circle as it undergoes two simple harmonic motions of the same amplitude in two perpendicular directions.

We can, on the other hand, look at simple harmonic motion as one component of circular motion. Imagine the handle of a wheel turning at a constant velocity. If this handle were illuminated from the side with parallel light, the shadow would sweep out simple harmonic motion on the wall—just as the conical pendulum might seem to be swinging only from side to side if you viewed its circular path from the edge in such a way that you could not tell that it was also swinging toward and away from you.

The fact that simple harmonic motion is one component of circular motion makes simple harmonic motion very common in the mechanical gadgets of our modern society. The horses on a merry-go-round, for example, go up and down in simple harmonic motion

FIGURE 11-4 *A pendulum swinging in a circle is undergoing two simple harmonic motions in perpendicular directions at the same time.*

FIGURE 11-5 *As a wheel turns at constant velocity, the shadow of the handle undergoes simple harmonic motion.*

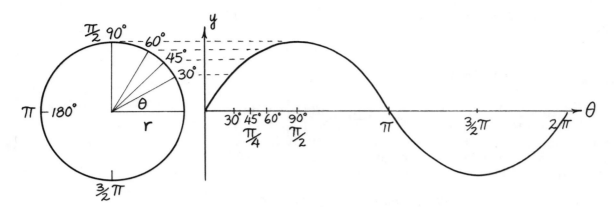

FIGURE 11-6 *Sinusoidal motion is a one-dimensional projection of circular motion.*

because they are raised and lowered by a crankshaft that turns at constant velocity. The pistons in an automobile engine also move in a near approximation to simple harmonic motion, since they are attached to a crankshaft that travels in circular motion.

We can use its relationship to circular motion to see why simple harmonic motion traces out a sine curve. Let us return to our example of the handle on the wheel and plot the height of the handle as a function of the angle through which the wheel turns.

The shadow of the handle starts off at zero distance from the equilibrium, or midpoint of its travel, when the angle θ is zero. This distance increases with the angle, at first rapidly and then more slowly, until the angle θ reaches 90° (which is the same thing as $\pi/2$ radians). The shadow then comes back down, at first slowly and then more rapidly, to cross the equilibrium position when θ reaches 180°, or π radians. It then does the same thing again in the opposite direction, returning to the equilibrium position to start the cycle over again at 2π radians.

The amplitude of the simple harmonic motion, or maximum distance that the shadow travels, is the radius of the circle involved. The distance from the equilibrium position only reaches this peak at $\pi/2$ and $3\pi/2$ radians. At all other angles, the distance is something smaller than the radius of the circle. In fact, you could find this distance by dropping a vertical line to the horizontal axis that passes through the equilibrium position. By so doing, you would form a right triangle whose hypotenuse is the radius of the circle. The distance from the equilibrium position, the length of the vertical line, is the side of the right triangle opposite the angle θ. The length of this line is, of course, related to the radius of the circle by the sine of the angle θ:

$$y = r \sin \theta$$

As the angle θ increases at a constant rate, the simple harmonic motion of the shadow naturally traces out the path of a sine curve.

The close relationship between simple harmonic and circular motion can also be used to derive a relationship between the acceleration and amplitude of simple harmonic motion and its period. We will find this relationship useful, because the equations we referred to earlier as giving the periods of a pendulum and a mass bouncing on a spring turn out to be special cases of this more general relationship.

Acceleration in simple harmonic motion may also be thought of as a one-dimensional projection of acceleration in circular motion.

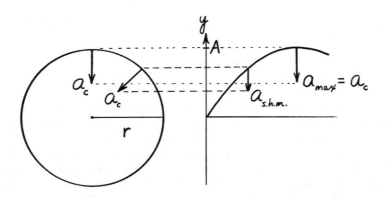

FIGURE 11-7 *Acceleration in simple harmonic motion is a projection of centripetal acceleration.*

As you recall, the acceleration in circular motion, centripetal acceleration, is given by the square of velocity divided by the radius of the path.

$$a = \frac{v^2}{r}$$

This acceleration is a vector quantity that is always pointed toward the center of the circular path. If we are only looking at one component of displacement, say the vertical component, then only that component of acceleration is of interest. This vertical component of acceleration is zero when the angle θ is zero. (The centripetal acceleration is then all in the horizontal direction.) As the angle increases, the vertical component of the acceleration grows to reach a maximum value as the displacement reaches the full amplitude, which is the same as the radius for circular motion.

$$a_{max} = a_c \qquad \text{when } y = r = A \text{ [amplitude]}$$

Velocity in the centripetal force expression can be expressed in terms of the period and amplitude of simple harmonic motion, using the definition of velocity:

$$v \equiv \frac{d}{t} = \frac{2\pi r}{T}$$

The distance, d, is equal to one circumference, $2\pi r$, when the time, t, is equal to one period, T. The centripetal acceleration then becomes

$$a_c = \frac{\left(\frac{2\pi r}{T}\right)^2}{r} = \frac{4\pi^2 r^2}{T^2 r} = \frac{4\pi^2 r}{T^2}$$

Replacing the radius r of circular motion with the amplitude, A, of simple harmonic motion, we get an expression for the maximum acceleration at the point where the simple harmonic motion goes through maximum displacement:

$$a_{max} = \frac{4\pi^2 A}{T^2}$$

This relationship is usually expressed in the form where it is solved for the period, T, in terms of the amplitude, A, and the maximum value of the acceleration, a_{max}, that occurs in simple harmonic motion.

$$T = 2\pi \sqrt{\frac{A}{a_{max}}}$$

This expression states that, in general, the period of oscillation of any object undergoing simple harmonic motion is directly proportional to the square root of the maximum displacement, or amplitude, of the motion and is inversely proportional to the square root of the maximum value of the acceleration that the object undergoes as it passes through the maximum displacement. The proportionality constant turns out to be 2π.

As you may remember, the form of the preceding expression is very similar to the equations mentioned earlier as giving the period of a mass bouncing up and down on a spring and the period of a pendulum. That is because the latter two expressions are just special applications of the general result. In both cases, the maximum acceleration is found in terms of the maximum force applied to the object in agreement with Newton's second law:

$$F_{max} = ma_{max}$$

In the case of a mass bouncing up and down on a spring, the force applied to the object is given by Hooke's law:

$$F = kx$$

The maximum value of the force is reached when the stretch of the spring reaches the maximum value, which is the amplitude, A, of the simple harmonic motion:

$$F_{max} = kA$$

Combining this expression with Newton's second law, we can see that the ratio of amplitude to maximum displacement is equal to the mass, m, over spring constant k in the case of a mass on a spring:

$$ma_{max} = kA$$
$$\frac{m}{k} = \frac{A}{a_{max}}$$

The period of oscillation is therefore 2π times the square root of the ratio of mass over spring constant:

$$T = 2\pi \sqrt{\frac{m}{k}}$$

A similar argument is used for the period of a pendulum, but the argument must be restricted to small angles. In this case, the force restoring the pendulum to its equilibrium position is one component of the weight, wt:

$$F = \text{wt} \sin \theta$$

The maximum acceleration is produced, in accordance with Newton's second law, when the pendulum has swung to its maximum displacement:

$$(ma_{max}) = (mg) \sin \theta_{max}$$

The sine of the angle can be expressed in terms of this maximum displacement, or amplitude A, using the small angle approximation. This approximation states that the sine of an angle and the tangent of the angle are nearly equal to the angle itself, expressed in radian measure, if the angle is sufficiently small. This approximation holds to three significant figures if the angle is less than about 0.1 radian or about 6°.

$$\sin \theta \cong \tan \theta \cong \theta$$

The angle measured in radians is equal to the arc length over the radius. In terms of the pendulum, this is the amplitude of the displacement over the length of the pendulum:

$$\theta = \frac{A}{l}$$

We can therefore use this small angle approximation to replace the sine of the angle in the earlier expression:

$$(ma_{max}) = (mg) \left(\frac{A}{l}\right)$$

Thus, we see that the ratio of the length of the pendulum to the acceleration due to gravity is equal to the ratio of amplitude to maximum acceleration in the case of the pendulum:

$$\frac{l}{g} = \frac{A}{a_{max}}$$

Using this ratio in the general expression for period in simple harmonic motion gives the period of the pendulum equal to 2π times the square root of length, l, of the pendulum divided by the acceleration due to gravity, g.

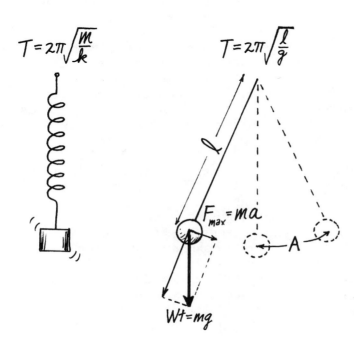

$$T = 2\pi \sqrt{\frac{m}{k}}$$

$$T = 2\pi \sqrt{\frac{l}{g}}$$

$$F_{max} = ma$$

$$A$$

$$Wt = mg$$

FIGURE 11-8 *The period of a mass on a spring is given by an expression similar to that which gives the period of a pendulum.*

▶ Complex Periodic Motion

The simple harmonic motion we have considered so far is the motion your eardrums experience when you hear a pure tone. It is possible for an object to undergo many simple harmonic motions at the same time, just as it is possible for you to hear several tones at the same time. Most sounds are, in fact, many tones mixed together. A harmony that is pleasing to the ear is a mixture in the right proportions of various tones that have frequencies related in a simple manner. A guitar note, for example, is made up mostly of the basic, or *fundamental,* tone plus the first *overtone,* a tone having exactly twice the frequency of the fundamental. It is a particular mix of these tones that gives a guitar its distinctive mellow sound. There are, moreover, tiny amounts of other overtones mixed in as well. The proportion of these vibrations is the factor that distinguishes an expensive guitar from a cheap one.

This same difference is more exaggerated when different instruments are involved. A piano sounds very different from a flute, even though they are both playing exactly the same note. One difference is that the piano note starts loud, as the hammer hits the string, and then dies out, while the flute note continues at an even level. That is called the *attack* of the note. But even if you could control the level of a flute note, you could never make it sound like a piano. The sounds are clearly distinguishable by the overtone mix you hear. A flute note is a much purer sound with almost all the vibration being in the fundamental tone. A piano note is made with two or three strings being struck simultaneously at a point which is one-seventh to one-ninth of their length from their end. These strings are thus made to vibrate in several ways at the same time. What is more, other strings on the piano pick up these vibrations to reinforce their sound in the mixture you hear. That is why a piano that is badly out of tune sounds all wrong from the first note that is played upon it.

The idea of adding vibrations on top of other vibrations can be visually understood by tying one pendulum on to another to make a double pendulum. The second pendulum will then add its influence to the first. They will both wobble back and forth under this mutual influence even as they swing back and forth together. This compound motion may be looked upon as the sum of several simple harmonic motions added together. Different lengths of string connecting the two pendulum bobs will change the frequency of the wobbling motion that is superimposed upon the swinging motion and produce different kinds of compound motion. The kinds of resulting motion that are associated with harmonies, in musical terms, are motions that repeat themselves in a noticeable pattern. Such a motion can be generated by the double pendulum by adjusting the lengths of the strings so that the wobbling motion is at a frequency that is some simple multiple of the general swinging motion. Then the two motions get back together and repeat their motion after a certain number of swings.

As you might well imagine, you can get even more complicated motion by adding more pendulum bobs. The more pendulums you string together, the more simple harmonic motions you add into the mix. In fact, you can get just about any motion you like, no matter how complex, if you are willing to add in enough simple harmonic motions with the right frequencies, amplitudes, and starting times.

There are many physical situations in which a number of independent influences are superimposed upon the same system. If these

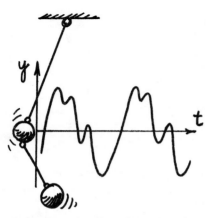

FIGURE 11-9 *A double pendulum illustrates the idea of adding several simple harmonic motions together.*

influences are linear in the sense that their effects add, a rule, called the **principle of superposition,** may be used to understand the result as simply the sum of the individual influences. One of the great values of this rule is that it can also be used in reverse. Not only can you add several simple harmonic motions together to get a more complex motion but you can look upon any complex motion as being made up of simple harmonic components.

As an example of the principle of superposition, let us consider the problems encountered in high fidelity sound systems. A funny-shaped electrical signal is generated by, say, a microphone or phonograph cartridge. Depending on the quality of the pickup device, this signal may be a more or less accurate representation of the sound you want to get out. The amplifier must be able to take this little funny-shaped signal and turn it into a big signal having nearly the same funny shape. To do this, it must be able to amplify a vast range of frequencies equally so that all of the frequency components in the original signal will remain in the finished product. The speaker must then take this big electrical signal and produce corresponding air pressure waves.

The demands of handling such a vast range of frequencies and giving them all equal treatment is just too great for any one speaker. Instead, high frequencies are split off and sent into a little speaker, called a tweeter, which has a low enough mass to respond to the rapidly changing signals associated with high frequency sound. The low frequencies go to a woofer, which is large enough to handle the long wavelengths associated with low frequency sound. Frequently, there is even a mid-range speaker to handle the middle frequencies. The point is to get an even response to all frequencies. But a single sound, if it is very complex, can have parts that come through on all three speakers.

One way to test a high fidelity sound system is to see what it does to a complex signal with lots of different frequency components. Such a signal need not necessarily look complex, however. One simple looking signal, which is easy to generate but which has a huge range of frequency components, is called a square wave. This signal just switches back and forth between a positive and a negative voltage at even intervals of time. To be a good square wave, the signal need only go along at a constant voltage until it suddenly changes to the negative voltage and remains there, going along until it suddenly changes back to repeat the cycle. This signal looks simple, as shown in Figure 11-10a. It has, however, a large number of frequency components, as is shown in the other parts of the figure.

Electronically, it is easy to generate a square wave signal, but just to see how the principle of superposition applies, let us show how you would go about making a square wave out of simple harmonic components. First you would need a sine wave having the same fundamental frequency as the square wave. Then you would need to add another sine wave having three times the frequency of the fundamental but one-third the amplitude. If you adjusted the starting time of this so-called third harmonic so that it goes positive and negative at the same times as the fundamental, the sum of these two would go positive extra fast and then wobble before going negative extra fast to wobble again. If you then add still another sine wave having five times the frequency, one-fifth the amplitude, and the same starting time, you can get the signal to go up still faster and

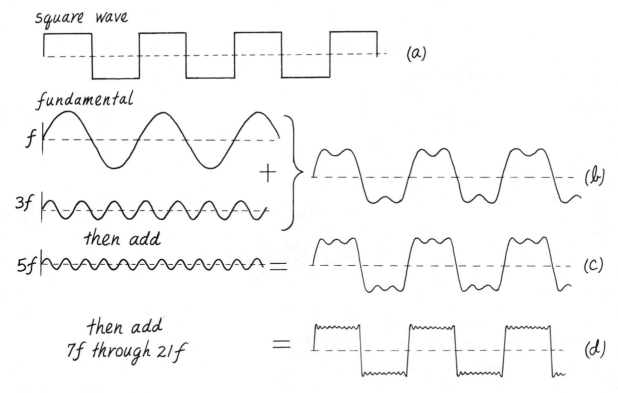

FIGURE 11-10 *A square wave (a) is made up of odd harmonics having the correct amplitudes and starting times. If only the fundamental signal and one having a frequency three times as great are added together, the result looks like (b). Add in a signal with five times the frequency, and (c) the result starts to look a little more like a square wave. The more frequency components you add (d), the more the result looks like a square wave.*

then "wibble-wobble" at the top before going down to wibble-wobble again at the bottom. The more odd-numbered harmonics of the right amplitude you add in, the flatter you can get the top and bottom and the faster you can make the signal change from positive to negative. A square wave therefore contains only odd harmonics, but it contains an infinite number of them—at least theoretically. The more of them that are in an actual signal, the sharper will be the corners and the flatter will be the tops and bottoms of the square wave shapes.

If you put a square wave signal into a decent high fidelity amplifier, you should find that you get a pretty good square wave out. If you put a square wave into even a fairly good speaker system, however, you could easily see the distortion in the output. Speakers, at their present state of development, are the weakest link in most home high fidelity sound systems.

► Waves

We have been discussing simple harmonic motion of a single object or of several objects coupled together, as in the double pendulum. If a larger number of objects are coupled together, energy can be transmitted along from one to the other in the form known as a **wave.** The wave may be thought of as something independent, in a sense, of the objects through which it moves. Ripples moving across the surface of a pond, for example, are thought of as going from one place to the other, although the water molecules do not themselves travel from one side of the pond to the other but merely bob more or less up and down as the wave passes through. As the wave moves along at what is known as the **velocity of propagation,** which is the

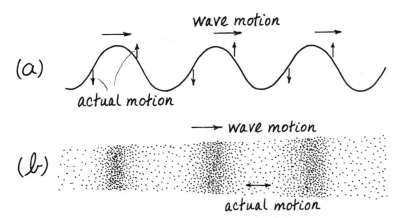

FIGURE 11-11 (a) *A transverse wave is one in which the actual motion is perpendicular to the direction of wave propagation.* (b) *The particles propagating a longitudinal wave move in the same or opposite direction as the wave itself.*

velocity at which the crests and troughs travel, the water actually travels in an elliptical path. Water that is in a trough of the wave at one moment moves up to form the crest of the oncoming wave at the next. While this is happening, nearby water that starts at a crest falls down to form the advancing trough.

This elliptic motion of the water molecules may be thought of as a composite of two simpler motions—one parallel to the direction of wave travel and the other perpendicular. To the extent that the actual motion of the water is back and forth in the same or opposite direction to that in which the wave is traveling, the wave motion is said to be **longitudinal.** A common example of a wave that is entirely longitudinal is a sound wave in air. The wave is made up of alternate regions of high and low pressure, which move through the air as the air molecules move back and forth in the same or opposite direction to form these regions of high and low density.

To the extent that a water wave involves motion perpendicular to the direction in which the wave travels, the wave motion is said to be **transverse.** One example of purely transverse wave motion is the propagation of a wave on a string. The string moves from side to side as the wave moves along its direction of extension. Next time you are cleaning house, give the vacuum cleaner cord a sudden sideward jerk and watch the wave move out across the room while the cord itself remains in your hand. Think "transverse wave" as you watch this fine demonstration.

An earthquake is usually made up of both kinds of wave motion propagating at different velocities. If you are any significant distance from the epicenter, or origin, of the earthquake, you feel two separate shock waves. The great rumbling you first feel is the longitudinal pressure wave, which seismologists call the P wave. This is followed by a swaying motion from side to side as the transverse shear wave passes through. This is called the S wave. Since they know the velocities of propagation for these two kinds of waves, seismologists can tell the distance to the epicenter of the earthquake by the separation in time between the P and S waves.

▶ Wavelength

The distance between adjacent crests, or between adjacent troughs, of a wave is aptly called the **wavelength** of the wave. Wavelength is to a wave what period is to simple harmonic motion. Both

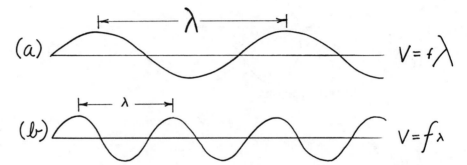

FIGURE 11-12 *The higher the frequency, the shorter the wavelength.*

indicate a separation, but period indicates a separation in time, while wavelength indicates an actual separation in space. The two are related by the velocity of propagation.

The relationship between wavelength, period, and the velocity of propagation results directly from the definition of velocity:

$$v \equiv \frac{d}{t}$$

For a wave, the appropriate distance and time are wavelength, λ, and period, T.

$$v = \frac{\lambda}{T}$$

That is all there really is to the relationship. It is usual, however, to express it in terms of frequency instead of period. Since frequency, f, is simply the reciprocal of period, the relationship becomes a product instead of a ratio:

$$\boxed{v = f\lambda}$$

Many people prefer not to memorize this relationship. They feel that they can be more reliable about getting it straight if they simply start with the definition of velocity and work it out each time they use it. If you prefer to rely on your memory, it is usually a good idea to write the relationship down and then stop for a moment to be sure that it seems reasonable. A high frequency should mean a short wavelength if the velocity of propagation remains unchanged.

▶ Reflection

Wave propagation requires a continuous medium of some sort. The medium can actually be lumpy, but it must be lumpy in a continuous manner. Air, for example, forms a continuous medium even though it is made up of hard little molecules. The molecules aren't even all the same size, but the mix of light molecules to heavy molecules is evenly distributed. It therefore behaves like a continuous medium in that wave energy is passed along from molecule to molecule without disruption.

For a medium to be continuous, each part of the medium must have the same amount of inertial give and take, at least on the average, as the rest of the medium. We saw an example of this when we considered a toy store demonstration of energy and momentum conservation. At the end of Chapter 9, you may recall, we considered the behavior of a string of metal balls all hung in a line from a wooden frame. If one ball is lifted and allowed to collide with the others, one ball will bounce out the other side. Each ball in the string has the same mass as the last and is therefore able to absorb all the energy and momentum of the ball colliding with it and pass it on to the next ball. The exception is the ball on the end. It has nothing to do with its energy and momentum but to swing out at the end. This end ball, however, does swing back and send its energy and momentum back down the line.

Whenever there is some interruption in a medium, at least part of the energy in a wave will be reflected at the boundary. If you give a sideward jerk to a vacuum cleaner cord, for example, and the other end of the cord is plugged into the electrical outlet, the wave energy will not disappear when it reaches the plug. Assuming the plug remains in the outlet, you will see the wave reflect at the boundary and come back toward you along the cord.

Even more fun than cleaning house, however, would be to build your own wave demonstration machine from a box of soda straws, paper clips, and a roll of cellophane tape.[1] Stretch a meter or so of the tape out on the floor with the sticky side up. Lay straws perpendicular to the tape with even spacing between them and with their centers stuck to the center line of the tape. An easy way to get the spacing even is to use spacers made by cutting a couple of straws into short sections. These spacers can be put between the straws and left there to add strength to the final assembly. Place another strip of tape, sticky side down, over the straws and spacers to hold the whole thing together. Insert a paper clip in both ends of each straw to give them a significant amount of inertia so that the kinetic energy stored in the wave is large compared to frictional losses. Hang the whole thing up by one end of the tape from a door frame and proceed to do experiments on waves.

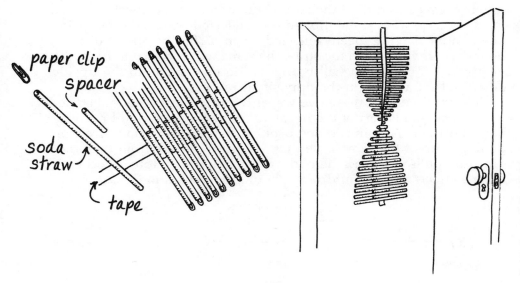

FIGURE 11-13 *A wave machine can be made from soda straws and tape. Spacers can be made by cutting a straw into short sections. Paper clips can be inserted in the ends to add mass. A door frame makes a convenient support.*

[1] A form of this type of machine using steel rods attached to a horizontally supported torsion wire was developed by John N. Shive of Bell Laboratories to demonstrate microwave transmission theory. A model suitable for classroom demonstrations is available through Ealing Corporation.

The first experiment is to twitch the top straw and watch the pulse travel along the machine until it gets to the bottom. Instead of getting lost, the pulse seems to turn around and return in the direction from which it came. One way to look at this is to say that the end straw (commonly called the *last straw*) must do something with its kinetic energy and momentum, these being conserved quantities. It therefore gives them back to the straw from which it got them, this being the only straw at hand. From a force point of view, the last straw is not restrained by another straw on down the line so that its motion continues under its own inertia until it is restrained by the very straw that gave it a shove in the first place. From a wave point of view, the motion of the last straw is greater than the motions of the other straws by virtue of the discontinuity, and this larger wave motion may be looked upon as an additional wave superimposed upon the original. The reflected wave is just the wave that came out of nowhere to make the last straw wiggle more than the others.

If you find it a bit hard to believe a wave coming from nowhere, you might find it a bit easier to understand the result of clamping the last straw in position (either with your fingers or by anchoring it to the door frame with a couple pieces of tape). Surprisingly, a wave coming into such a boundary will also be reflected, but the reflected wave will be upside down relative to the incoming wave. From a force point of view, the fixed boundary applies whatever force is necessary to hold the last straw stationary. This force is greater than the force between straws, since the straws up the line have some give to them. The reflected wave is then seen as a result of this reaction force. From a wave point of view, the fixed boundary supplies a wave that exactly cancels the wave coming in. The reflected wave is then seen as this canceling wave supplied by the fixed boundary.

This picture of the reflected wave as a new wave that comes out of nowhere to satisfy the boundary conditions agrees with what happens when two waves traveling in opposite directions run into each other. You can do this experiment by sending a second wave into the wave machine at the moment that the first is reflected from a fixed end. As we just mentioned, the reflected wave will be upside down after having been reflected from a firmly attached fixed boundary. If you twitch the straw the same way the second time, the new wave you send in will be right side up relative to the upside down pulse traveling in the opposite direction.

You might expect these two waves to cancel each other out somehow, since one is sort of the negative of the other. A point midway between them is going to see a pulse coming from one direction pulling it up at the same time that the pulse coming from the other direction pulls it down. The two waves should therefore cancel each other's effects and the midpoint should remain stationary.

FIGURE 11-14 (a) *Two pulses approach each other from opposite directions. (b) The two pulses just seem to cross over each other and depart from the midpoint, although their joint effect on that midpoint may have been simply to hold it stationary.*

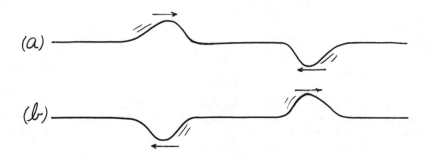

Instead, what you actually observe is that these two waves seem to cross right over each other and continue in the same directions as they have been going. The poor straw at the midpoint may indeed get opposite orders from opposite directions and remain stationary as a result, but the energy in the two conflicting wave pulses cannot be so easily destroyed. For one brief moment in time, the straws on both sides of the midpoint may all line up in a straight line so that you might think both pulses have disappeared, looking only at displacement. That would be to forget, however, the kinetic energy of the straws. The straws on one side of the midpoint are traveling downward as they pass through their equilibrium position, while those on the other side are traveling upward. Both sets of straws coast on through their equilibrium positions to form pulses in the other direction. It is exactly as though the pulse coming in from the right has crossed over the one coming in from the left.

As far as the straws on one side of the midpoint are concerned, the result is the same if you hold the midpoint stationary by attaching it to some solid object. Those straws don't know what is holding that midpoint straw stationary, and they don't particularly care. It's all the same to them whether there is some upside down pulse coming in from the other side of the boundary or whether the same thing is being accomplished by a clamp. The main thing is that the stationary boundary condition is satisfied.

This argument can be repeated for a wave reflected from a free end of the wave machine. The last straw has nothing to restrict its motion and therefore moves farther than its neighbors. The situation is the same as if a positive pulse, one pulling in the same direction, were coming in from the opposite direction and superimposed on the first pulse. The straws down the line respond to the extra large motion of the last straw by passing this reflected pulse back in the other direction with no regard as to whether it was caused by a boundary or by an actual pulse coming from beyond the boundary. The only difference is that a pulse reflected by a free-end boundary is right side up relative to the incoming pulse.

▶ Standing Waves and Interference

A rather pleasant thing happens when the pulses fed into the wave machine are replaced by simple harmonic or sine wave motions. Instead of jerking the straw at one end of the chain from one side to the other, try pushing it back and forth in simple harmonic motion. You will see this sinusoidal wave travel down to the other end and get reflected. As the reflected wave comes back along the machine and mixes with the wave you are sending in, the progression of the waves seems to come to a halt. The waves just seem to stand there. This phenomenon is known as a **standing wave.**

A standing wave is really two waves moving in opposite directions. At certain places, called **nodes,** the wave coming from one direction cancels out the wave coming from the other direction. At one moment in time, one wave tells the straw to go up while the other tells it to go down. Later, when the first wave gets around to telling the straw to go down, the second wave is telling it to go up. The result is called **destructive interference** of the waves. The straw remains stationary as the two waves cross over. At other points the

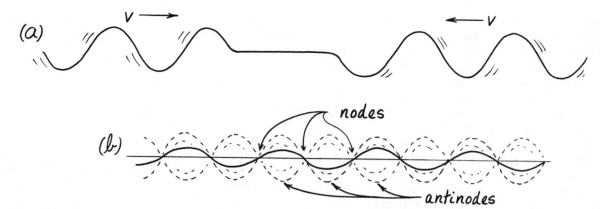

nodes

antinodes

FIGURE 11-15 *Two waves approaching from opposite directions (a) will combine to produce a standing wave, (b) which has nodes and antinodes.*

two waves **constructively interfere,** or combine to produce a motion greater than that which would be produced by either wave independently. These places are distinctly the opposite of nodes. They are called **antinodes.**

The wavelength of the two waves that make up a standing wave is, of course, closely related to the distance between nodes. In fact, you might think at first that this distance is the wavelength itself. The standing wave does seem to repeat itself after going from a node to an antinode and back to a node again. Actually, this distance is only half a wavelength. When one antinode is moving in a positive displacement, in one direction, the next is moving in a negative displacement, in the other. Just as a full cycle of simple harmonic motion involves both positive and negative peaks, a full wavelength includes two antinodes. A wavelength can therefore be measured between *alternate* nodes.

Not just any frequency of oscillation will produce standing waves when applied to the top straw of the wave machine. Generally, you can get short wavelength standing waves with a high frequency wiggle and long wavelength standing waves with a low frequency wiggle, but these wavelengths have to fit on the wave machine if they are to produce stable standing waves. If the end straw is left free, the only waves that will produce standing waves on the machine are ones that produce an antinode at that point. If you wiggle the top straw with just the right amount of force, you can force an antinode at that point also. Depending on the frequency of the wiggle, you could produce standing waves having one, two, three, or any other integral number of nodes in between, but the ends must be antinodes

If you attach the end straw to something solid so that it can't move, you will find that a different set of frequencies produce standing waves. By holding the end straw stationary, you force that end to be a node. You thereby change the boundary conditions that waves must satisfy if they are to produce standing waves.

▶ Resonance

Nearly everything around you has its own sound. If you tap your fist on the chair you are sitting in, it makes a different sound than if you reach over and bang on the wall. If someone in the other room starts banging on the wall in response, you would probably recognize right away what is happening. The sound of someone pounding on the wall is much different from, say, the sound of that person banging on the radiator pipes. The difference comes from the ability of different objects to support different kinds of standing waves.

Your chair, for instance, is made up of various lengths of different materials attached together at certain points and free at others. Any standing wave induced in your chair will be forced to have nodes at certain points and antinodes at others. Depending on the velocities of sound in the various materials, only certain frequencies of vibration are capable of producing standing waves that fit these boundary conditions. They are called the **resonate frequencies** of the object. Any vibration close enough to one of the resonate frequencies of an object will excite the corresponding standing wave vibration in the object.

A shock wave, such as you might make by rapping your chair with your fist, is really a complex sound having a vast number of simple harmonic components. Only those components that are near a natural resonate frequency will be picked up by the chair and reinforced to produce the chair's distinctive sound. If you reach over and pound on the wall, the shock wave will excite a whole different set of resonate frequencies, depending on the boundary conditions and the velocity of wave propagation in the wall.

There is also a difference in how fast friction dampens out a particular vibration in different material. The wall is likely to go "thud" while the radiator pipe is likely to "ring" for a longer time. The characteristic sound of an object is not only recognized by its frequency components but by how long they last.

▶ Musical Instruments

Musical instruments produce their characteristic sounds by means of various types of standing waves. String instruments such as pianos and guitars use strings of various lengths and under various tensions to support standing waves having the proper frequency. Wind instruments, on the other hand, use a column of air. In both cases, different notes are played by setting boundary conditions to select the length of the proper standing wave. The wavelength, together with the velocity of propagation of the wave, determines the frequency in accordance with the relationship we derived when we first started talking about wavelength:

$$v = f\lambda$$

(As you remember, this is really the definition of velocity, v, applied to the case of a wave where distance is wavelength, λ, and time is the period, which is the reciprocal of frequency, f.)

String instruments are first tuned by adjusting the velocity v of wave propagation along the strings. Two factors determine the velocity of a transverse wave on a string. One is the mass of the string and the other is its tension.

Imagine a wave pulse traveling along a string. As the pulse passes a certain point on the string, it jerks the string up and then down again. The force that causes this accelerated motion is really nothing more than the tension in the string, since the motion of the pulse along the string results from the way that the string pulls on it-

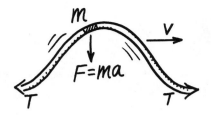

FIGURE 11-16 *The velocity of a wave pulse depends on the tension in the string as well as on the mass of the string.*

self. It therefore seems reasonable that the wave would move along more quickly if the tension in the string were increased. That is exactly what happens. Piano and guitar strings are both tuned by adjusting their tension. Increasing the tension raises the pitch, or frequency, f, of the standing waves by increasing the velocity, v, with which waves propagate along the string.

Some of the tuning is built into the string itself, however. Bass strings on both guitars and pianos are big and massive compared to the strings for higher notes. They are even wrapped with wire to increase their mass. The purpose of this is to slow down the waves. The greater the mass of a piece of string, the less will be its acceleration under a given force. Increased mass of the string therefore has the opposite effect of increased tension. The bass strings vibrate at a low frequency, f, because they have a slow velocity, v, of wave propagation.

Once a guitar string is tuned by adjusting its tension, it can still be used to play different notes by changing its effective length. In playing different chords, the guitarist holds the strings down on different frets to change the boundary conditions, which, in turn, changes the allowable wavelengths that can produce standing waves. Clamping the string between the finger and the fret forces that point to be a node. Another node is forced on the string by the bridge. The longest distance between nodes is therefore the distance from the bridge of the guitar to the guitarist's finger. Since a wavelength is the distance between alternate nodes, the distance from the bridge to the finger is half a wavelength of the fundamental frequency being played.

The guitar string can also have one or more nodes between the two that are forced by the boundary conditions. If there is one node in the middle of the string, the frequency is called the first overtone. If there are two nodes between the ones on the end, it is called the second overtone, and so forth. These different ways of supporting standing waves are called **modes** of vibration. The same string can actually vibrate in several modes at the same time. In a guitar the first overtone is strongly excited by strumming the string at a point one-fourth of the way along the string's length. That induces an antinode at that point, a node in the middle, and an antinode one-fourth of the way from the other end of the string. The string also vibrates in its fundamental mode with an antinode in the middle. The actual motion is a superposition of these two main modes of vibration with much smaller amounts of other vibrations thrown in.

One technique of guitar playing, called playing the harmonics, is to dampen out the fundamental mode by touching the string lightly in the middle and letting it vibrate in the higher frequency modes, mainly the first overtone. This is actually known as the "second" harmonic because, as it happens, the "first" harmonic is a name assigned to the fundamental.

Boundary conditions also determine which note is played by a wind instrument. Instead of forcing a node at a particular point as in a string instrument, however, the boundary condition is frequently such as to force an antinode. A flute, for example, is a hollow pipe that has a mouthpiece at one end and is open at the other. Both ends force an antinode in the standing wave that is generated when the lowest note is played. Blowing across the mouthpiece produces a swirling flow of vortices that make a soft whispering sound. The air

FIGURE 11-17 *The distance from the bridge of a guitar to the finger holding the string on a fret is one-half the wavelength of the fundamental mode of vibration and equal to the wavelength of the first overtone.*

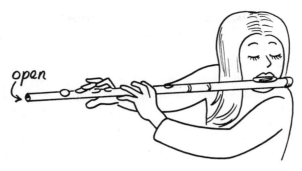

open

FIGURE 11-18 *A flute is played by directing a stream of air across the opening in the mouthpiece to reinforce standing waves in the tube.*

molecules are displaced in a complex waving motion that disturbs the molecules next to them, causing them to be displaced, and so forth as the wave moves on down the tube. Much the same as a pulse is reflected by a free end on the wave machine, the displacement wave in the air molecules is reflected when it gets to the free end of the flute. Just as the inertia of the last straw on the wave machine leads to an extra large motion, so the inertia of the air in the end of the tube leads to an extra large displacement as the displacement wave comes to the sudden freedom of the open end of the flute. This extra large motion leads to a reflected wave that travels back toward the mouthpiece.

When the reflected wave gets back to the mouthpiece, it not only is reflected again but is amplified. An outward displacement wave deflects the air stream away from the opening to form a decreased pressure that is even greater than that caused by the inertia of the air at a simple open end. An inward displacement wave deflects the air stream into the opening to form an increase in pressure that is even greater than that caused by the inertia of the air alone. Thus the waves move back and forth along the tube, reflected at the open end and both reflected and driven at the mouthpiece end. The resulting standing wave has an antinode at each end since the displacement wave motion is forced to be extra large at these points.

The note that the flute is playing can be controlled by opening various holes on the side of the tube to force antinodes at the proper points. The flute can easily be shifted into the upper register by simply blowing harder to excite the second harmonic. Interestingly, the overtones that are present in the lower register almost completely disappear in the higher octave and the notes turn out to be almost pure sine waves.

We have examined resonance in musical instruments in which boundary conditions force both ends of the standing wave to be either a node or an antinode. In both cases, the distance between the boundaries is half a wavelength for the fundamental mode of vibration. Depending on the wave velocity in the instrument, that puts a lower limit on the frequency that the instrument can play. The lowest note is one with a wavelength equal to twice the distance between the boundaries. There is a way, however, to set boundary conditions so that the fundamental has a wavelength equal to four times the length of the instrument. This trick is sometimes used to make very low notes in organ pipes. The pipe is closed at one end and open at the other. The fundamental mode of vibration then has a node at one end and an antinode at the other. The length of the pipe is therefore equal to one-fourth of a wavelength.

A good amount of musical instrument design has been done by

people who knew very little about the physics of sound. People just tried different things and used the things that seemed to work. In fact, much of what we know about the physics of sound has come from people who were interested in studying musical instruments. Once our understanding of theory caught up to our practical knowledge, however, substantial improvement in instrument design was possible. The transverse flute, the instrument we know today, was designed by Theobold Boehm in the 1860s, after he first studied physics for two years to better understand musical sounds.

The problems in this chapter are related to vibration and sound. **Simple harmonic motion,** the kind of vibration associated with a pure tone of a single frequency, is the sinusoidal motion that results from a mass oscillating on a spring or a pendulum bob swinging back and forth on a string. The time required for the motion to repeat itself is called the **period,** T, of the motion, but people often speak of the **frequency,** f, of the motion to convey the same information, the frequency simply being the reciprocal of the period:

$$f \equiv \frac{1}{T}$$

Amplitude, A, of the simple harmonic motion is defined as the distance from the equilibrium position to the maximum displacement. It is equal to one-half the distance from one peak to the other

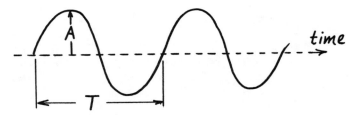

along the displacement axis. The period, T, is related to the amplitude, A, of the simple harmonic motion by

$$T = 2\pi \sqrt{\frac{A}{a_{max}}}$$

where a_{max} is the maximum acceleration associated with the maximum displacement of the motion. The ratio of A to a_{max} turns out to be equal to the ratio of mass, m, to the spring constant k for a spring and the ratio of the length, l, to the acceleration due to gravity, g, for a pendulum.

$$T_{spring} = 2\pi \sqrt{\frac{m}{k}}$$

$$T_{pendulum} = 2\pi \sqrt{\frac{l}{g}}$$

The **principle of superposition** holds that any motion can be made up of the sum of simple harmonic motions of different amplitudes, frequencies, and starting times.

Sound and other wave motion results from a vibration moving through a medium. The **wavelength,** λ, associated with such motion is just the length corresponding to one period, T, of the motion as it moves along at the velocity, v, of propagation through the medium:

$$v \equiv \frac{d}{t} = \frac{\lambda}{T}$$

This relationship is usually stated in terms of frequency: velocity equals frequency times wavelength, since frequency, f, is the reciprocal of period, T.

$$v = f\lambda$$

This relationship has frequent application to resonant systems, where a half wavelength refers to the distance between nodes or antinodes of a standing wave oscillating in the system. The boundary conditions of the system determine the positions of various nodes or antinodes and thereby determine the possible wavelengths that will fit. The only frequencies at which the system will resonate are those that produce the proper wavelengths to fit these boundary conditions.

11-1 The pistons in an automobile engine go up and down each time the crankshaft revolves one time, but they only produce power on alternate down strokes. The other down strokes between power strokes are used to draw air-gas mixture into the cylinder in preparation for the next power stroke.

Given: A four-cylinder automobile engine is turning at 2,000 revolutions per minute when the car is going 40 mph in top gear.

Find: (a) What is the period, in seconds, of revolution of the crankshaft?

(b) What is the frequency of the power strokes for one cylinder?

(c) What is the frequency of the power strokes for all four cylinders?

Your Solution

Discussion

This problem illustrates the relationship between period and frequency. The size of a cycle along the time axis can be described either in terms of the amount of time for the cycle to complete itself and start over again or in terms of the number of complete cycles that take place in a given amount of time.

Sample Solution 11-1

$$f \equiv \frac{1}{T}$$

(a) $T = \frac{1}{f}$

$$= \frac{1}{(2,000 \ rev/min)} \left(\frac{60 \ s}{1 \ min}\right)$$

$$= \boxed{3.0 \times 10^{-2} \ s/rev}$$

(b) $f_{\substack{power \\ stroke}} = \frac{1}{2} f_{piston \ stroke}$

$$= \frac{1}{2T} = \frac{1}{2(3.0 \times 10^{-2} \frac{s}{rev})} \left(\frac{1 \ Hz}{cycle/s}\right) \left(\frac{1 \ cycle}{1 \ rev}\right)$$

$$= 16.7 \ Hz \cong \boxed{17 \ Hz}$$

(c) $f_{4 \ cylinder} = 4 \ f_{1 \ cylinder}$

$$= 4 \ (16.7 \ Hz) = \boxed{67 \ Hz}$$

Discussion

Frequency is the reciprocal of period. Therefore, the period of a cycle can be found, as in part (a), as the reciprocal of frequency. Time units may be converted from minutes to seconds, since the time interval under consideration is so short.

The frequency of a power stroke is only one-half the frequency of total piston strokes, because there are twice as many piston strokes as power strokes, every other piston stroke being a power stroke.

The frequency of all four pistons combined is found in part (c) as being simply four times the frequency of one of the four pistons.

piston

cylinder

crankshaft

11-2 A mother is feeding her baby in a rocking chair. The gentle acceleration in alternate directions soothes both mother and child and makes them sleepy.

Given: The baby's head moves back and forth over a distance of 7.3 cm, making the trip 45 times in 1 min. The motion is simple harmonic.

Find: (a) What is the amplitude of the simple harmonic motion?
(b) What is the period of the motion?
(c) What is the maximum acceleration of the baby's head?

Your Solution

Discussion

This problem illustrates the relationship between the amplitude and period of simple harmonic motion and the maximum acceleration. If you find yourself wanting to take a little nap, think instead of the acceleration of students' heads as they shake themselves awake.

Sample Solution 11-2

(a) $2A = d_{total}$

$A = \dfrac{d_{total}}{2}$

$= \dfrac{(7.3\,cm)}{2} = 3.65\,cm \cong \boxed{3.7\,cm}$

(b) $f \equiv \dfrac{1}{T}$

$T = \dfrac{1}{f}$

$= \dfrac{1}{(45\ cycles/m)}\left(\dfrac{60s}{1\,m}\right)$

$= \boxed{1.3\ s}$

(c) $T = 2\pi\sqrt{\dfrac{A}{a_{max}}}$

$a_{max} = A\,\dfrac{4\pi^2}{T^2}$

$= (3.65\,cm)\,\dfrac{4\pi^2}{(1.3s)^2}\left(\dfrac{1\,m}{10^2\,cm}\right)$

$= \boxed{.81\ m/s^2}$

Discussion

The amplitude of simple harmonic motion is measured from the equilibrium position to one peak. It is therefore equal to one-half the peak-to-peak displacement. Period is found, as in part (b), as the reciprocal of the given frequency.

The maximum acceleration can be found by solving the relationship between period, amplitude, and maximum acceleration. The general form of this relationship is useful to remember, because it works as well for the period of a pendulum and a spring oscillator if you just change the variables under the square root sign to be reasonable in light of the physics of the situation. You might expect the maximum acceleration, a_{max}, to be in the denominator, for example, since greater acceleration would reasonably produce a shorter period of oscillation. The resulting acceleration is a bit less than a tenth of that due to gravity.

11-3 A radio speaker has a small coil of wire mounted in a magnetic field and attached to a flexible paper cone. A changing electric current through the coil shoves it back and forth against the cone to set the air in motion and make sound. The loudness of the sound is greater, however, if the frequency of the electric signal is near the natural frequency of the speaker, determined by the mass of the coil and cone and the springiness of the cone.

Given: A speaker has a peak response at 25 Hz. The mass of the coil and cone is 45 g.

Find: (a) What is the natural period of oscillation of the speaker?
(b) What is the spring constant of the flexible paper cone?

Your Solution

Discussion

This problem is an application of the relationship between the natural period and the mass and spring constant of a spring oscillator. This natural resonance of a speaker cone causes the speaker to sound "boomy" because of its exaggerated response to base notes at near the resonance frequency. Acoustic suspension speaker systems attack this problem by weakening the spring constant of the speaker cone and allowing the air pressure in the speaker enclosure to hold the speaker cone generally in place.

Sample Solution 11-3

(a) $f = \frac{1}{T}$

$T = \frac{1}{f}$

$= \frac{1}{(25\,Hz)}\left(\frac{1\,Hz}{cycle/s}\right)$

$= \boxed{4 \times 10^{-2}\,s}$

(b) $T = 2\pi\sqrt{\frac{m}{k}}$

$\frac{T^2}{4\pi^2} = \frac{m}{k}$

$k = \frac{4\pi^2 m}{T^2}$

$= \frac{4\pi^2(45\,gm)}{(4\times10^{-2}\,s)^2}\left(\frac{1\,kg}{10^3\,gm}\right)\left(\frac{N}{kg\cdot m/s^2}\right)$

$= \boxed{1.1 \times 10^3\,N/m}$

Discussion

The period is found in part (a) as simply the reciprocal of the frequency. The standard unit of frequency is the hertz, defined as a cycle per second. The resulting period is in seconds per cycle, but the units of cycles in the denominator may be left out as they are implicit in the meaning of period. Units such as cycles or revolutions or other angular measure can often be inserted or left out as the occasion demands. They are sort of unitless units, whose presence or absence is implied by the thing being measured.

11-4 A boy takes his sister to the park to use the swings. Although they usually use the little swing set, they decide to use the big swings today.

Given: They are interested in what the period will be on the big 20-ft swings. The period of the little swings was 2.8 sec when the boy was swinging. The boy weighs 65 lb, and his younger sister weighs 72 lb.

Find: (a) What will be the period of the boy on the big swings?
(b) What will be the sister's frequency on the big swings?
(c) What was the length of the little swings?

Your Solution

Discussion

This problem illustrates the relationship between the period of a pendulum and its length and the acceleration due to gravity. The condition for simple harmonic motion is that the restoring force be proportional to the displacement. This condition is met for a pendulum only for small angles of oscillation, much smaller angles than are common for small children on a swing set. The assumption of simple harmonic motion gives, however, a reasonable approximation to reality, which would be good enough for most purposes in this situation.

Sample Solution 11-4

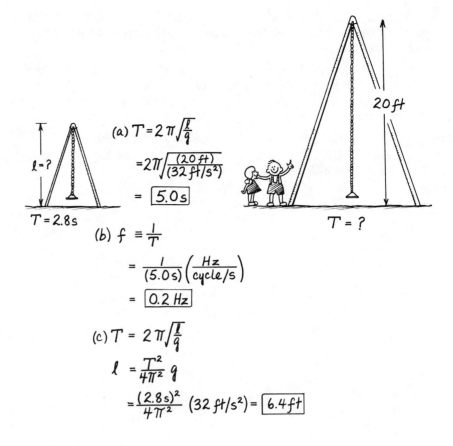

(a) $T = 2\pi\sqrt{\dfrac{\ell}{g}}$

$\quad = 2\pi\sqrt{\dfrac{(20\,ft)}{(32\,ft/s^2)}}$

$\quad = \boxed{5.0\,s}$

(b) $f \equiv \dfrac{1}{T}$

$\quad = \dfrac{1}{(5.0\,s)}\left(\dfrac{Hz}{cycle/s}\right)$

$\quad = \boxed{0.2\,Hz}$

(c) $T = 2\pi\sqrt{\dfrac{\ell}{g}}$

$\quad \ell = \dfrac{T^2}{4\pi^2}\,g$

$\quad = \dfrac{(2.8\,s)^2}{4\pi^2}\,(32\,ft/s^2) = \boxed{6.4\,ft}$

Discussion

The period for the little girl in part (b) is the same as that of the little boy in part (a), in spite of the difference in their masses, because mass doesn't enter into the relationship for the period of a pendulum. If the lengths of two pendulums are the same, they will have the same period in the same gravitational field, since the weight effects of mass balance the inertial effects for the same reason that light and heavy objects in free fall have the same acceleration.

11-5 The velocity of an ocean wave depends on its wavelength compared to the depth of the ocean floor. A wave generated by a seismic disturbance, called a tsunami (soo-nah-mee), generally moves with a very high velocity, because its extremely long wavelength makes it a shallow water wave even in ocean depths of four kilometers.

Given: A typical wave generated by wind action has a period of 8.0 s and a wavelength of 85 m. A tsunami, on the other hand, might have a period of 15 min and move with a velocity of up to 200 m/s (400 mph) because of its wavelength.

Find: (a) What is the velocity of the wave generated by wind action?

(b) What is the wavelength of the tsunami?

Your Solution

Discussion

This problem illustrates the relationship between velocity, frequency, and wavelength of wave propagation. The difference in velocity for waves of different wavelength is of interest to the surfer, who is waiting for just the right wave to ride on a surfboard. The change in the ocean depth relative to the wavelength of the wave as it comes into the beach also changes the wave's velocity and causes it to break.

Sample Solution 11-5

$$V = f\lambda$$

(a) $V = \left(\dfrac{1}{T}\right)\lambda$

$= \dfrac{(85\,m)}{(8.0\,s)} = \boxed{10.6\,m/s}$

(b) $\lambda = \dfrac{V}{f} = \dfrac{V}{\left(\frac{1}{T}\right)}$

$= VT$

$= (200\,m/s)(15\,m)\left(\dfrac{60\,s}{1\,m}\right)\left(\dfrac{1\,km}{10^3\,m}\right)$

$= \boxed{180\,km}$

Discussion

The basic relationship, velocity is equal to frequency times wavelength, is easy to derive if you aren't quite certain that you have it right. The definition of velocity is distance over time:

$$v \equiv \frac{d}{t}$$

The appropriate distance in this case is a wavelength, and the appropriate time is a period. The relationship in this form is useful in part (a) just the way it is. It may also be solved for wavelength to be used in part (b).

The wavelength of a tsunami as found in part (b) turns out to be a bit over 100 miles.

11-6   A guitar is tuned by adjusting the tension to change the velocity of waves along the strings so that the ends are nodes for the proper fundamental frequency. Different notes are then played on the same string by effectively shortening it by pressing down on a fret.

Given: The E string of a guitar is 61.7 cm long and vibrates at 330 Hz when properly tuned. The A note, at a frequency of 440 Hz, can be played on this string by holding it on the proper fret.

Find: (a) What is the proper velocity of the wave along the E string such that the distance between nodes corresponds to the length of the string?
   (b) What string length at this same velocity will produce an A note?

Your Solution

Discussion

This problem illustrates resonance on a string where both ends are clamped and thereby forced to be nodes. A guitar note is made up mostly of the fundamental and first overtone. The fundamental has a wavelength equal to twice the length of the string, and the first overtone has a wavelength equal to one times the length of the string.

Sample Solution 11-6

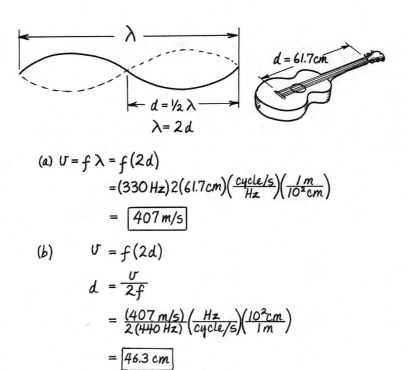

$$d = 61.7 cm$$

$$\lambda = 2d$$

(a) $\quad v = f\lambda = f(2d)$

$$= (330\,Hz)2(61.7cm)\left(\frac{cycle/s}{Hz}\right)\left(\frac{1\,m}{10^2 cm}\right)$$

$$= \boxed{407\,m/s}$$

(b) $\quad v = f(2d)$

$$d = \frac{v}{2f}$$

$$= \frac{(407\,m/s)}{2(440\,Hz)}\left(\frac{Hz}{cycle/s}\right)\left(\frac{10^2 cm}{1\,m}\right)$$

$$= \boxed{46.3\,cm}$$

Discussion

The distance between nodes of a standing wave is equal to half of a wavelength, since a full wave contains part of the string above the equilibrium position as well as part below at any one time. Part (a) gives the fundamental frequency times the wavelength as the velocity of the wave along the string, the wavelength of the fundamental being expressed as twice the length of the string.

Given this velocity for waves on the string, the string can be shortened to produce a higher fundamental frequency, as in part (b). The regular spacing of frets on a guitar results from the proportional relationship between notes on our musical scale.

11-7   Some organ pipes are open at one end and closed at the other. The length of such a pipe is the distance from a node to the first antinode of a wave having the proper frequency and traveling at the velocity of sound in air.

Given: A certain organ pipe is 11 inches long. The velocity of sound in air is 1,100 ft/s at a certain temperature.

Find: (a) What is the wavelength, in feet, of the note produced by this organ pipe?
(b) What frequency does this note correspond to?

Your Solution

Discussion

This problem illustrates resonance in a pipe closed at one end and open at the other. This boundary condition forces a node at one end and an antinode at the other.

Sample Solution 11-7

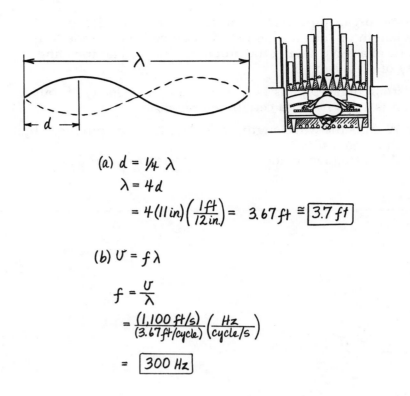

(a) $d = \frac{1}{4} \lambda$

$\lambda = 4d$

$= 4(11 in)\left(\frac{1 ft}{12 in.}\right) = 3.67 ft \cong \boxed{3.7 ft}$

(b) $v = f \lambda$

$f = \frac{v}{\lambda}$

$= \frac{(1,100 ft/s)}{(3.67 ft/cycle)}\left(\frac{Hz}{cycle/s}\right)$

$= \boxed{300 Hz}$

Discussion

The distance between a node and the first antinode is one-fourth of a wavelength, since it is half the distance between nodes, which in turn is half a wavelength. A wavelength of the sound generated by this organ pipe is therefore four times the length of the pipe.

The frequency of this sound can then be calculated from the velocity of sound in air, using the relationship between velocity, frequency, and wavelength of wave propagation.

11-8   The horses on a merry-go-round go up and down as they go round and round. They are attached to a crank shaft that is connected to the turning mechanism of the merry-go-round.

Given: The horses go up and down 12 times each minute and 5 times each time the merry-go-round goes around once.

Find: (a) What is the frequency of the horse?
(b) What is the frequency of the merry-go-round?
(c) What is the period of the merry-go-round?

11-9   A gambler is reciting an incantation about his baby needing a new pair of shoes as he shakes a dice cup with simple harmonic motion.

Given: The gambler shakes the dice cup back and forth over a distance of 20 cm, from one end to the other, 7 times in 4.0 s.

Find: (a) What is the amplitude of the motion?
(b) What is the period of the simple harmonic motion?
(c) What is the maximum acceleration of the dice cup?

11-10   A little girl playing with her kitten ties a ball to a rubber band and allows it to dangle near the floor. The kitten bats it with her paw and discovers that it bounces up and down at a given frequency.

Given: The ball has a mass of 125 g, and the rubber band has a spring constant of 12 N/m.

Find: (a) What is the period of oscillation of the ball on the rubber band?
(b) What is the frequency?

11-11   The heroine of a Japanese monster movie is dangling at the end of a long rope that is held in one hand by the monster. A close-up shot showing the heroine screaming, with the monster's face in the background, is followed by a long shot showing the whole monster holding a doll on the end of a string. A later shot shows the hero screaming at the end of the same rope, with the monster's face in the background.

Given: The length of the rope in the two close-up shots was 5.0 m long. The heroine looked as if she had a mass of about 65 kg, while the hero seemed to have a mass of 85 kg. The doll on the end of the string in the long shots swings back and forth with a period of 1.0 sec, because the producer did not bother to change the speed of the motion to correspond to the change in scale. It all turned out happily in the end.

Find: (a) What was the period of the heroine's swing at the end of the rope?
(b) What was the frequency of the hero's swing in the same situation?
(c) What length of string did the movie producers use in filming the long shot?

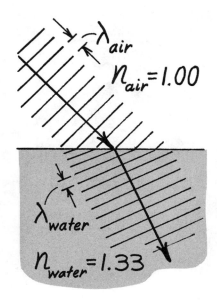

λ_{air}

$n_{air} = 1.00$

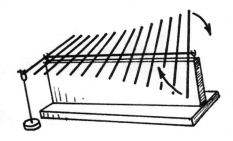

λ_{water}

$n_{water} = 1.33$

11-12 Light waves bend when they pass from one medium into another having a larger index of refraction, because the wavelength for the same frequency of light must be smaller when the light travels slower in the second medium.

Given: Light travels at 3.00×10^8 m/s in air but only 2.25×10^8 m/s in water. The wavelength of the yellow light from a sodium arc lamp is 5.89×10^{-7} m in air.

Find: (a) What is the frequency of this light?
(b) What is the wavelength of this same frequency of light as it travels at a slower velocity in water?

11-13 A xylophone is a percussion instrument consisting of a series of wooden or metal bars that are graduated in length to sound the musical scale. These bars are generally supported on felt pads so that their ends are free to vibrate. This boundary condition forces an antinode at both ends and thereby fixes the fundamental frequency at which a bar will ring when struck with a small wooden hammer.

Given: A xylophone has a middle-C bar that is 28.0 cm long and a fundamental frequency of 261.6 Hz. The G bar of the same octave has a fundamental frequency of 392 Hz (American Standard pitch).

Find: (a) With what velocity does the wave in the C bar travel?
(b) Assuming the same velocity for the wave in the G bar, what is the length of the G bar?

11-14 A physics instructor is using a Shive wave machine to demonstrate standing waves. He has one end clamped and is driving the other end up and down in simple harmonic motion to produce the longest possible wave on the machine. The clamped end is a node and the other end is the first antinode.

Given: The wave machine is 1.0 m long. The velocity of waves along the machine is 2.5 m/s.

Find: (a) What is the wavelength of the fundamental mode of oscillation where one end is clamped and the other is free to move?
(b) With what frequency should the physics instructor push the other end up and down to excite this fundamental mode?

MORE INTERESTING PROBLEMS

11-15 An automobile is a mass on springs and has a natural frequency of oscillation. The function of the shock absorbers is to dampen this motion so that the car does not bounce up and down as it goes down the road. You may have noticed a car with worn-out shock absorbers doing just that.

Given: An automobile having a mass of 1,700 kg must have worn-out shock absorbers, because you notice that it is bouncing up and down with a period of about 2.5 s as you follow it down the freeway.

Find: (a) What is the effective spring constant of this automobile's suspension system?

(b) How much would a person having a weight of 800 N make such an automobile settle when they got in?

Hooke's law: $F = kx$

11-16 A swimmer is poised on the end of a diving board about to start bouncing up and down to gain altitude for a back flip. You estimate her weight and notice how much she makes the diving board sag when she is just standing there.

Given: Her weight seems to be about 120 lb, and this weight makes the diving board bend a distance of about 1.2 ft.

Find: (a) What is the spring constant of the diving board?
(b) What is the diver's mass?
(c) With what period would you expect the diver to start bouncing when she first starts and is moving in simple harmonic action?

11-17 The limiting factor in determining the "red line," or maximum safe rotational velocity, of an automobile engine is the forces that develop between parts that go up and down, such as the pistons, and other parts that go around, such as the crankshaft. A common failure resulting from over-revving an engine is to break the connecting rod that connects the piston to the crankshaft.

Given: The piston in an experimental engine has a mass of 1.00 kg and is operated almost in simple harmonic motion with an amplitude of 4.00 cm. The engine is red-lined at 6,000 revolutions per minute (rpm).

Find: (a) What is the period of the simple harmonic motion at the red-line rotational velocity?
(b) What is the maximum acceleration of the piston at red-line?
(c) With what maximum force does the connecting rod push on the piston when the engine is operated at its red-lined rpm?
(d) What would be the force acting on the piston if the engine were over-revved to 8,000 rpm?

11-18 An FM radio receiver must be able to receive radio signals having wavelengths anywhere within the FM broadcast band.

Given: The FM band extends from 88 to 108 Mc on your dial. Mc stands for megacycles per second, or 10^6 Hz. The velocity of radio waves is the same as that of light, 3.00×10^8 m/s.

Find: (a) What is the length of the longest radio wave your FM receiver must be able to handle?
(b) What is the length of the shortest radio wave your FM receiver must be able to handle?

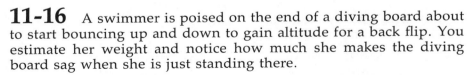

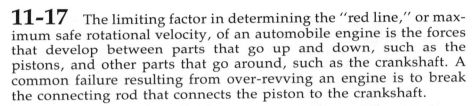

11-19 Bats hunt and navigate by listening for the echos of high-pitched chirps they emit from several times to hundreds of times a second, depending on how hard they are "looking." The sounds must be of a very high frequency to have a wavelength of about the same size as the insects the bats feed upon. The long-eared bat is even able to distinguish its prey when it is sitting on the surface of a leaf.

Given: The speed of sound in air is about 1,100 ft/s. A bat directs a beam of high-pitched sound toward a mosquito that is 0.4 in. long when flying in the air but that stands only 0.25 in. high when sitting on a leaf. The bat warbles the frequency of the sound so he can distinguish size.

Find: (a) At what frequency will the wavelength of the sound emitted by the bat correspond to the length of the mosquito in flight?

(b) At what frequency will the wavelength of the sound match the height of the mosquito sitting on the leaf?

11-20 The speed of sound in air increases with temperature, but the effect of this change on the tune of an organ pipe is compensated in part by the thermal expansion of the metal from which the pipe is made.

Given: The speed of sound in air at 0 °C is 331.4 m/s, but it increases by 0.61 m/s for each degree Celsius above zero. A brass organ pipe is open at both ends and has a length of 2.3 m at 0 °C and a coefficient of linear expansion of 1.8×10^{-5}/°C.

Find: (a) What is the fundamental resonant frequency of the pipe at 0 °C.

(b) What is the velocity of sound at 25 °C?

(c) What is the resonant frequency of the pipe at 25 °C?

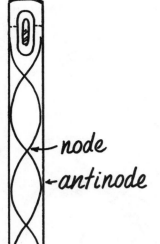

node
antinode

11-21 The transverse flute is a woodwind instrument developed by Theobold Boehm in the 1860s, after he had studied physics for two years to better understand musical sound. It is a hollow tube closed at one end and open at the other, but the placement of the mouthpiece makes it act as if it were open at both ends.

Given: The velocity of sound in air is 343.6 m/s at 20 °C. The lowest note that can be played on older flutes is C, at 261.6 Hz, and the highest note is C three octaves higher, at 2,093 Hz.

Find: (a) What is the wavelength of the lowest C that can be played on a flute?

(b) What is the distance from a node to an antinode for the lowest C on this instrument?

(c) What is the wavelength of the highest C that can be played on the flute?

ANSWERS

11-8   (a) 0.20 Hz; (b) 0.040 Hz; (c) 25 s.

11-9   (a) 10 cm; (b) 0.57 s; (c) 12 m/s².

11-10 (a) 0.64 s; (b) 1.56 Hz $\cong$ 1.6 Hz.

11-11 (a) 4.5 s; (b) 0.22 Hz; (c) 25 cm.

11-12 (a) 5.09×10^{14} Hz; (b) 4.42×10^{-7} m.

11-13 (a) 146 m/s; (b) 18.7 cm.

11-14 (a) 4.0 m; (b) 0.625 Hz $\cong$ 0.63 Hz.

11-15 (a) 1.07×10^4 N/m $\cong 1.1 \times 10^4$ N/m; (b) 7.5 cm.

11-16 (a) 100 lb/ft; (b) 3.75 slug $\cong$ 3.8 slug; (c) 1.2 s.

11-17 (a) 1.00×10^{-2}s; (b) 1.58×10^4 m/s²; (c) 1.58×10^4 N;
           (d) 2.81×10^4 N.

11-18 (a) 3.4 m; (b) 2.8 m.

11-19 (a) 33,000 Hz; (b) 52,800 Hz $\cong 5.3 \times 10^4$ Hz.

11-20 (a) 72 Hz; (b) 346.7 m/s; (c) 75 Hz.

11-21 (a) 1.313 m; (b) 32.84 cm; (c) 16.42 cm.

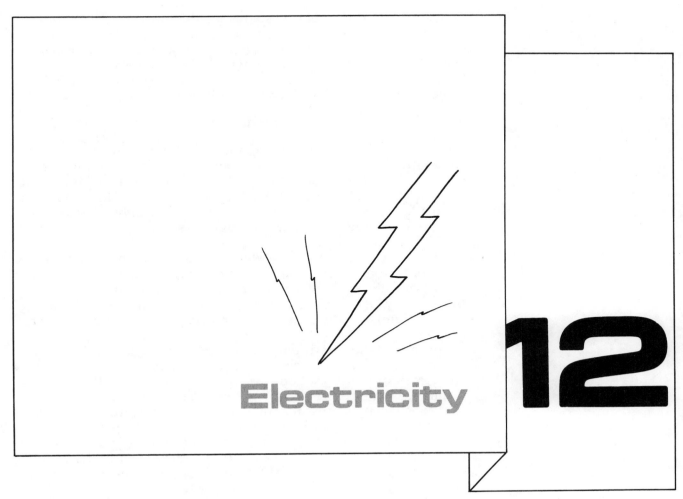

► Statics

Only two kinds of forces account for the majority of everyday phenomena—forces of a gravitational nature and forces of an electrical nature. We have already discussed the gravitational force between two objects. It is directly proportional to the product of the masses and inversely proportional to the square of the distance between the two objects. As you recall from our discussion of Newton's laws of motion, gravitational force is given by

$$F = G\frac{m_1 m_2}{d^2}$$

The electrical force between any two stationary objects obeys a similar inverse square relationship with distance. The thing about an object that produces an electrical force, however, is something called **charge.** Charge plays a role in electrical phenomena much like the role that mass plays in gravitational phenomena. The electrical force between any two objects is directly proportional to the product of their charges q_1 and q_2 and inversely proportional to the square of the distance between them.

$$F = k\frac{q_1 q_2}{d^2}$$

The letter k in this equation stands for the proportionality constant which in the electrical force law is similar to the universal gravitational constant G in Newton's law of gravitation. Instead of being a very small number, as in the gravitational case, however, the electrical proportionality constant k turns out to be a very large number if the charges are of ordinary size in human terms.

$$k = 9 \times 10^9 \text{ N} \cdot \text{m}^2/\text{C}^2$$

where C stands for the usual unit for measuring electrical charge, the **coulomb.** We may think of it as the amount of charge that is on a certain number of electrons—or on the same number of protons, for that matter. That number turns out to be 6.25×10^{18}. This might seem like a great number of electrons, but it only represents the amount of charge that passes through a 100-W light bulb in a little over a second.

If two charges of 1 C each were 1 m apart, the previously stated value of k would mean that the force of repulsion between the two charges would be 9 billion newtons. That would be about twelve times the weight of the *Queen Elizabeth* ocean liner. Obviously, such amounts of pure charge do not exist in room-sized objects around us.

In spite of the vast difference in size between electrical and gravitational forces, they still look a lot alike. They are both inverse square law forces in that they both decrease inversely with the square of distance. The striking similarities between these two forces have made some people think that they might really be different aspects of the same thing. Albert Einstein was one of those people. He spent the latter part of his life working with little success on this problem. Although there are similarities between these two kinds of force, there are also some very basic differences between them—and the vast difference in size is only one of these.

Another difference is that electrical forces can be balanced out. They can be either attractive or repulsive, whereas gravitational forces can only be attractive.

The old saying that opposites attract, usually referring to people, was first popularized by public lecturers who traveled about the countryside on horseback to entertain people by demonstrating the scientific marvels of electricity. An essential part of these demonstrations was the charging and discharging of pith balls. Pith is a light spongy plant tissue resembling Styrofoam, and balls made of it were coated with silver paint so that their outside surfaces would conduct electricity. Such a ball, hung from a silken thread, would exhibit a strong attraction for a rubber rod that had been rubbed on cat's fur, but this force of attraction would suddenly change to a force of repulsion as soon as the ball touched the rod. Another ball charged in the same manner would exhibit the same behavior. The two balls charged the same way would be found to repel each other. Pith balls would also be first attracted and then repelled by a glass rod, but a ball charged with a glass rod would be attracted to a ball charged with a rubber rod. The lecturer could make much of the fact that nature provides two kinds of charge just as it provides two sexes.

Benjamin Franklin named these two kinds of charge positive and negative in accordance with his feeling that an electrical charge represented either an abundance or lack of some kind of electrical

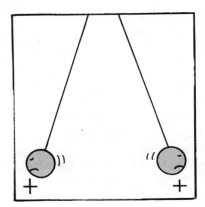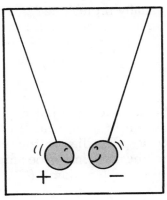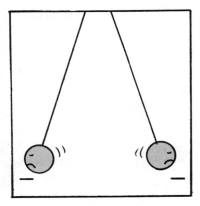

FIGURE 12-1 *Opposites attract.*

fluid. For some reason, he decided to call the rubber negative and the glass positive. Benjamin Franklin's fluid theory was remarkably like our present understanding of electrical charge in terms of the excess or lack of electrons. He missed, however, when it came to guessing which was which. It turns out that the rubber rod gains electrons in the rubbing process, while the glass rod loses them. We compensate for Franklin's wrong guess by saying that electrons have a negative charge and protons have a positive charge. Thus, a positive charge does wind up representing an excess of *something* even if that thing is not the electrons that normally flow through a wire. The positively charged protons are tightly bound in the nuclei of the atoms making up the wire.

▶ Current

The fluid theory of electricity was based on the observation that some things, such as the silver paint on the pith balls, allow electrical charge to flow from one place to another in much the same way as a pipe allows water to flow from one place to another. These things, usually metallic objects such as watch chains or wires, were called electrical **conductors**. Other things, which did not allow electrical charge to flow, such as the silken thread attached to the pith balls, were called **insulators**.

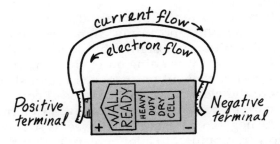

FIGURE 12-2 *Current flows from positive to negative, although electrons go the other way.*

The idea of an electrical current arose naturally from the fluid theory of electricity as an analogy to water flow in a pipe. Just as the rate at which water flows through a garden hose might be measured in gallons per minute, the electrical flow rate, or current, is the ratio of the amount of charge that passes a particular point in a wire to the amount of time that it takes to pass.

$$\text{Current} \equiv \frac{\text{charge}}{\text{time}}$$

Current is usually represented by I, while the symbol we will use for charge is q. Time is, as usual, represented by t.

$$I \equiv \frac{q}{t}$$

Anyone who has had experience with changing fuses is probably familiar with the unit for current. It is called the **ampere,** or "amp" for short. Much less familiar to the average person is the unit for electrical charge, the coulomb. One ampere (A) of electrical current is defined as the flow of 1 coulomb (C) of charge per second.

$$1 \text{ A} \equiv \frac{1 \text{ C}}{1 \text{ s}}$$

You might wonder at this point where the coulomb unit of charge came from. We mentioned earlier that this is the charge of about $6\frac{1}{4}$ billion billion electrons. That's a lot of electrons, and the electrostatic forces between charges of this size are, as we also mentioned, huge. Still, this quantity of charge was originally defined in terms of about the smallest amount that could, in a sense, be weighed. When an electric current is passed through an electrolytic cell containing silver nitrate, pure silver is plated out on one electrode. The amount of silver is directly proportional to the charge that passes through the cell, since individual electrons deposit individual atoms in the plating process. One coulomb of charge was defined as the amount that would deposit about 1 mg of silver—a small amount, but one that could be measured on an analytic balance.

While it is true that this unit of charge is too large for use in most electrostatic force problems, it turns out to be just right for problems involving the flow of current through household appliances. Just a little less than a coulomb of charge flows, for example, through a 100-W light bulb each second. An electric toaster typically passes 11 C/s, or 11 A, when it is in operation. These units of charge and current have therefore been widely adopted, because they are so convenient for common use.

The direction of current flow has also been established by common usage. The plus and minus sign convention was, as we have mentioned, arbitrarily assigned by Benjamin Franklin on the basis of what he thought was going on in charging rubber and glass rods by friction. This convention was well established and widely used long before the electron was discovered. People just naturally assumed that current flows out of the positive (+) terminal of a battery and into the negative (−) terminal. It seems so sensible that current should flow from a high to a low potential that this convention has stuck. In fact, positively charged ions do move in this direction when they are free to move. In a wire, however, the positive charges are tightly bound in the nuclei of the metal atoms and therefore don't move. The charges that do move, the electrons, actually travel in the other direction. Sometimes the term *electron current* is used to indicate the direction of electron flow. It is from negative to positive. The word *current* alone, however, is simply defined as something that is imagined to flow from positive to negative.

▶ Batteries and Bulbs

Some people find the subject of electricity somewhat abstract because they have had little direct experience with making electrical things work. Mechanical things seem more concrete because, as children, most people have had experience playing with blocks and mechanical toys. If you find yourself among the many who have had far less direct experience with the inner workings of electrical devices, as compared to mechanical gadgets, you are encouraged to do the following experiment with your own hands.

An ordinary flashlight has a simple electrical circuit, or hookup, but many people never stop to figure out how it operates. It is instructive to take one apart and make the bulb light up in your fingers. The wires you will need can be made either from paper clips or strips of aluminum foil cut from that roll you probably keep in the kitchen for wrapping food. Remove the bulb from the flashlight and try to make it light up first with one battery and then with two.

One pleasant diversion is to take the bulb, batteries, and wires in hand and seek out children of various ages to see how long it takes them to light the bulb.

After some smart kid makes the bulb light with one battery in less time than it took you, remind yourself that children come equipped with less intellectual baggage than adults and that it is therefore easier for them to experiment with random combinations. In order to save face, produce the other battery from your pocket and suggest that there are two ways to light the same bulb with two batteries and that one of these ways will make the bulb light brighter. If the wise kid beats you at those combinations, start pulling out additional bulbs. Eventually your superior experience with mechanical things will prevail as you start producing, from the other pocket, rubber bands with which to hold things together.

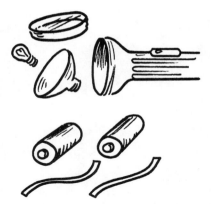

FIGURE 12-3 *An experiment using a flashlight bulb and batteries and strips of aluminum foil.*

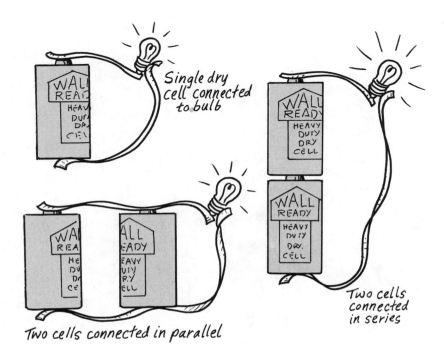

Single dry cell connected to bulb

Two cells connected in parallel

Two cells connected in series

FIGURE 12-4 *Various configurations of one and two batteries lighting a flashlight bulb. With the batteries in series, the bulb is significantly brighter.*

► Voltage

The whole point of fooling around with flashlight bulbs and batteries is to gain a practical feeling for what pushes electricity around a circuit—or, to say it another way, to gain a practical understanding of something called voltage. We have been using the term *current* to refer to the flow of electrical charge. Voltage is the thing that causes current; it is the thing that pushes electrical charge around and makes it flow.

You might like to think of voltage as something a little like force in mechanical situations or like pressure in fluid flow situations. Actually, voltage is more like an energy concept. Voltage is formally defined as the electrical potential energy per unit charge.

$$1 \text{ V} \equiv \frac{1 \text{ J}}{1 \text{ C}}$$

Where V stands for volt, J for joule, and C for coulomb. Still, the analogy with fluid pressure is very useful in understanding what is going on in an electrical circuit. This mental model, even with certain limitations, allows you to gain a qualitative understanding, for example, of series and parallel circuits.

Thinking of a battery as a kind of pump that pushes current through a bulb by developing some sort of electrical pressure across it, we can see how two batteries connected in series would light the bulb brighter than a single battery acting alone. Two water pumps would, after all, develop more pressure than one acting alone if the output of one were connected to the input of the other. Picture in your mind a water fountain with a stream of water squirting up in the air because of the pressure developed by a pump under the fountain. If you connected two pumps in series, the water would certainly squirt higher in the air than before, because the second pump

FIGURE 12-5 *Two water pumps connected in series to one figure in a water fountain and another two pumps connected in parallel to another. The series connection develops more pressure, and the parallel connection develops more volume.*

would take the pressurized water from the first pump and boost its pressure still more. The volume of water would not be greatly increased by this arrangement but the pressure would be. All the water that passes through one pump in this arrangement must also pass through the other, and so the volume of water is restricted by the capacity of a single pump. Therefore, you wouldn't necessarily get a lot more water, but the water you would get would squirt higher.

Much the same thing happens when two batteries are connected in series to a flashlight bulb. The voltage of one battery is boosted still higher by the second battery; there is more voltage across the bulb, so it lights brighter. The current would be limited by the capacity of a single battery cell, since all the current that passes through one must pass through the other. Although the series connection of batteries doesn't necessarily increase the current, it puts more energy into that current.

If you want more volume in a water flow system, or more current in an electrical system, the better connection would be parallel. Suppose, for example, that you had a double figure in a water fountain and you needed a combined volume of water greater than could be carried by a single pump. You could obtain this volume of water by connecting two pumps in parallel so that they both draw from the same source and deliver to the same figure. You wouldn't get much increase in pressure with this arrangement, but you would get a lot more water.

Much the same thing happens when you connect two batteries in parallel to the flashlight bulb. Both negative terminals of the batteries are connected to one side of the bulb, and both positive terminals are connected to the other. This parallel connection does not seem to light the bulb much brighter, as long as the batteries are fresh, because voltage is not much increased by this arrangement, but the batteries will stay fresh longer. Only part of the current flowing through the bulb comes from each battery, so that the current drain on each battery would be less. The parallel connection therefore increases the capacity of the batteries for delivering current.

The analogy between electrical voltage and fluid pressure can also be used to understand the relative nature of voltage. Since voltage is an energy concept, it has meaning only in terms of a reference level. As you know from our discussion of energy, the potential energy of this book you are holding has meaning only if you specify whether you are talking about its height relative to the floor, to the ground outside, or to the center of the earth. It is even negative if you are talking about the ceiling over your head as the reference level. That is why you need to connect both terminals of the battery to the bulb if you are to get it to light. The voltage of the battery has meaning only in reference to the other terminal. For this reason, voltage is also known as **potential difference.** Anybody who tries the experiment with the flashlight bulb and batteries will soon discover that the key to lighting the bulb is to provide a difference in electrical potential across the bulb by making a closed circuit from the battery, through the bulb, and back to the other terminal of the battery. This is very much like needing to provide a difference in pressure to get fluid to flow. No amount of pressure from a pump will make water squirt out of a fountain if that same pressure exists at the opening of the fountain. A stopped-up water fountain doesn't squirt water. Unstop the fountain and water flows because of the difference in pressure between the output of the pump and the atmosphere.

FIGURE 12-6 *Voltage is relative. A bird sitting comfortably on a high tension wire will not get shocked because both legs are at the same potential.*

Have you ever wondered how a bird can perch comfortably on a high tension wire without getting shocked? Power lines for trolley cars, for example, are typically at 400 V relative to ground. This is nearly four times the voltage considered safe for house use. A bird can land on this wire, however, without even noticing the high voltage. That is because this high voltage is relative to the potential of the ground, and the bird does not touch the ground. Both legs of the bird are on the wire and are therefore at the same potential. So long as there is no potential difference, or voltage, across the bird, no current will flow and the bird will not feel anything.

Perhaps you might wonder what would happen if the bird should happen to land on the wire with one foot on either side of an insulator. The overhead trolley power wires are supported by other wires, which are attached to lamp posts. Since the lamp posts are at ground potential, these support wires are interrupted by porcelain or glass insulators to keep them from conducting current to ground. In fact, a close examination will show that each wire is double insulated. It is interrupted with an insulator in two different places just in case something should happen to one of the insulators. If the bird should happen to get across one of the insulators, nothing would happen. The bird would still be protected by the other insulator. It would take two birds to get across both insulators simultaneously to produce Kentucky fried bird.

Ground is a very common reference potential for use in power supplies like the large 400-V generators used to power the trolleys, particularly streetcars that run on metal rails set in the street. These rails are used as a return conductor to complete the circuit from the streetcar back to the power generator. They are said to be *grounded* since they are electrically at the same potential as the earth as a whole. That is very convenient so that you don't get shocked if you should step on one of these tracks or board a streetcar. It is a good idea to have electrical appliances you use grounded. Otherwise, you might become a little bit Kentucky fried yourself.

▶ Ohm's Law

If we think of voltage as something that pushes current through a conductor like pressure pushes water through a pipe, it seems reasonable that a greater voltage would push a greater current through the conductor. That is exactly what happens. The voltage even turns out to be directly proportional to the current for many conductors, including most metals, as long as the temperature and everything else is kept constant.

$$V \propto I$$

This proportionality was first noticed experimentally by George Simpson Ohm in 1825. It is therefore called **Ohm's law.** The proportionality constant between the voltage, V, and the current, I, is called the resistance, R, of the conductor.

$$V = IR$$

The resistance of an electrical conductor depends upon its length and thickness as well as what it is made of. Silver is the best electrical conductor at ordinary temperatures, in that the resistance of a silver wire of certain length and thickness will be less than the same size wire made of another material. You might reason that the electrical wiring in your house would be better if the wires were made out of silver. Copper, however, comes in as a close second. A copper wire of a certain length and thickness has only about 6% more resistance than the same size silver wire. Gold and aluminum would be next, in that order. Gold, of course, would be too expensive for house use, although it does seem a shame for it to be lying around Fort Knox not doing anything. Actually, most of the gold in industrial use is being used for conducting electricity. Not only does it conduct electricity with low resistance but it is highly resistant to corrosion. That combination makes gold an excellent choice for electrical contacts. It is only used in very small quantities for this purpose, however, because of its expense.

Aluminum, on the other hand, is cheaper than copper. For that reason, it is sometimes used in house wiring, even though it has over one and a half times as much resistance for the same size wire. From Ohm's law, a higher resistance means that a higher voltage is needed to push a certain amount of current through the wire.[1]

$$V = RI$$

In house wiring, this effect is countered by using thicker wire. A thick wire has less electrical resistance than a thin wire of the same length and composition. The usual copper wire used in a 15-amp house circuit is No. 14 gauge. Aluminum wire for the same circuit would be No. 12 gauge, which has about 1.6 times as much cross-sectional area. The greater thickness brings the resistance back down to about the same value as the thinner copper wire has.

To be specific about the numbers involved, we must define units for this thing we call resistance. The common unit is the **ohm,** commonly represented by the Greek letter omega, Ω. One ohm of resistance is that amount of resistance that requires 1 volt of potential difference to cause a current flow of 1 ampere:

$$1 \text{ V} = (1 \text{ } \Omega)(1 \text{ A})$$

$$1 \text{ } \Omega = \frac{1 \text{ V}}{1 \text{ A}}$$

A piece of No. 14 gauge copper wire would be about 400 ft long to have a resistance of 1 Ω. An aluminum wire of the same diameter would only be about 240 ft long to have the same resistance.

CHECK QUESTION

An electric toaster draws 11.25 A at 120 V. What is its resistance?

Answer: $R = \dfrac{V}{I} = \dfrac{(120 \text{ V})}{(11.25 \text{ A})} = 10.7 \text{ } \Omega$

---

[1] Notice that here we have chosen to place the R next to the equality sign rather than in the reversed order shown in the proceding equation, because we wish to emphasize that it is a proportionality constant. Ohm's law is usually written $V = IR$. It makes no difference algebraically which way you write it.

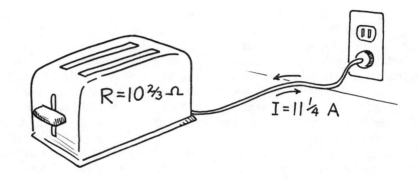

FIGURE 12-7 *The resistance of an electric toaster is the ratio of voltage, or potential difference, across it to the current that the voltage causes to flow through the toaster.*

Our discussion of Ohm's law assumed that the resistance of a wire, or of anything else, is constant. It is actually a variable that can depend on several factors. The resistance of your skin, for example, can change over a wide range depending on moisture. Should you tell a little lie, the tiny amount of sweat your body produces as an emotional response can be easily measured by the change in the resistance of your skin. Skin resistance is, as a matter of fact, one of the three parameters commonly measured in polygraph lie detector tests.

The electrical resistance of something can even depend, directly or indirectly, on the current. The resistance of a wire, for example, depends on its temperature. The hotter the wire gets, the more the thermal agitation of the atoms disrupts the flow of electrons. The flow of current, in turn, tends to heat the wire and thus raise its resistance. This effect is quite pronounced in the case of a light bulb, where the resistance of the tungsten wire filament is several times greater when the lamp is lit than when the wire is cold. A graph of voltage against current would therefore produce more of a curve than the straight-line characteristic of a direct proportion.

There are many cases, however, in which the departure from a strict proportionality is of little consequence in a given situation. Ohm's law may then be used as a close approximation to reality. As with other mental models, what is close enough depends on the accuracy you need. If the situation demands it, you can easily obtain an electrical device, called a precision **resistor,** whose resistance is constant to five or six significant figures over a wide range of current. The resistance of the common radio resistor is constant to within two to four significant figures, depending upon its quality. Such a device may be nothing more than a little chunk of carbon with electrical leads attached to opposite ends, or it may be a length of special alloy wire wrapped into a coil in a special way so as to eliminate magnetic effects. A resistor may be defined as any electrical device whose resistance may be considered constant. Even a light bulb may be thought of as a resistor for some purposes.

The idea of a resistor is a useful mental tool for understanding electrical circuits, although, if you stop to worry about such things, it is a simplification of real life situations. The electric toaster, for example, could be thought of as a resistor connected to the power lines by resistanceless wires. This simplified picture would ignore any voltage drop at the toaster due to the resistance that must be present in even a good heavy toaster cord. It also ignores any voltage drop in the power lines themselves. The voltage drop is there, and you might notice it by the slight dimming of a light bulb connected to the

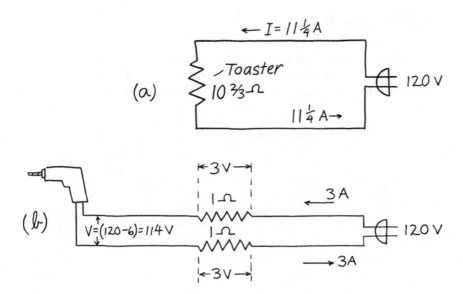

FIGURE 12-8 (a) A diagram showing a toaster as a pure resistor ignoring the resistance of the power cord. (b) A diagram showing the resistances in long extension cords for an electric drill.

same circuit when you start the toaster; but these are only secondary effects and may be safely ignored for many purposes.

It is easy to imagine a situation, however, in which the voltage drop in the power cord assumes major importance. Let us see how the idea of a resistor can be used to upgrade our mental model of what is going on in such a situation. Suppose you want to drill a hole in the dashboard of your automobile parked in the alley behind your house. You string out extension cords from your house so you can use your electric drill. You happen to own three 50-ft light-duty extension cords. Actually, you don't need that much length, but you decide to go ahead and use all three just in case. (If you've got them, you might as well use them.) Besides, you find a little label on the cords that says they are rated to carry 5 A of current, and the label on the drill says that it draws only 3.0 A. It seems that you should be okay, but when you try it you find that the electric drill just doesn't have as much power as when you plug it directly into the outlet in the house. Obviously, there must be a significant loss of some kind in the extension cords.

One way to picture the electrical circuit is to think of the two wires in the extension cord as two resistors connected in series with the drill. Just to make the numbers easy to work with, let us assume that each wire has a resistance of 1 Ω. That would correspond to 150 ft of light-duty extension cord of No. 18 gauge.

If the drill draws 3.0 A, this amount of current must pass through each of the two wires. It requires 3.0 V to push 3.0 A through 1.0 Ω of resistance, in accordance with Ohm's law.

$$V = IR$$
$$= (3.0 \text{ A})(1 \ \Omega)$$
$$= 3 \text{ V}$$

The same voltage, or potential difference, is needed to push the current through the other wire. The voltage available to the electric drill is thereby reduced by a total of 6.0 V below that available at the wall outlet. Even though the extension cords will handle the current, you shouldn't use longer extension cords than you need to reach the project.

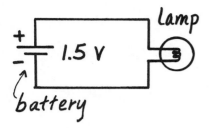

FIGURE 12-9 *A schematic diagram of the circuit in which a flashlight bulb is lit with a single battery.*

► Schematic Diagrams

Electrical circuits are frequently described in terms of simplified diagrams, using a few common symbols to represent ideal wires, ideal resistors, and other idealized electrical components to approximate what really happens in a practical situation. The preceding diagrams for a toaster and an electric drill are examples of this kind of diagram, called a **schematic diagram.** Resistance is shown as a zig-zag line, and ideal resistanceless wires are shown with straight solid lines.

Although real wires have resistance, and real resistors may not follow Ohm's law precisely, you can more easily understand how current flows in a circuit if you ignore these little difficulties. You can always go back and add in the complications if they turn out to matter. Let us return to the flashlight example to see how a schematic diagram may be used to describe a simple circuit. The simplest way to light the bulb, as you remember, was to connect one side of the bulb to one terminal of the battery at the same time that the other side is connected to the other terminal. This would be represented by showing ruler-drawn straight lines connecting the bulb with a single-cell battery, represented by long and short parallel lines. The convention is to represent the positive terminal of the battery with a long line and the negative terminal with a short line.

Let us now see how this circuit is similar to what is actually used in a two-cell flashlight. The two cells would be represented by two pairs of long and short parallel lines. One side of the bulb is connected, as before, to one terminal of the battery. In most flashlights one terminal of the bulb actually rests upon one terminal of one battery; but this fact is not important in understanding how the current flows. This connection is still shown as a ruler-straight line drawn from one side of the bulb to one terminal of the batteries. The other side of the bulb is connected to the other terminal of the batteries through a switch.

The purpose of the switch is, of course, to interrupt the circuit to prevent current from flowing when the flashlight is turned off. For our purposes, we may think of this as an ideal switch, which either has zero or infinite resistance. A switch is said to be closed when the contacts touch each other and open when the contacts do not touch. In this respect, the commonly accepted terminology is exactly the opposite as that used for a water valve. When you close a switch, you allow current to flow in much the same way as you allow water to flow by opening a faucet. Opening a switch should give it infinite resistance to electrical flow in much the same way that closing a faucet should shut off the flow of water.

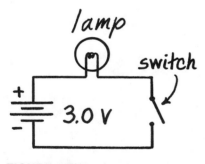

FIGURE 12-10 *A schematic diagram of a flashlight.*

CHECK QUESTIONS

1. What is the voltage across the *bulb* when the switch is closed?

   Answer: 3 V

2. What is it when the switch is open?

   Answer: 0

3.   What is the voltage across the *switch* when the switch is closed?

   Answer: 0

4.   What is it when it is open?

   Answer: 3 V

The fact that the resistance of the light bulb changes when the bulb is lit has little effect on our understanding the operation of the circuit. For the purposes of comparing the voltage across the bulb with that across the switch, we would do as well to draw the circuit using a resistor to represent the bulb. This kind of simplification allows you to ignore unimportant factors.

One complication frequently ignored in a simplified analysis of a circuit is the fact that the voltage of a battery decreases as the current delivered by the battery increases. This effect is called the **internal resistance** of the battery. If it turns out to be important in a given situation, it may be taken into account by representing the battery as a constant voltage source, called the **electromotive force,** or emf, of the battery, in series with a resistor. The emf of a battery is defined as the maximum voltage the battery can deliver when there is no load connected to it. A battery that has run down can usually still provide a voltage in the absence of a load, but the internal resistance has become so high that the voltage available at the terminals drops to a low value when you attempt to draw current from it. You might look upon this high internal resistance of a dead battery as having resulted from most of the available current paths inside the battery having been exhausted. This is just another example of creating a mental model to account for additional factors as they become important.

The value of a schematic diagram is that it allows us to focus our attention on essential matters. The actual structure of the flashlight, for example, may well be more complicated than our diagram would indicate. Parts of the flashlight may serve both to hold things together and to act as wires to conduct current. Just as in other mental models of reality, we need only be aware of simplifying assumptions so that we are capable of taking them into account should our judgment indicate that a more sophisticated model is warranted.

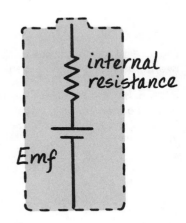

FIGURE 12-11 *The internal resistance of a battery may be pictured as a resistor in series with the emf of the battery.*

▶ Resistors in Series and Parallel

A battery can light two flashlight bulbs in two ways. The bulbs are said to be connected in **series** if they are attached to the battery in such a way that all of the current from the battery passes through each of the bulbs. The bulbs are said to be connected in **parallel** if they are attached in a branching circuit such that part of the current passes through one bulb and the rest passes through the other.

If you get a spare bulb for your flashlight and try this experiment, you will find that the bulbs connected in series will not be as bright as the two bulbs connected in parallel. The two bulbs connected in series have more resistance than either taken independently and will not draw as much current as would either one if it were connected

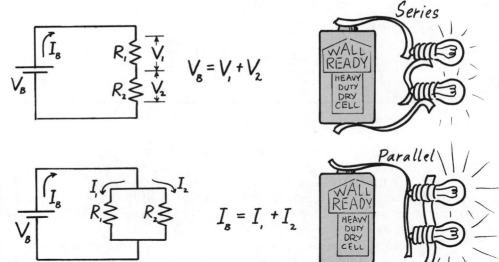

FIGURE 12-12 *Two ways to light two bulbs with the same battery. The bulbs connected in series (a) will not light as brightly as the bulbs connected in parallel (b), because they will not draw as much current.*

across the battery by itself. Only enough current can flow, in fact, for the sum of the voltages across these two bulbs to add up to the voltage of the battery. This kind of series arrangement of resistors is sometimes called a **voltage divider,** because it can divide a voltage into two parts. It is useful when you need a smaller voltage than that of the source.

The two bulbs connected in parallel, on the other hand, are equivalent to one, having less resistance than either taken independently. The entire voltage of the battery is across each bulb so that more current flows than if only one bulb were connected. The current produced by the battery is, in fact, the sum of the currents passing through each bulb.

We can generalize this argument to find the **equivalent resistance** of any two resistors connected in series or in parallel. The rule for adding resistance in series is that the equivalent resistance is simply the sum of the resistance of each of the separate resistors. This rule follows from Ohm's law by solving for resistance:

$$V = IR$$
$$R \equiv \frac{V}{I}$$

The total voltage V_t across both resistors is the sum of the voltages V_1 and V_2 across each of the resistors. Since the same current flows

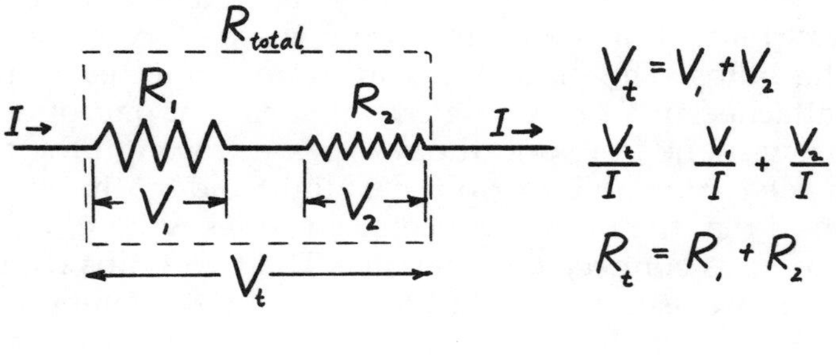

$$V_t = V_1 + V_2$$
$$\frac{V_t}{I} = \frac{V_1}{I} + \frac{V_2}{I}$$
$$R_t = R_1 + R_2$$

FIGURE 12-13 *The resistance of a series circuit is the sum of the resistances.*

$$I_t = I_1 + I_2$$

$$\frac{I_t}{V} = \frac{I_1}{V} + \frac{I_2}{V}$$

$$\frac{1}{R_t} = \frac{1}{R_1} + \frac{1}{R_2}$$

FIGURE 12-14 *The reciprocal of the total resistance of a parallel circuit is equal to the sum of the reciprocals of the resistances in the separate branches.*

through both resistors, we can divide the total voltage, V_t, on one side of the equation and the sum of the voltages V_1 and V_2 on the other with this current and we have proved the rule.

$$R_{total} = R_1 + R_2$$

The rule for adding resistance in parallel, on the other hand, to take the *reciprocal* of the sum of the *reciprocals* of each of the resistances. This rule comes from the fact that the reciprocal of resistance, again from Ohm's law, becomes

$$V = IR$$

$$\frac{1}{R} = \frac{I}{V}$$

In the parallel circuit, the current I_t passing through the total circuit is the sum of the currents I_1 and I_2 passing through the separate branches. Since the voltage across the total circuit is the same as across each of the resistors, we can divide $I_t, I_1,$ and I_2 by this voltage and we will have proved the rule.

$$\frac{1}{R_{total}} = \frac{1}{R_1} + \frac{1}{R_2}$$

CHECK QUESTION

Find the equivalent resistance for each circuit.

Answer: (a) 2 Ω, (b) $\frac{1}{2}$ Ω, (c) 9 Ω, (d) 2 Ω

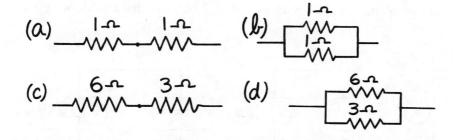

▶ Electric Power

When discussing work and energy, we defined power as the rate at which work is done, or the ratio of energy to time.

$$\text{Power} \equiv \frac{\text{energy}}{\text{time}}$$

It is easy to see that the power consumed by an electrical device depends on the voltage across it. We might, for example, connect two or three flashlight bulbs in series and try to make them light exactly as bright as one bulb connected to one battery. Since voltage in a series circuit is divided between resistances, we would find it necessary to boost the voltage of our power source in proportion to the number of bulbs we wish to light. Increasing the voltage of the power source, by such means as putting more batteries in series, would bring up the brightness of the bulbs, so that each one would dissipate power at the same rate as one bulb connected across one

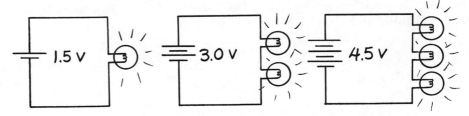

FIGURE 12-15 *The power dissipated by light bulbs connected in a series circuit is proportional to the voltage across them.*

battery. We can see from this argument that power dissipated in an electrical circuit is proportional to the voltage across a circuit.

$$P \propto V$$

We could have connected the bulbs in parallel, of course, and we would not have had to raise the voltage to make them all light as bright as one. The current drain on the battery would be greater, however, and the battery would wear out more quickly. Additional batteries could be connected in parallel to the original battery to keep the current drain per battery constant, but this would not be essential to our argument except to emphasize the fact that the power dissipated by an electrical circuit is also proportional to current passing through it.

$$P \propto I$$

Since power is proportional to both voltage and current, it is also proportional to their product.

$$P \propto VI$$

FIGURE 12-16 *The power dissipated by light bulbs connected in parallel is proportional to the current they draw.*

It happens that by a very clever choice of units for both voltage and

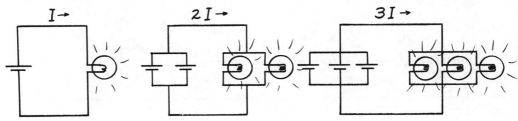

current our forefathers have managed to make the proportionality constant come out to be unity (1), so we can say,

$$\boxed{P = VI}$$

As mentioned earlier, voltage is formally defined as potential energy per unit charge. All our forefathers had to do to make things come out right was to define a volt as 1 joule per coulomb, since 1 ampere (the unit for current or flow rate of charge) is already defined as 1 coulomb/second. One volt times 1 ampere then works out to be 1 watt, which is indeed appropriate for power.

$$P = (1\ V)(1\ A)$$
$$= \left(1\ \frac{J}{C}\right)\left(1\ \frac{C}{s}\right)$$
$$= 1\ \frac{J}{s}$$
$$= 1\ W$$

CHECK QUESTION

A 60-W light bulb is rated at 120 V. How much current must flow at this voltage to dissipate 60 W of power?

Answer: 0.5 A

► Compound Circuits

The rules for adding resistors in series and parallel, as well as the rule for power dissipated in an electrical circuit, can be applied to a whole network of electrical components. Just start small and use the rules to build up simpler and simpler equivalent circuits until you can find the equivalent resistance of the whole circuit. Ohm's law applied at various stages of simplification will tell you the current flowing through each component as well as the voltage across it. The power rule will then yield the power dissipated by each component.

Suppose, for example, that you wish to connect two 8-Ω speakers in series to your radio so that they will produce nice loud music in your livingroom.[2] Suppose further that your neighbors have asked you to place small auxiliary speakers in their bedrooms so that they can listen to the high notes as well as the booming bass notes that inevitably come through the walls. You decide to connect these two auxiliary speakers, also 8-Ω impedance, in parallel, but you connect that circuit in series with a 12-Ω power resistor to reduce the output of the little speakers before connecting the whole mess in parallel with your livingroom speakers. You wonder how much power will be dissipated by one of the little speakers when your radio is putting out 40 W.

The first thing to do is draw a picture of the whole circuit, as in Figure 12-17a. We could then draw an equivalent circuit, Figure

---

[2] Audio speakers are rated in ohms of **impedance.** We can think of this impedance as the same thing as resistance, although it is really a generalization of the concept of resistance for circuits involving alternating current.

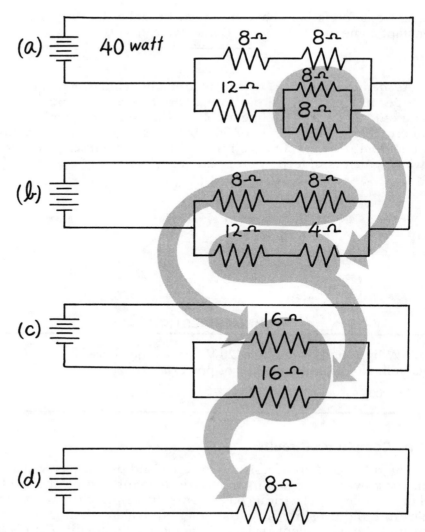

FIGURE 12-17 *Successive equivalent circuits used to analyze an arrangement of 8-Ω speakers and a 12-Ω power resistor.*

12-17*b*, in which the two 8-Ω auxiliary speakers (in parallel) are represented as a single 4-Ω resistor.

$$\frac{1}{8\ \Omega} + \frac{1}{8\ \Omega} = \frac{1}{4\ \Omega}$$

Of course, this equivalent 4-Ω resistor is still connected in series with the 12-Ω power resistor. This circuit, shown in Figure 12-17*b*, can in turn be represented by a still simpler equivalent circuit, as in Figure 12-17*c*. In this figure the rule for adding resistors in series is used to represent the two 8-Ω main speakers as a single 16-Ω resistor, and the 12-Ω power resistor and the 4-Ω equivalent resistor are represented as another 16-Ω resistor. This circuit can be further simplified by looking at the two 16-Ω branches as a single 8-Ω resistor connected to the radio, since by the rule for adding resistors in parallel

$$\frac{1}{R_t} = \frac{1}{R_1} + \frac{1}{R_2}$$

$$= \frac{1}{16\ \Omega} + \frac{1}{16\ \Omega} = \frac{2}{16\ \Omega}$$

$$R_t = \frac{16\ \Omega}{2} = 8\ \Omega$$

We can now apply the rule for power to find the current flowing through the speaker network. We recall that power is voltage times current:

$$P = VI$$

Although we don't know either the voltage or current at this point, we do know their ratio. Ohm's law tells us that voltage is current times resistance.

$$V = IR$$

We can use this relationship to eliminate voltage from our power relationship:

$$P = (IR)I$$

$$\boxed{P = I^2R}$$

Solving for current,

$$I = \sqrt{\frac{P}{R}}$$

and inserting out data, we find that the total current passing through the speaker circuit is 2.24 A.

$$
\begin{aligned}
I_t &= \sqrt{\frac{40 \text{ W}}{8 \text{ }\Omega}} \\
&= \sqrt{\frac{40 \text{ V} \cdot \text{A}}{8 \text{ V/A}}} \\
&= \sqrt{5 \text{ A}^2} \\
&= 2.24 \text{ A} \cong 2 \text{ A}
\end{aligned}
$$

At this point, we could apply Ohm's law to find the voltage across the speaker circuit and then deduce the current through each branch, but in this special case where the branches are of equal resistance it is easy to see that the current will split up into two equal 1.12-A currents in each branch. The current in the lower branch will then again split up into two 0.56-A currents through the two 8-Ω auxiliary speakers. We can then put this result back into the general solution for power we just derived in terms of current squared and resistance to get the result we were looking for, namely the power dissipated by each little speaker:

$$
\begin{aligned}
P &= I^2R \\
&= (0.56 \text{ A})^2(8 \text{ }\Omega) \\
&= (0.31)(8)(\text{A}^2)\left(\frac{\text{V}}{\text{A}}\right) \\
&= 2.5 \text{ W}
\end{aligned}
$$

The power dissipated in your neighbors' bedrooms would be somewhat less than in your livingroom. Their speakers would dissipate 5 W of sound energy, and 15 W would go into the 12-Ω resistor—20 W altogether. From the symmetry of your speaker circuit, the other half of the 40 W from the radio will go into your main speakers.

The problems in this chapter are related to electrical forces, voltage, current, resistance, and power.

The electrical force between two charges q_1 and q_2 is directly proportional to the product of the charges and inversely proportional to the square of the distance, d, between them.

$$F = k\frac{q_1 q_2}{d^2}$$

The proportionality constant k is equal to 9.0×10^9 N-m²/C².

The unit of charge, the **coulomb,** may be thought of as the charge on a very large number of electrons, namely 6.25×10^{18} electrons.

The rate of charge flow, called **current,** is measured in coulombs per second, which is the definition of the **ampere:**

$$I \equiv \frac{q}{t}$$

$$1\text{ A} \equiv \frac{1\text{ C}}{1\text{ s}}$$

Voltage may be thought of as the electrical pressure that causes current to flow through a wire or resistor. For many wires, the voltage, V, is directly proportional to the current, I:

$$V = IR$$

The proportionality constant R is called the **resistance** of the wire. The resistance of two wires connected in series is the sum of the resistances of the two separate wires:

$$R_{\text{total}} = R_1 + R_2$$

The resistance of two wires connected in parallel is the reciprocal of the sum of the reciprocals of the two separate resistances.

$$\frac{1}{R_{\text{total}}} = \frac{1}{R_1} + \frac{1}{R_2}$$

Electric power, which like all power is defined as the ratio of energy to time or the rate of doing work, turns out to be in the product of voltage, V, and current, I:

$$P = VI$$

12-1 In demonstrating the concept of electrical attraction to a friend, you inflate a rubber balloon and charge it by rubbing it against your shirt and show how the force of attraction pulls your shirt away from your body. The patter developed around this demonstration has endless variations.

Given: A force of 0.12 N is produced when the balloon is held 1.0 cm from the shirt, but this force is shown to decrease when the distance is increased to 2.0 cm to simulate a heart throb. ($k = 9.0 \times 10^9$ N-m²/C²)

Find: (a) How much electrical charge was transferred between the balloon and the shirt to produce the known force at the smaller distance.

(b) What force is produced at the larger distance?

Your Solution

Discussion

This problem illustrates the force law between electrical charges, sometimes called the coulomb force law. This force law is similar in form to Newton's law of universal gravitation. Assume that the charges are located at two points that are separated by the given distances. Charges that are distributed over the surface of the balloon and shirt would require more sophisticated mathematical techniques.

Sample Solution 12-1

(a)

$$F = k \frac{q_1 q_2}{r^2}$$

$$F = k \frac{-q^2}{r^2} \quad \text{assuming that } q = q_1 = -q_2$$

$$q = \sqrt{\frac{-Fr^2}{k}}$$

$$= \sqrt{\frac{(0.12\,N)(1\,cm)^2}{(9.0 \times 10^9\,N \cdot m^2/\,C^2)} \left(\frac{1\,m}{10^2\,cm}\right)^2}$$

$$= 3.65 \times 10^{-8}\,C \cong \boxed{3.7 \times 10^{-8}\,C}$$

(b)

$$F = k \frac{q_1 q_2}{r^2}$$

$$= (9.0 \times 10^9\,N \cdot m^2/coul) \frac{-(3.65 \times 10^{-8}\,coul)^2}{(2.0\,cm)^2} \left(\frac{10^2\,cm}{m}\right)^2$$

$$= \boxed{-3.0 \times 10^{-2}\,N}$$

Discussion

The force between two charges is directly proportional to the product of the charges and inversely proportional to the square of the distance r between them. Assuming that the two charges are equal and opposite, since the positive charge on the shirt was formed by transferring a certain number of electrons from the balloon to the shirt, the product of the two charges is $-q^2$. Part (a) is worked out by solving for q and inserting the data. Part (b) is solved by inserting the data into the force law, which happens to be expressed in terms of force, F, as stated. The negative result means that the force is attractive, since the force law is set up for the force between two positive charges, and like charges repel.

12-2 An automobile starter draws a heavy current and will soon discharge your battery if something is wrong with your engine and it doesn't start right away. The generator will recharge the battery, once the engine does start, by passing the same total amount of charge through the other direction but at a much slower rate.

Given: Your automobile is hard to start. You operate your starter for a total of 30 s (at ten-second intervals), drawing an average of 300 A. The engine finally starts and the generator replaces this charge at the rate of 15 A.

Find: (a) How much total charge flows through the starter?
(b) How long does it take the generator to replace this charge?

Your Solution

Discussion

This problem illustrates the definition of current. Most people encounter the unit of charge flow, the ampere, in their daily lives, but they do not usually think of it in terms of the charge transferred. One time when the actual amount of charge becomes important is when the charge must be replaced—as in recharging a battery.

Sample Solution 12-2

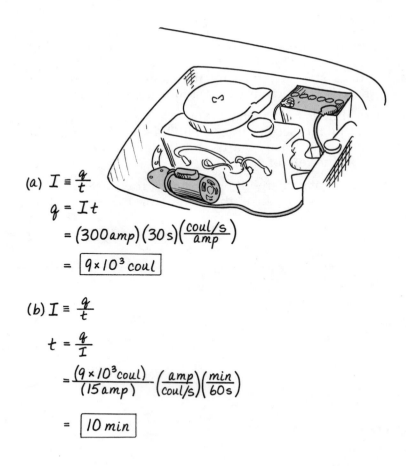

(a) $I \equiv \dfrac{q}{t}$

$q = I t$

$= (300 \, amp)(30 \, s)\left(\dfrac{coul/s}{amp}\right)$

$= \boxed{9 \times 10^3 \, coul}$

(b) $I \equiv \dfrac{q}{t}$

$t = \dfrac{q}{I}$

$= \dfrac{(9 \times 10^3 \, coul)}{(15 \, amp)}\left(\dfrac{amp}{coul/s}\right)\left(\dfrac{min}{60 \, s}\right)$

$= \boxed{10 \, min}$

Discussion

The amount of charge drawn by the starter motor in 30 s is found in part (a) by solving the definition of current for charge and inserting the given current and time. The units are straightened out by using a conversion factor based on the fact that an ampere of current is defined as a coulomb of charge per second.

Part (b) is found by solving the definition of current for time. The result is that it takes 20 times as long to charge the battery as it took to discharge it, since the generator only produces one-twentieth of the amount of current that the starter motor draws.

12-3   An unwise workman is operating an electric drill without a three-wire grounded extension cord. Fortunately, he gets a little electric shock as a warning before he works up a sweat.

Given: Dirt on an insulating surface inside the electric drill permits enough leakage current to flow so that the metal casing is at 90 V above ground. The workman's body has a resistance of $2 \times 10^4 \ \Omega$ before he works up a sweat. The current through his body is less than the bad range, which starts at about 10 mA.

Find: (a) How much current flows through his body, no sweat?
(b) At what electrical resistance does the current through his body enter the bad range?

Your Solution

Discussion

This problem is an example of Ohm's law. The voltage across the workman's body is proportional to the current passing through.

Leakage currents across dirty insulators in electric appliances are a major problem. They have killed many people. Even new appliances with three-wire cords and plugs, which provide an extra ground, can be lethal if the safety ground is removed so that the plug can be used with an older outlet. People who remove that third prong probably don't understand the magnitude of the problem.

Sample Solution 12-3

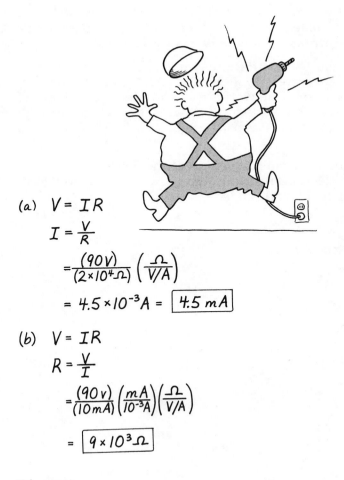

(a) $V = IR$

$I = \dfrac{V}{R}$

$= \dfrac{(90V)}{(2 \times 10^4 \Omega)} \left(\dfrac{\Omega}{V/A} \right)$

$= 4.5 \times 10^{-3} A = \boxed{4.5 \, mA}$

(b) $V = IR$

$R = \dfrac{V}{I}$

$= \dfrac{(90v)}{(10 mA)} \left(\dfrac{mA}{10^{-3}A} \right) \left(\dfrac{\Omega}{V/A} \right)$

$= \boxed{9 \times 10^3 \, \Omega}$

Discussion

The current is found in part (a) by solving Ohm's law and inserting the data. The resulting current, 4.5 mA, will give a good stout shock. Solving for resistance, as in part (b), gives a value several times greater than the actual resistance of a person's body after they have become moist with perspiration. That third wire is a life saver.

12-4 The ignition coil in an automobile is connected in series with the points, which act like a switch to interrupt the current through the coil at the proper time. Even when closed, however, the points have considerable resistance since they are made of tungsten.

Given: In a twelve-volt auto ignition system, the coil is designed to pass 3.0 A of current when the voltage across it is 10 V, with the assumption that the voltage drop across the points will be about 2 V.

Find: (a) What is the resistance of the points?
(b) What is the resistance of the coil?
(c) What is the total resistance of the circuit, including coil and points?

Your Solution

Discussion

This problem is an example of two electrical resistors connected in a series circuit. The electrical resistance of ignition points increases with age as the points become pitted by the tiny electric spark that takes place each time they are opened. The points should be changed every 20,000 miles or so for this reason.

Sample Solution 12-4

$$V = IR$$

(a) $R_p = \dfrac{V_p}{I}$

$= \dfrac{(2V)}{(3.0A)}\left(\dfrac{\Omega}{V/A}\right)$

$= \boxed{0.67\,\Omega}$

(b) $R_c = \dfrac{V_c}{I}$

$= \dfrac{(10V)}{(3.0A)}\left(\dfrac{\Omega}{V/A}\right) = \boxed{3.3\,\Omega}$

(c) $R_{total} = R_p + R_c$

$= (0.67\,\Omega) + (3.3\,\Omega) = \boxed{4\,\Omega}$

(circuit diagram: 3.0 A →, 12 V source, $V_{points} = 2V$, $R_{points} = ?$, $V_{coil} = 10V$, $R_{coil} = ?$)

Discussion

Resistance in parts (a) and (b) is found from the voltage and current, using Ohm's law. The resistance of the series circuit can be found in part (c) by using the rule for adding resistance in series. The total resistance of two resistors connected in series is the sum of the two resistances. An alternative approach would be to use the total voltage and current in the same general solution as found from part (a) and part (b).

$$R_{total} = \frac{V_{total}}{I} = \frac{12\ V}{3\ A} = 4\ \Omega$$

The result is the same.

12-5 A light fixture holds two light bulbs and is wired so that they are connected in parallel to the line voltage.

Given: A sixty watt lamp and a seventy-five-watt lamp are connected in parallel to 115 V. The sixty watt lamp passes a current of 0.52 A, and the seventy-five-watt lamp passes 0.65 A.

Find: (a) How much current is passed by both lamps together?
(b) What is the resistance of the sixty-watt lamp?
(c) What is the resistance of the seventy-five-watt lamp?
(d) What is the total resistance of both lamps together?

Your Solution

Discussion

This problem illustrates the rule for adding resistances in parallel. Instead of the voltage adding as for a series circuit, the current flowing through the two parallel paths is the sum of the current through each separate path.

Sample Solution 12-5

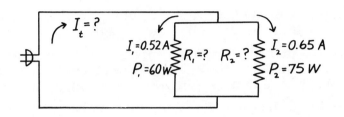

(a) $I_t = I_1 + I_2$
$= (0.52 A) + (0.65 A)$
$= \boxed{1.17 A}$

(b) $R_1 = \dfrac{V_1}{I_1}$
$= \dfrac{(115 V)}{(0.52 A)}$
$= \boxed{221 \,\Omega}$

(c) $R_2 = \dfrac{V_2}{I_2}$
$= \dfrac{(115 V)}{(0.65 A)}$
$= \boxed{177 \,\Omega}$

(d) $\dfrac{1}{R_t} = \dfrac{1}{R_1} + \dfrac{1}{R_2}$
$= \dfrac{1}{(221\,\Omega)} + \dfrac{1}{(177\,\Omega)} = 0.0102 \,\Omega^{-1}$
$R_t = \boxed{98 \,\Omega}$

Note that the answer to part (d) is the same would be given by a calculation from Ohm's Law using the total current found in part (a) $V = IR$

$$R = \frac{V}{I} = \frac{(115 V)}{(1.17A)} = 98 \,\Omega$$

Discussion

The current in part (a) is the sum of the currents in the separate branches. The resistance of the two separate paths can be calculated from the voltage and current, using Ohm's law as shown in parts (b) and (c).

The total resistance of the combined parallel circuit can be found from the rule for adding resistors in parallel. The reciprocal of the total resistance is the sum of the reciprocals of the resistances of the branches. Instead of inserting the data as the second step, as shown in part (d), a general solution could be obtained by solving for R_{total} instead of its reciprocal. The result is the sum divided by the product of the branch resistances. An alternative to using the parallel resistance rule would be to calculate the resistance from Ohm's law, using the total current as found in part (a).

$$R_{total} = \frac{V}{I_{total}} = \frac{115 \text{ V}}{1.17 \text{ A}} = 98 \ \Omega$$

12-6   You decide to have a coffee party in a dormitory room and invite as many friends as you can without blowing out a fuse when they all plug in their immersion heaters at the same time. Everyone has the same kind of immersion heater for boiling water because the campus bookstore only handles one kind.

Given: You look on your immersion heater and discover that it is rated for 300 W at 115 V. The fuse panel shows that the wall outlets in your room can handle 15 A altogether.

Find: (a) How much current does your immersion heater draw?
(b) How many friends can you invite?

Your Solution

Discussion

This problem is an example of electric power expressed in terms of voltage and current. From the power rating of an electric appliance, you can figure out the current it draws and estimate the number of appliances you can use at the same time on a particular circuit.

Sample Solution 12-6

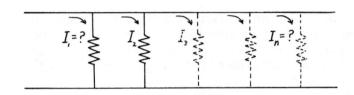

(a) $P = VI$

$I_1 = \dfrac{P_1}{V}$

$= \dfrac{(300 \text{ Watt})}{(115 \text{ V})} \left(\dfrac{V \cdot A}{\text{Watt}} \right)$

$= \boxed{2.61 \text{ A}}$

(b) $I_t = I_1 + I_2 + \cdots + I_n$

$I_t = n I_1 \quad \text{since } I_1 = I_2 = I_3 = \cdots = I_n$

$n = \dfrac{I_t}{I_1}$

$= \dfrac{(15 \text{ A})}{(2.61 \text{ A})} = 5.75$

$= \boxed{5}$

Discussion

The electric power rule states that power is the product of voltage and current. Solving for current, as in part (a), the power is divided by the voltage and the units are handled with a conversion factor based on the fact that a watt is a volt times an amp.

The total current in a parallel circuit is the sum of the currents in each branch. This parallel circuit rule is used in part (b) to calculate the number of branches, each carrying the same current, which will add up to current limit of the fuse. The total current is just the number of branches times the average current flowing in each branch. Solving for the number of branches and inserting the data shows that something over five heaters could be connected. The result must be rounded to the lowest integer, however, since six heaters will blow the fuse.

12-7   A Wheatstone bridge is a common circuit used in electrical measurements. In this bridge, two branches consisting of two series connected resistors each are connected in parallel to a current limiting resistor.

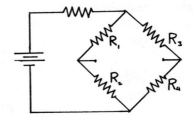

Given: $R_1 = 6\ \Omega$, $R_2 = 8\ \Omega$, $R_3 = 12\ \Omega$, $R_4 = 16\ \Omega$, $R_L = 10\ \Omega$.

Find: (a) What is the resistance of the branch containing R_1 and R_2?
   (b) What is the resistance of the branch containing R_3 and R_4?
   (c) What is the resistance of the whole circuit?

Your Solution

Discussion

This problem illustrates a compound electrical circuit in which resistors are connected in series and parallel.

In the usual application, the Wheatstone bridge is balanced by adjusting the values of resistances in the two arms so that there is no voltage between the two corners, which represent the junctions between R_1 and R_2 on the one hand and between R_3 and R_4 on the other hand. It is easy to show that the ratio of R_1 to R_2 is equal to the ratio of R_3 to R_4 if there is a null reading, or zero voltage, between these two junctions. This circuit can be used to measure an unknown resistance by putting it into a bridge with known resistances or to detect the change of resistance of a temperature-, pressure-, or light-sensitive resistor.

Sample Solution 12-7

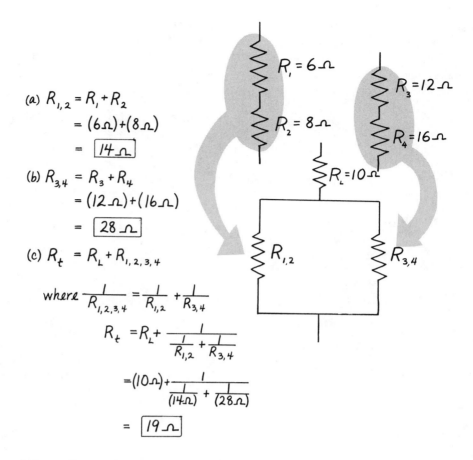

(a) $R_{1,2} = R_1 + R_2$

$\quad = (6\Omega)+(8\Omega)$

$\quad = \boxed{14\,\Omega}$

(b) $R_{3,4} = R_3 + R_4$

$\quad = (12\Omega)+(16\Omega)$

$\quad = \boxed{28\,\Omega}$

(c) $R_t = R_L + R_{1,2,3,4}$

where $\dfrac{1}{R_{1,2,3,4}} = \dfrac{1}{R_{1,2}} + \dfrac{1}{R_{3,4}}$

$R_t = R_L + \dfrac{1}{\dfrac{1}{R_{1,2}} + \dfrac{1}{R_{3,4}}}$

$\quad = (10\Omega) + \dfrac{1}{\dfrac{1}{(14\Omega)} + \dfrac{1}{(28\Omega)}}$

$\quad = \boxed{19\,\Omega}$

Discussion

The resistances of the two branches are found by using the series resistance rule, as shown in parts (a) and (b).

These two combined resistances are used in a simplified parallel circuit having as its two branches the results of parts (a) and (b). The reciprocal of the total resistance representing all four resistances, R_1, R_2, R_3, and R_4, is the sum of the reciprocals of the two branches according to the parallel resistance rule. Then this resistance, which is the reciprocal of the sum of the reciprocals, is added to the load resistance, R_L by the series resistance rule.

Even the most complicated circuit can be handled with the series and parallel resistance rule, since those are the only two ways of connecting two resistors together. Just start small and build up like building blocks.

12-8   As you are taking clothes out of the dryer, you notice a force of attraction between a cotton shirt and a polyester blouse. In fact the darn things stick together and you find it a bit of a hassle to pull them apart. You get to wondering how much force would be involved if the charge transfer between the shirt and blouse were on the order of coulombs.

Given: You are holding the shirt in one hand and the blouse in the other, both wadded up into a ball and held 1.8 m apart, which is as far apart as you can get them. You imagine that one has a charge of $+1.0$ C and the other has a charge of -1.0 C. You decide that the maximum force you could hold apart would be about 500 N. ($k = 9 \times 10^9$ N-m^2/C^2)

Find: (a) What would be the force if the charge were plus and minus a coulomb?
　　　(b) What charge, one plus and the other minus, would produce the maximum force you could withstand at that same separation?

12-9   Automobile batteries are rated in terms of the total amount of charge, flowing from one terminal to the other, that will completely discharge them. The unit used to measure this charge is the amp-hour.

Given: A typical battery, rated at 60 A-h, is charged with a trickle charger that charges at the rate of 1.0 A.

Find: (a) How many coulombs of charge can the fully charged battery cause to flow before it becomes completely discharged?
　　　(b) How long would it take the trickle charger to charge the battery back to the fully charged state?

12-10   A physics teacher is trying to demonstrate the magnetic field around a current-carrying wire. He connects a length of No. 14 gauge wire to a low voltage power supply and turns up the voltage until the wire starts to get hot.

Given: It requires 1.14 V to push 30 A through a 15-ft-long piece of No. 14 gauge wire at a temperature of 20 °C. When the wire heats up to 50 °C, its resistance increases by 4×10^{-3} Ω.

Find: (a) What is the resistance of the wire at 20 °C?
　　　(b) What is the resistance of the wire at 50 °C?
　　　(c) What will be the current passing through the wire at 50 °C assuming the voltage remains constant?

12-11   A string of Christmas tree lights is made up of a large number of low-voltage light bulbs connected in series to the power line so that the 110 V available at the wall outlet is evenly divided between the bulbs.

Given: Eleven (11) light bulbs, each designed to operate at 10 V, are connected in series to a 110-V power source. The current passing through the string is 0.15 A.

Find: (a) What is the resistance of each lamp?
(b) What is the resistance of the whole string of lamps?

12-12 The headlights in an automobile are connected in parallel. You turn on the lights on an old sports car, equipped with two headlights and an ammeter, so that you can observe the current drain on the battery.

Given: The current through both headlights together is 9.0 A when the battery is fully charged and provides a full 12 V. You estimate that the current is evenly divided between the two lights from the fact that they both seem to be about the same brightness.

Find: (a) How much current flows through each headlight?
(b) What is the resistance of each headlight?
(c) What is the resistance of the parallel circuit comprised of both headlights taken together?

12-13 Ellen Wonderful likes to watch TV while fixing breakfast. She has a TV set and toaster plugged into the same wall outlet (in parallel).

Given: Ellen looks on the toaster and notices that it draws 10.5 A. She becomes concerned that she has been operating it while the TV set is on. She therefore looks on the TV and finds that it is rated at 120 W. Both the TV and the toaster are rated for 120 V.

Find: (a) How much power does the toaster use?
(b) How much current does the TV set use?
(c) What is the current drain of the toaster and TV set together?

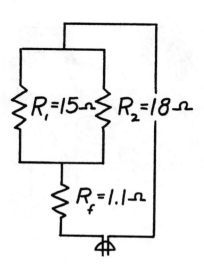

12-14 Two heating coils in an electric room heater are connected in parallel to a thermal delay fuse, which has a small but noticeable resistance.

Given: The resistance R_f of the fuse is 1.1 Ω. The current passing through the fuse is almost evenly divided between the two coils, because their resistances, 15 Ω and 18 Ω, are nearly the same.

Find: (a) What is the combined resistance of the two heating coils connected in parallel?
(b) What is the total resistance of the heating coils and the thermal delay fuse?

12-15 Jane Coolly is trying to help solve the air pollution problem by building an electric motorcycle.

Given: Jane's motorcycle runs on a 12-V battery rated at 80 A-h. Her motor draws current at 32 A. (1 kilowatt = 1.34 horsepower)

Find: (a) How long can Jane ride her motorcycle on one charging?
(b) What is the horsepower equivalent of her motor?
(c) How many joules of energy are stored in her battery?

12-16 Chlorine gas is manufactured by passing an electric current through molten salt in an electrolysis cell. It takes two electrons to produce one molecule of this diatomic gas.

Given: A current of 100 A is passed through an electrolysis cell in the production of chlorine. One mole of this gas occupies 2.24 liters at atmospheric pressure and contains Avogadro's number, 6.02×10^{23} molecules per mole, of molecules. Each electron of the current passed through the cell contains a charge of 1.60×10^{-19} C.

Find: (a) How many molecules are in 1 liter of chlorine at atmospheric pressure?
(b) How much electrical charge must pass through the electrolysis cell to produce 1 liter of gas?
(c) How long does it take to produce 1 liter of chlorine?

12-17 One deterrent to burglary is to leave your front porch light on all the time. Some people don't like to do this because of the cost of the electrical energy.

Given: Your front porch light contains a 60-W bulb. You look on your electric bill and find that you were charged $10.53 for 356 kWh of electrical energy one month (30 days).

Find: (a) How much did 1 kWh of energy cost you that month?
(b) How many kilowatt-hours would a 60-W bulb use in a month?
(c) How much would it cost you to leave the bulb on for the whole month?

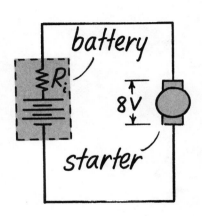

12-18 The voltage provided by an automobile battery decreases when the starter is operated because of the internal resistance of the battery. The more the battery is run down, the greater is the internal resistance. Sometimes the voltage drop across the internal resistance of the battery is so great that there is not enough left to make the ignition coil give a good spark, even though there is enough to turn the engine over.

Given: The voltage of a particular automobile battery drops from 12 to 8 V while the starter is drawing 200 A of current.

Find: (a) What is the effective resistance of the starter motor?
(b) What is the internal resistance of the battery?

12-19 A standard electric light bulb rated at 100 W costs 45 cents when purchased in a package of four. The package lists the average light output and the average life of the bulb to help you decide if you should purchase the more expensive long-life bulbs.

Given: The standard bulbs are rated at 100 W, assuming a voltage of 120 V. The average life of 750 hours listed on the package assumes the same voltages and is considerably extended if your line voltage is lower, say 115 V. The cost of the electrical energy to run the bulb is 3.0 cents/kWh.

Find: (a) What is the resistance of the bulb when operated at 120 V?
  (b) If the resistance of the bulb did not decrease, what would be the wattage of the bulb when operated at 115 V?
  (c) How much will it cost to operate the bulb for its rated life?

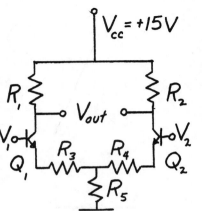

12-20 A differential amplifier is a circuit that looks a lot like a Wheatstone bridge (see Problem 12-7) except that two transistors, Q_1 and Q_2, form the resistances in two arms of the bridge. The resistance of these transistors depends on the voltages V_1 and V_2 applied to their bases at the terminals as shown. A small difference in these voltages causes a larger voltage, V_{out}, to appear between the output terminals.

Given: $R_1 = R_2 = 440\ \Omega$, $R_3 = R_4 = 50\ \Omega$, $R_5 = 120\ \Omega$. The voltages applied to the bases of the transistors give them the following resistances: $R_{Q1} = 200\ \Omega$, $R_{Q2} = 300\ \Omega$. The voltage between the supply terminal V_{cc} and ground (shown as three horizontal lines) is 15 V.

Find: (a) What is the resistance of the whole circuit? (*Hint:* Redraw diagram combining resistances to simplify according to the rules for adding resistors in series and parallel.)
  (b) How much voltage is across R_5? (*Hint:* Calculate current through the whole circuit.)
  (c) What is the output voltage, V_{out}? (*Hint:* Calculate the current through each branch and use this to find the voltage across the two resistors, R_1 and R_2.)

ANSWERS

12-8 (a) -2.8×10^9 N; (b) 4.2×10^{-4} C.

12-9 (a) 2.16×10^5 C $\cong 2.2 \times 10^5$ C; (b) 60 hours.

12-10 (a) 3.8×10^{-2} Ω; (b) 4.2×10^{-2} Ω; (c) 27 A.

12-11 (a) 67 Ω; (b) 7.3×10^2.

12-12 (a) 4.5 A; (b) 2.7 Ω; (c) 1.3 Ω.

12-13 (a) 1,260 W; (b) 1.0 A; (c) 11.5 A.

12-14 (a) 8.2 Ω; (b) 9.3 Ω.

12-15 (a) 2.5 hours; (b) 0.51 hp; (c) 3.5×10^6 J,

12-16 (a) 2.68×10^{22} molecules; (b) 8.60×10^3 C; (c) 86 s.

12-17 (a) 2.96 cents; (b) 43.2 kWh; (c) $1.28.

12-18 (a) 4×10^{-2} Ω; (b) 2×10^{-2} Ω.

12-19 (a) 144 Ω; (b) 92 W; (c) $2.25.

12-20 (a) 488 Ω; (b) 3.7 V; (c) 0.91 V.

Modern Physics

▶ Something Old and Something New

Classical physics is frequently called Newtonian physics, because it is based on the laws of motion laid down by Newton. It generally refers to that body of physics that was fairly well understood by the turn of this century. At about that time, some new developments came to light that led to relativity on the one hand and atomic and nuclear physics on the other. These topics are generally referred to as *modern physics,* which is somewhat misleading, since it implies that there is a replacement for the old-fashioned kind. This is not the case. Modern physics supplements our classical understanding, adding depth and extending it into the realms of the very fast and the very tiny. At the boundaries, however, the old and the new match perfectly, leaving both intact in their own right.

We shall end this book by probing these boundaries—first in the realm of the very fast, to see how the rules of relativity are consistent with classical physics, and then in the realm of the very small, to take a look at the atom and its nucleus. Should the physics of these realms seem strange, it is because they are not a part of your everyday experience. You must stretch your imagination to understand what is right and proper in these realms of existence.

467

► The Special Theory of Relativity

Our present understanding of things that move at velocities near the speed of light is based on a theory called the *special theory of relativity*. Some things don't happen exactly as you might expect at these speeds. The laws of classical physics, which work so well at ordinary velocities, clearly need some modification as one enters the realm of the very fast. For one thing, the laws of motion must be modified to account for nature's speed limit. We have found that nothing can be accelerated through the speed of light. Try as we may, we just can't get anything to go any faster. In fact, we cannot even get bits of matter, regardless of how tiny, up to a speed quite as great as the speed of light. The laws of classical physics do not predict this speed limit, and they need to be modified to account for it.

To understand how things should behave at high velocities, imagine that you are going nearly that fast yourself. The theory of relativity is largely based on how things look from a moving **frame of reference.** A frame of reference is an arbitrary set of axes that you use to judge the position or velocity of something. Let us examine this concept of a frame of reference to see how you already use it in your thinking.

Most people usually select a frame of reference in which most of the things around them seem stationary. Their minds make this selection so automatically that they seldom stop to question the process. For example, you might be sitting at home at your desk, and your frame of reference would be selected so that your chair, desk, bookcase, walls, ceiling, as well as most of the other objects in the room, would seem stationary. Even if you got up and walked across the room, you would probably keep using the same frame of reference. You would likely think of yourself as moving and everything else as stationary. The idea that they are actually stationary would probably be so firmly established in your mind that you would not even give it a second thought. When an earthquake strikes, people frequently panic because of the sudden and unexpected loss of their stationary frame of reference. Most people find it very disquieting to have everything upon which they judge their bearings suddenly start to move.

On the other hand, people sometimes select a frame of reference that they know full well is moving but that still makes things around them seem stationary. Commuters gathered around the bar in the club car of a train are aware of the fact that the bar, their drinks, and the people standing around them are all moving, but they still judge position and velocity relative to this moving frame of reference. Two friends standing at the bar might match coins to see who pay for the next round of drinks. They flip the coins into the air, catch them, and slap the coins onto the backs of their hands in the time-honored gambling ritual. If you asked them to stop and think about it, they would agree that the "true path" of the coins was an arc, as the train carries both them and the coins down the track at the same time that the coins go up into the air and back down to their hands. If you don't remind them of the motion of the train, however, they would intuitively use a moving frame of reference in which the bar, their drinks, and the club car furniture all seem to be stationary.

The special theory of relativity is based on the assumption that the laws of physics all work the same in a moving frame of reference without regard to the velocity of that frame of reference as long as the

FIGURE 13-1 *Commuters in a club car understand relativity, at least in a limited sense.*

velocity is constant. That is just what the commuters in the club car have learned to expect, at least insofar as it applies to mechanical things around them. The coins flipped into the air go through the same motion, as far as the gamblers are concerned, without regard to the speed with which the world is flashing by outside. The bartender pours liquor the same without glancing out the window to check on the velocity of the train. Only when the train lurches or sways, would the motion of the train affect the difficulty with which people stagger across the floor.

Albert Einstein was the first to suggest that the same principle of relativity that applies to mechanical things, such as coins flipped into the air and people staggering across the floor, would apply as well to electrical and magnetic phenomena, even including the behavior of light. He saw no reason why some laws of physics, some forms of experience, would look different in a moving frame of reference if others do not. He therefore took it as a basic postulate that all forms of experience, even experiments to measure the speed of light, are relative to the frame of reference you use and would work the same in any uniformly moving frame of reference. This means that the laws of physics are no better suited to one frame of reference than to any other.

This assumption forms the basis of the *special* theory of relativity, so called because it is restricted to the special case of frames of reference moving at a constant velocity—cases in which there is no acceleration. This is the principle of relativity that applies to the train moving smoothly down the track without lurching or swaying. In this special case, the assumption is that the motion of the reference frame has *no effect* on the laws of physics as long as there is no acceleration.

After you have fully understood the special theory of relativity, you will be in a position to appreciate the **general theory of relativity,** which applies even to frames of reference that accelerate. This is the theory that applies to the train as it lurches or sways. The general theory of relativity is based on the assumption that the effects of acceleration are the same as a change in the gravitational field. For now, however, we will not get into this theory of gravitation. It will be enough to understand something about the structure of space and time implied by the special theory. All we will do in this book will be to introduce some of the basic ideas of relativity, so that you can start to think about some of the puzzles they contain.

▶ The Structure of Space and Time

Einstein was the first to realize that the relativity experienced by the commuters gathered around the bar in the club car of the moving train is even more important than the mental picture they have formed of space and time, based on what they probably feel is common sense. Common sense tells them, for example, that they are really moving and that the world outside their window is really stationary. If they wanted to go even faster, they could take a walk toward the front of the train. Common sense tells them that their resultant velocity would be the sum of a person's walking speed and the speed of the train. This simple rule for adding velocities assumes that everybody agrees on their measurements of space and time. Common sense tells them that time is independent of space and that

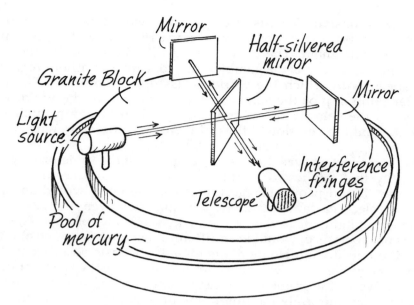

FIGURE 13-2 *The Michelson interferometer splits up a beam of light and compares the light paths by recombining the light to produce fringes of constructive and destructive interference of the light waves.*

any two observers who see two events simultaneously will agree that they happened at the same time. These common-sense assumptions are based on common experience with things moving at ordinary velocities. Einstein realized that these common-sense assumptions would be modified if we lived our lives at velocities nearer to the speed of light.

Take the simple rule of adding velocities, for example. If the train were going at near the speed of light, say nine-tenths the speed of light, and you were to walk forward in the train at two-tenths the speed of light, the common-sense rule for adding velocities would state that your actual velocity would be eleven-tenths the speed of light. Common sense therefore seems to tell you that you can go faster than the speed of light. Since you can't, common sense must be in need of some fixing up.

Even the notion that you have an "actual" velocity assumes that one frame of reference is better than all the rest. You may like to think that the world outside your train window is stationary and that you are moving. Of course, the world isn't really stationary; it is circling the sun. Even the sun isn't stationary. We know that the sun is a member of a galaxy that is spinning on its axis. Even the galaxy is moving through space. Common sense would seem to suggest that the sum of all of these motions should add up to something we could measure. People have looked long and hard for some sort of absolute frame of reference by which we could measure how fast we are "really" going. There just doesn't seem to be any.

One of the most famous experiments in the history of physics was an experiment to measure the velocity of the earth through the universe by using the speed of light as a reference. In 1887, Albert Michelson teamed up with a young chemist named Edward Morley to build a big interferometer, which could compare the velocity of light in two perpendicular paths. Michelson had invented this interferometer and found it useful for measuring tiny distances, even to a fraction of a wavelength of light.[1] Light is split into two perpendic-

---

[1] The Michelson interferometer is still used as one of the finest measures of distance. Our standards of length, in fact, are now based on the wavelength of light as measured by this instrument.

ular paths by a half-silvered mirror and then reflected back on itself by two mirrors at the end of each path. When the light in these two paths gets back to the half-silvered mirror, it forms light and dark fringes of constructive and destructive interference, depending on whether the light waves reinforce or cancel each other. Michelson realized that the wavelength in these two paths depends on the velocity of light. He wondered if the motion of the earth around the sun wouldn't affect the apparent velocity of light in one direction more than the other. It was a simple matter to point one arm of his interferometer in the direction he knew the earth was going, note the position of the fringes, and then turn the apparatus around so that the other arm of the interferometer was pointed in that direction. Any shift in the fringe pattern would indicate a difference in the apparent velocity of light in these two perpendicular directions. But if there was any shift in the fringe pattern due to the velocity of the earth, he discovered, it was too small to measure with any of the interferometers he had built to that date. That was when he teamed up with Morley to build a bigger interferometer. They mounted the mirrors and everything on a large granite block, so that the distances would be as stable as possible, and then they floated the whole thing in a large pool of mercury so that there wouldn't be any chance that the granite block would bend in the slightest as they turned it about a vertical axis.

Michelson and Morley thought that they would get at least some change in the fringe pattern due to the motion of the earth. It seemed to them that light in one path would have to chase after the mirror at the end of the path, which would be receding at the velocity of the earth, until it finally is reflected and returns to the half-silvered mirror, which the light beam would now see as coming toward it—also at the velocity of the earth. Light in the other path, meanwhile, would have to take a slightly diagonal path to get back to the presumably moving half-silvered mirror. The net effect should have been small but measurable if only the motion of the earth around the sun was to be taken into account, but they thought they might see other motions as well.

In order to get the path lengths long enough to measure the tiny change they predicted, Michelson and Morley's interferometer became so sensitive that you couldn't even stand in the same room with it. The heat of your body would warm one side of the granite block and cause a noticeable shift in the fringe pattern due to thermal expansion. Michelson and Morley had to stand outside the room and manipulate the block with strings and pulleys to turn it on its axis so as to exchange the roles of the two paths. But they finally got all of these problems worked out, and then they tried it.

Much to their surprise, nothing happened. Turning the block on its axis didn't change the fringe pattern even a tiny bit. It was just as though the earth weren't moving. This was one of the most famous experiments in the history of physics, and its importance lies in the fact that it failed.

For years, people tried to explain the failure of the Michelson-Morley experiment in terms of some change in the length of the apparatus. The motion of the granite block through space was supposed to push the electron orbits out of shape or something, and that would make the granite block get smaller in the direction it was going. This change in length was supposed to be just the right amount to compensate for the change in the apparent velocity of light.

All sorts of people started getting interested in this problem about the turn of the century, and they all tried different ways to measure the velocity of the earth through space. All attempts failed. The theories became more complicated as they tried to explain away the failure of each fresh approach. It began to look as if all the laws of nature had entered into some kind of conspiracy to prevent these people from measuring the absolute velocity of the earth.

Then in 1905, a young unknown Swiss patent examiner, Albert Einstein, published a paper that swept away all of these contrived difficulties by suggesting a whole new theoretical approach. Instead of assuming that the motion of the earth made Michelson and Morley's granite block contract, Einstein suggested that the contraction was a property of space itself. Before that, no one had thought to question the common-sense view that space and time were absolute quantities, somehow fundamental constants in all the universe. Einstein saw space and time as things to be measured. He reasoned that if every method you can think of for measuring the distance between two points gives the same result, then that is the distance that actually exists as far as you are concerned. Similarly, if you measure the time between two events, then that is the time that actually exists.

It is true that someone who was stationary with respect to the sun would get a different result if he were to measure the distance between the mirrors mounted on Michelson's granite block. This hypothetical observer would see the earth and everything on it as being in motion. To that person, it would seem that any distance measured by Michelson and Morley along the forward direction of their travel would be foreshortened by a certain factor, depending on their velocity. As we shall see, that factor turns out to be the square root of 1 minus the ratio of velocity, v, squared over the velocity of light, c, squared:

$$\sqrt{1 - \frac{v^2}{c^2}}$$

This factor is always less than 1 as long as v is something between zero and c. Thus, the person who is stationary with respect to the sun would see the length l', measured in Michelson and Morley's moving frame of reference, as being shorter than his own length, l, by an amount that would be found by multiplying by this factor:

$$l' = l\sqrt{1 - \frac{v^2}{c^2}}$$

This does not mean that the hypothetical observer's measurements are any better than Michelson's. In fact, Michelson would see the hypothetical person's measurements as foreshortened in exactly the same way. From Michelson and Morley's point of view, the hypothetical observer is the one who is moving. The motion of the earth would make a person who is stationary with respect to the sun seem to be moving in the opposite direction. Instead of agreeing with that person's view of space, the Michelson and Morley team would see any distances measured by him as being shorter by exactly the same factor. Strange as it may seem, each sees the other's distances as foreshortened—and they are both right.

The resolution of this seeming paradox comes from the fact that both parties also measure time differently. The person who is sta-

FIGURE 13-3 *A hypothetical observer who is stationary with respect to the sun would see the distance between the interferometer mirrors mounted on Michelson and Morley's granite block as being foreshortened by the motion of the apparatus.*

tionary with respect to the sun sees Michelson's clocks as running slow. If that person measures the time, t, between any two events, he will see Michelson as measuring a longer time t', between the same two events. He will see Michelson's time measurements as lengthened or dilated by the same factor by which he sees Michelson's space measurements foreshortened. Just as you multiply by this factor to get a shorter distance, you divide by it to get a longer time:

$$t' = \frac{t}{\sqrt{1 - \dfrac{v^2}{c^2}}}$$

Michelson would also see the hypothetical person's time measurements as dilated in exactly the same way. To Michelson, that person is moving, and moving clocks, as we will see, always seem to run slow. Just as each person sees the other's distances as foreshortened, they also see the other's time as expanded.

These two changes work together to bring each observer's view of things back into agreement. If only one effect occurred, one frame of reference would have to be better than the other. As we will see by considering several examples, however, the fact that both effects occur allows both observers to see the same events happening and agree on such things as cause and effect even if they see the events from different perspectives of space and time.

The truly remarkable feature of the theory of relativity is that it points out a whole new structure of space and time. Instead of being absolute things, tied to some fundamental frame of reference, they are relative things, which can be seen equally well from different perspectives. Instead of being independent things, as time and space might seem to be if viewed only from one frame of reference, they are closely related concepts, so that a new perspective from a different frame of reference shows you more of one and less of the other.

This relativistic structure of space and time differs in several respects from our common-sense notion of the way things are. Consider, for example, our common-sense notion of simultaneity. We think of two events as being simultaneous when we see them happen at the same time. Just as it seems natural to think in terms of an absolute frame of reference in space, it is also natural to think of that frame of reference as being absolute in time. We might like to think that our idea of "at the same time" is an absolute. Just as we might like to think of ourselves as being the center of the universe in space, we might also like to believe our zero point in time is better than everyone else's. From the relativistic point of view, however, the idea of any absolute frame of reference is only a simplified mental model, too simple, in fact, to deal with reality. Just as our spatial reference frame is a mental contrivance so is our notion of simultaneity. If two things happen at different places but at the same time as you see it, they might well seem to happen at different times as another observer would see it. If he is moving relative to you, he will have a different perspective of what is meant by "at the same time."

Let us return to the Michelson-Morley experiment as an illustration. As far as Michelson and Morley were concerned—or at least as far as they could tell from their experiment—the light beam was split up to travel two equal paths and be reflected back simultaneously by the mirrors at the ends of their paths. The interference pattern produced when the two beams returned indicated that this is exactly what did happen. To our hypothetical observer, who is stationary with respect to the sun, however, the two light beams would not bounce off the two mirrors simultaneously. The two paths would not be the same length, and the speed of light would be the same in both directions only in *his* frame of reference. He would, however, agree with Michelson and Morley on the final result. He would see these two effects as ganging up to produce the same fringe pattern.

At this point, you might well wonder who is "correct." As we consider examples in relativity, you will find that you have a strong tendency to look at the observers in moving frames of reference as being somehow mistaken in interpreting what they see. It might seem that these people should realize that *they* are the ones who are moving and that they should take that into account when they measure time. The fact is that Michelson and Morley thought that the hypothetical observer had a better view of reality than they did, since he would be more stationary. When they tried to partly compensate in their calculations for their own motion, however, they came out with a result that failed to agree with the experiment. That was because they didn't understand the unified structure of space and time. Einstein's contribution to our understanding of that structure made it unnecessary for either observer to compensate for his own motion.

If you find it difficult to put aside your common-sense view of space and time, think what it must have been like for people in Einstein's day. When his paper was published in 1905, there was a huge wave of protest. If you look under "relativity" in the card catalog of the Library of Congress, you will find that nearly half the books written on the topic are intended to show why it would never work. All these books now sit forgotten on the shelf, of historical interest only. Even in 1921, 16 years after Einstein's original paper on relativity was published, the topic was so controversial that it wasn't directly mentioned in Einstein's Nobel prize presentation.

One reason that our common-sense view of space and time has such a strong grasp on our thinking is that we have little direct experience with frames of reference that go fast enough for relativity to make a noticeable difference. You have to be going at near the speed of light relative to another observer before what you see differs much from what the observer sees. These velocities lie so far beyond what is technically possible for people to travel, even today, that it requires a stretch of the imagination even to think about going that fast.

▶ A Thought Experiment at Near the Speed of Light

People of today have a cultural advantage over Einstein's contemporaries when it comes to thinking about frames of reference moving at high velocities. People of Einstein's day had plenty of experience with moving frames of reference on trains and the like, but the intelligentsia had a low regard for science fiction. Science fiction had not yet developed into an accepted art form. Today, however, sci-fi fans have little trouble imagining what it would be like to go at speeds near the speed of light.

To get an idea of why length is contracted in a moving frame of reference, let us imagine that a spaceship has been assigned the task of making some space surveys at velocities below the speed of light. Two special survey ships have been outfitted to make careful angular measurements on far-distant objects. Captain Epstein puts Officer Stern in charge of one of these vessels and Officer Bower in charge of the other. These officers are directed to keep their survey craft exactly 1 light-second in front of and behind the mother craft. (One light-second, the distance traveled by light in 1 s, is less than the distance from the earth to the moon.) Mr. Stern and Ms. Bower position their ships and set their instruments to measure the far-distant objects of interest. While the computer on board the mainship is comparing the readings reported from the two survey craft, Captain Epstein directs the helm officer to get underway at one-tenth the velocity of light, $0.1c$. Lieutenant Antonelli, the communications officer, signals Stern and Bower to get underway at this velocity. They both get the message at the same time and simultaneously increase their speed to $0.1c$.

Let us use some advanced science fiction technology to simplify for our imagination the problems involved in this little speed adjustment. Given the physics we understand today, it would take a little over a month to attain the velocity we are talking about at the comfortable acceleration of 9.8 m/s^2. Even at ten g of acceleration, it would take over 3 days. Using sci-fi technology, however, we can pretend that someone has been clever enough to figure out a way to do it in a few moments. It is important to ignore what goes on during the acceleration, because we are now considering only the special theory of relativity, the theory of reference frames moving at a uniform constant velocity.

Suppose we are observing this whole operation from a space station that remains in the original frame of reference. We make a graph with the position of the mainship and the two survey craft on one axis and time on the other. If the mainship started out from our starbase, the graph of its position before it got underway would be a straight line along the time axis, representing no change in position.

If we could use the vertical axis to represent time, Mr. Stern and Ms. Bower will both have positions that are shown as vertical lines before they get their first signal to speed up.

We could choose any convenient scale to represent distance and time; let us measure distance in light-seconds, so that light signals travel upward and outward at a 45° line, covering equal distances on both scales of our graph. This choice of scale tends to put time and distance on an equal footing in our thinking.

We can see that Bower and Stern get the signal to get underway simultaneously by the fact that the first two 45° lines cross Ms. Bower's and Mr. Stern's paths at the same height on the time axis. At that point, the paths bend to be inclined at a certain angle to the space axis, so that their velocity can be represented as progress along the space axis with increasing time.

We need to assume that the helm officer on the mainship delays execution of the captain's command for 1 s to allow that command to reach the survey craft so that they can all start up simultaneously. The path representing the mainship also moves in the vertical direction and bends at the same angle at the same time as the other two paths. That way, the mainship stays halfway between the two survey craft.

We can now see how distances get foreshortened in a moving frame of reference. If Captain Epstein orders both craft to speed up

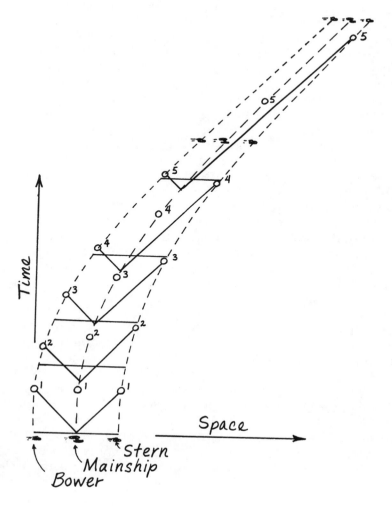

FIGURE 13-4 *A space-time diagram shows that the distance between Ms. Bower and Stern gets smaller, because Bower always gets the signal to speed up sooner than Mr. Stern.*

again, his order will reach Bower before it reaches Stern, at least in our frame of reference. We see Ms. Bower traveling toward the command ship and thus intercepting the signal before it has a chance to catch up to Mr. Stern. On our graph this would be represented by the 45° signal path intercepting Ms. Bower's path earlier, or lower on the time axis, than Mr. Stern's, since both craft's paths are inclined to the right. Each time the captain orders an increase in speed, Bower gets the order first and speeds up sooner than Stern. The distance between them therefore decreases the faster they go.

The same argument for the distance between ships can be made for the ships themselves and any material object on board the ships. Everything is squeezed down in the direction of travel. We can stand outside on our space station and say that this squeezing down is caused by the length of time it takes to get information from one end of an object to the other about how fast it is supposed to be going, but we could not expect people on board the moving spacecraft even to notice this squeezing effect, since it happens uniformly on everything around them. Everything, even a meter stick that they might use to measure things with, is contracted in length by the same amount.

The people on the spacecraft seem to us to have a warped perspective of time as well as space. Notice that the helmsman on the mainship must always wait for a certain length of time before increasing the speed of the mother craft if he is to keep the command ship halfway between the two survey craft. If we mark with numbered circles the points on our graph where the survey craft receive their orders and speed up, we can see that the proper time for the helmsman to increase the speed of the mainship is at a point halfway between the other two points along a straight line connecting them. The fact is that the people on the spacecraft see these points as simultaneous in their frame of reference. The helmsman just increases the speed of the mother craft at the time when he thinks both Ms. Bower and Mr. Stern are increasing theirs. The distance between ships may seem shorter to us; but to those people in the moving frame of reference, it seems that the two survey craft are exactly 1 light-second from the mother craft. The helmsman therefore waits for exactly 1 s of his time and then increases speed. It may seem longer to us, but that is because we think his watch is running slow.

▶ The Time Dilation

It is easy to see why a moving timepiece would run slow. The usual trick is to think in terms of a **light clock,** a clock that uses the speed of light as a standard for measuring time. Put two parallel mirrors on the ends of a meter stick and send a pulse of light bouncing back and forth between them. If we assume that the speed of light, 3.00×10^8 m/s, is a constant in any frame of reference, the amount of time for light to travel from one mirror to the other 1 m away would be $(1/3.00) \times 10^{-8}$ s, at least in that frame of reference. We might even fix the apparatus up to be used as a clock by putting only a thin layer of silver on the mirrors, so that some of the light gets through to photocells mounted behind the mirrors. One photocell would be connected to an amplifier that says "tick" every time the light pulse bounces from the mirror. The other photocell would, of course, be connected to an amplifier that says "tock."

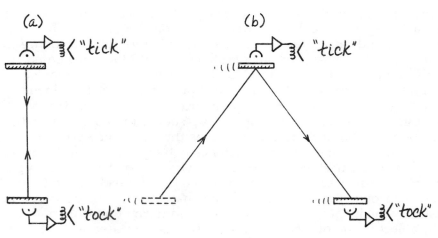

FIGURE 13-5 (a) A stationary light clock would measure time based on the speed of light and the distance between two mirrors. (b) A moving light clock would run slower, because the distance traveled by the light would need to be longer.

We could now define a standard time interval as the time between a "tick" and a "tock" of our light clock. Using the definition of velocity,

$$v \equiv \frac{d}{t}$$

solved for distance,

$$d = vt$$

we can relate the lengths of the light paths, as seen from two different points of view, to the time intervals as seen in those frames of reference. For a stationary clock, or at least one that you see as stationary, the time interval, t, would be defined in terms of the perpendicular distance between the mirrors and the velocity of light, c, as it travels between these mirrors. If the clock were moving, however, you would see the time interval as being longer because the distance the light would have to travel would be longer, as you see it. This longer time interval, t', would be defined in terms of the velocity of light, c, and the hypotenuse of a right triangle having as one leg the distance you originally used for a stationary clock. In fact, the moving clock runs slow by exactly the same proportion that the hypotenuse of this right triangle is greater than the leg that represents the distance between the two mirrors.

Let us label the sides of a right triangle in Figure 13-6, using a prime to indicate the time t' as measured by the moving clock and no prime to indicate the time t as measured by the stationary clock. In this diagram, the hypotenuse is the velocity of light times the primed time:

$$\text{Hypotenuse} = ct'$$

while the side leg is the velocity of light times the unprimed time:

$$\text{Side leg} = ct$$

As you see it, the hypotenuse is longer than the side leg because the mirrors moved a distance that is represented by the top leg of the triangle:

$$\text{Top leg} = vt'$$

The distance the mirrors move is their velocity times the primed time, t', which is the time as measured in the same frame of reference in which the light seems to travel along the hypotenuse of the right triangle.

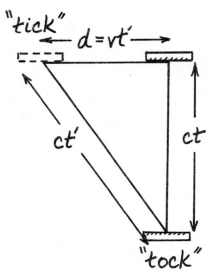

FIGURE 13-6 We can relate the time measured by a moving clock to that measured by a stationary clock by using the Pythagorean theorem to relate the distance between the mirrors of a light clock to the distance we see the light travel while we see the mirrors move in our frame of reference.

We can now apply the Pythagorean theorem to show that the times as measured in these two frames of reference are related by the square of the hypotenuse being equal to the sum of the squares of the two sides:

$$(ct')^2 = (ct)^2 + (vt')^2$$

We can get the primed values of time on one side of the equation by subtracting $(vt')^2 = v^2(t')^2$ from both sides of the equation:

$$c^2(t')^2 - v^2(t')^2 = c^2t^2$$

This can be solved for t' by factoring $(t')^2$ out of the two terms on the left and then isolating it on one side of the equation and taking the square root. It is traditional, however, to divide both sides of the equation by c^2 first;

$$\frac{(t')^2(c^2 - v^2)}{c^2} = t^2$$

simplify by dividing c^2 into both terms of the numerator, and then take the square root of both sides

$$t'\sqrt{1 - \frac{v^2}{c^2}} = t$$

before solving for t'

$$t' = \frac{t}{\sqrt{1 - \frac{v^2}{c^2}}}$$

This relationship is called the **relativistic time dilation.** The theory of relativity is based on the assumption that not only do moving clocks run slow but that time itself runs slow in a moving frame of reference and that length is foreshortened by the same factor.

▶ The Relativistic Point of View

We have seen how the measurement of length might be distorted by the foreshortening of a moving meter stick, and we have seen how the measurement of time might be distorted by the slow running of a moving light clock. Neither of these two effects taken alone require the theory of relativity to understand. In fact, the relationships for length foreshortening and time dilation were known before Einstein proposed the theory of relativity, but people thought that these relationships applied only to the distortion of physical objects. They thought that the distortions of their measuring instruments seemed to conspire to make it difficult to measure their velocity relative to some fundamental frame of reference. Einstein was the first to suggest that these were more than experimental difficulties, that they were actually theoretical impossibilities. He suggested that the real reason that you couldn't measure your velocity relative to some fundamental frame of reference was that no fundamental frame of reference exists. Einstein realized that if the absence of a fundamental frame of reference were a basic fact of nature, the length foreshortening and time dilation are more than independent effects resulting from the distortion of the physical objects used to make the measurements. They become two related effects, which apply to the structure of space and time itself. If there is no such thing as an absolute frame of reference, then the structure of space and time must be such that all measurements of any physical phenomena, including the velocity of light, must come out the same in any frame of reference. Einstein's genius was to realize that the length contraction and time dilation were two aspects of a single phenomenon and that they fit together in just the right way to suggest this new view of the structure of space and time.

It is possible to get pretty well into the study of relativity and still cling unconsciously to the idea of a fundamental frame of reference and an absolute time permeating the universe. You might, for example, agree that Mr. Stern's survey craft is closer to Ms. Bower's and that they have difficulty measuring this fact because their clocks are running slow, but it might still seem that Stern should know that this is the case. Mr. Stern is a very smart guy; he should know he is the one who is moving. It is true that Stern would remember having accelerated from our frame of reference into the one he is in now. Also, he would know that his frame of reference is moving relative to us. But Mr. Stern might think of *us* as moving relative to *him*. As long as he remains at a constant velocity, his point of view is just as valid as ours. From his frame of reference, *our* lengths are foreshortened and *our* clocks run slow.

It seems a bit strange at first that observers moving relative to each other would both observe the same changes in space and time in the other's frame of reference. If we could look out into the universe and find another race of people like ourselves, except that they are moving at near the speed of light, we would see them living their lives in slow motion. If they were moving fast enough, their life spans might well be several times our own. They, on the other hand, would see us living in slow motion. We would live, as they see it, several times as long as they. Each race would see the other as having a longer life span. How could they both be right? If we considered only the time dilation, or only the length foreshortening, this would be a paradox. Both time and distance must be taken into account if we are to understand how people moving relative to us make measurements that differ in both respects but agree in the final result.

To see how a different view of space in a moving frame of reference must be accompanied by a different view of time, let us return to our example of the mainship and the two survey craft. Suppose the two survey craft are to fly past two space buoys in the course of their mission, and suppose they are to maintain a constant velocity of $0.8c$ (eight-tenths the speed of light). The buoys are 1 light-hour apart, as we see them. From our point of view, the mainship would take a little longer than 1 hour to travel this distance at less than the speed of light. In fact, we could easily calculate the time from the definition of velocity. Solving the definition of velocity for time,

$$v \equiv \frac{d}{t}$$

$$t = \frac{d}{v}$$

and setting the velocity equal to $0.8c$ while the distance is c times 1.00 hour,

$$t = \frac{(c)(1.00 \text{ hour})}{0.8c}$$

we see the mainship taking 1.25 hours to travel from one buoy to the next:

$$t = 1.25 \text{ hours}$$

Although we will measure 1.25 hours for the starship to travel 1.00 light-hour at eight-tenths the speed of light, we do not expect its crew to measure that much time. Their clocks are running slow. The time dilation in their frame of reference makes them measure less time than we measure for the same events. Their time as we see it is t', while their time as they see it is t. Solving the time dilation relationship for the time as they see it,

$$t' = \frac{t}{\sqrt{1 - \dfrac{v^2}{c^2}}}$$

$$t = t' \sqrt{1 - \frac{v^2}{c^2}}$$

At eight-tenths the speed of light, their clocks are running so slow that they only measure 1.25 hours, as we see it, as three-quarters of an hour:

$$t = (1.25 \text{ hours}) \sqrt{1 - \frac{(0.8c)^2}{c^2}}$$
$$= (1.25 \text{ hours})\sqrt{1 - 0.64}$$
$$= (1.25 \text{ hours})\sqrt{0.36}$$
$$= (1.25 \text{ hours})(0.6)$$
$$= 0.75 \text{ hour}$$

As it turns out, the crew on the mainship agrees that the time they measure between passing one buoy and the next is 0.75 hour, but they do not see this time measurement as resulting from their clock running slow. To them, the shorter time is a natural result of a shorter distance. In their frame of reference, they are stationary and the buoys approach and pass them at $0.8c$. The distance between the buoys, however, is foreshortened from their point of view. In their frame of reference, the distance between the space buoys is only 0.6 light-hour.

$$l = l'\sqrt{1 - \frac{v^2}{c^2}}$$
$$= (1.0c \cdot \text{hour}) \sqrt{1 - \frac{(0.8c)^2}{c^2}} = 0.6c \cdot \text{hour}$$

Given this shorter distance, it seems quite natural that the time necessary to travel it at $0.8c$ would be 0.75 hour. The definition of velocity, solved for time,

$$v \equiv \frac{d}{t}$$

$$t = \frac{d}{v}$$

gives 0.75 hour as a natural result.

$$t = \frac{0.6c \cdot \text{hour}}{0.8c} = 0.75 \text{ hour}$$

It is no accident that these numbers come out the same. The special theory of relativity is built on the premise that people in different frames of reference will agree on the same experimental results. Since the laws of physics work the same in any frame of reference, the same result should occur no matter how you look at it. Since one of the laws of physics is that the speed of light is the same universal constant in any frame of reference, people in different frames of reference will not see things happening in the same way, but they will still see the same things happening. The nice thing about the special theory of relativity is that people can not only agree on the final result but they can understand each other's points of view.

Suppose that Mr. Stern has been directed to leave a record of his distance standards on the space buoys as he passes. That way, Ms. Bower can compare her distance standards against those left by Stern when she passes the same buoys a few moments later. Mr. Stern therefore arranges two marking lasers (or other marking devices) to burn images of both ends of a meter stick at the same time on the hull of the space buoy. To make certain that the two lasers go off simultaneously, he actuates them both from a button located at the 50-cm mark halfway between the ends of the meter stick. Since the signal from the button travels the same distance to the two phasors at the ends of the stick, they both make their marks at the same time.

As long as Ms. Bower looks at these marks while she is traveling at the same speed Mr. Stern was going when he made them, there is no doubt that she will see them as 1 m apart. Bower and Stern, after all, are not moving relative to each other. But let us look at this from the point of view of a workman who has stopped by one of the space buoys on his regularly scheduled service rounds. Suppose the workman has made whatever adjustments are necessary and has paused for a lunch break. As he sits back to enjoy his cup of coffee and watch the maneuvers of the mainship, Mr. Stern comes along in the first survey craft.

Much to the workman's distress, Mr. Stern fires his marking lasers at the exact point on the hull of the space buoy where the workman has left his meter stick. No harm is done, however, as the workman's meter stick is foreshortened in Mr. Stern's frame of reference. From Mr. Stern's point of view, the space buoy and everything on it are zipping past at eight-tenths the speed of light and are shorter by a factor of $\sqrt{1 - v^2/c^2}$, which turns out to be six-tenths at this velocity. Since Mr. Stern sees the workman's meter stick as only 60 cm long, he is easily able to miss it by putting his marks beyond both ends of it with plenty of room to spare at either end.

This might all seem very logical from Mr. Stern's point of view, but you might ask how the workman sees this result. When he goes

over to pick up his meter stick, he will agree that Mr. Stern has been kind enough to miss it. He finds burn marks on the hull of the space buoy, but these marks are clearly more than 1 m apart. The strange thing is that the workman sees Mr. Stern's marking lasers as being closer than 1 m apart. From his point of view, Mr. Stern is zipping along at $0.8c$; everything on Mr. Stern's survey craft must be foreshortened by a factor of six-tenths. He is therefore left with the problem as to how marking lasers that are closer than 1 m apart somehow leave marks that are further than 1 m apart.

This problem is resolved by the other side of the space-time distortion. The only way that marking lasers can leave burn marks that are significantly further apart than the separation between phasors is if they go off at different times. The result observed by the workman could only have occurred if the phasor toward the rear of Mr. Stern's craft went off first and then, after it had traveled for a considerable distance, the laser toward the front of the craft went off. This, in fact, is exactly the way things look in the workman's frame of reference. He sees Mr. Stern reach out and touch the button halfway between the two marking lasers, but he does not see the signal from the button reach the two lasers at the same time. The laser traveling toward the button (the one toward the rear) gets the signal first and goes off before the one that is traveling away.

The fact that the signal from the button travels in both directions with the same velocity in the workman's frame of reference makes the two events, which Mr. Stern sees as simultaneous, happen at different times. The workman can, however, see how Mr. Stern might think that the two lasers went off simultaneously. Light scattered from the first laser that goes off must travel a longer distance to catch up with Mr. Stern than light from the second. Mr. Stern sees the flashes from the two lasers at the same time and, as far as he is concerned, they went off simultaneously.

People in both frames of reference agree on the final result. Both see the burn marks miss the ends of the workman's meter stick. Mr. Stern thinks it is because the workman's meter stick is foreshortened; the workman thinks it is because Mr. Stern's marks were made at different times. Which is right depends on your preference. The Einsteinian point of view is that they are both right.

▶ Relativistic Mass

Einstein's theory of relativity requires some modification of Newton's second law of motion. As objects approach the speed of light, their acceleration is no longer strictly proportional to the force applied, as stated in the second law.

$$F = ma$$

At least, F and a are not proportional in the sense that m is constant. Newton's second law can be fixed up, however, so that it continues to work even at velocities near the speed of light if we simply think of the mass, m, as something that increases by the same relativistic factor that seems to affect space and time. That is, we simply define a **relativistic mass,** m, which increases as the square root of 1 minus v squared over c squared decreases:

$$m = \frac{m_0}{\sqrt{1 - \dfrac{v^2}{c^2}}}$$

As long as the velocity, v, is very small compared to the velocity of light, the square root factor is not much different from 1, and the relativistic mass is not much different from the **rest mass,** m_0, which is the good old constant inertial mass that we have become familiar with in classical physics. Thus, Newton's second law is little affected by this redefinition of mass until the velocity approaches that of light. When the velocity does approach the velocity of light, both the mass and the acceleration become a function of velocity. Newton's second law still holds, but only in the form that refers to the change of momentum with respect to time.

$$F = \frac{\Delta(mv)}{\Delta t}$$

At the one extreme where the velocity is small, mass is constant and we have the classical $F = ma$. At the other extreme where v approaches c, the square root factor in the denominator of the relativistic mass expression approaches zero and the mass becomes infinite. We find that this is actually consistent with our experience.

You might wonder what experience we could possibly have had that would make this relativistic mass more than a theoretical premise. It is true that our present technology is far short of being able to get the velocity of objects as large as people and spaceships up to any value where the relativistic nature of mass would have any noticeable effect on Newton's second law. It is quite common, however, for us to get little particles of mass up to relativistic speeds.

One device in which relativistic particle speeds are quite common is the ordinary television set. The electrons in the picture tube may easily be accelerated up to about half the speed of light before they crash into the phosphore screen. The light you see as you watch television is therefore energy that is released by a sudden change in the relativistic mass of these electrons.

The relativistic nature of mass has a substantial effect on the technology of particle accelerators. It is no trick at all to get electrons or other particles up to 99% of the speed of light. They do that in the first 10 ft of the 2-mile-long accelerator at Stanford University. For the rest of the 2 miles the velocity increases by only tiny amounts. First they get the particles up to 99.9% of the speed of light, then

99.99%, and then 99.999%. The accelerating force is exerted on the particles for the whole 2 miles, dumping more and more energy into the particles, but the only effect on their velocity is to add nines after the decimal point. The closer the speed of the particles comes to the speed of light, the harder it is to make them go any faster.

What happens to the energy they keep dumping into the particles? If the particles don't go much faster, where does all the energy go? The energy, it turns out, goes into increasing the mass of the particles instead of their velocity, and this increased mass has a very real effect on their inertia.

Particles that come out of an accelerator not only have increased mass in terms of how hard they smash into the target but they also have increased mass in terms of how hard they are to deflect. It is common, for example, to deflect the particles from a high-energy accelerator from one experiment to the other so that several experiments can be set up at the same time. People who operate the accelerator find that they need a stronger magnetic force to deflect the beam on a good day when everything is working just right and the particles in the beam have the maximum energy the accelerator can deliver. Thus, the relativistic mass is just as substantial as the rest mass in terms of the inertial effects it has on Newton's second law.

The idea that energy can show up in terms of increased mass led Einstein to what some people feel is one of the most important concepts of the twentieth century—the celebrated equivalence between mass and energy. The idea is that energy is proportional to mass, with the proportionality constant being the square of the speed of light.

$$E = mc^2$$

This relationship suggests more than that equal increases of energy will be accompanied by equal increases in mass. It also suggests that there is energy associated with rest mass and that the energy is huge. The speed of light is a very big number; the speed of light squared is that much bigger again.

Some feel that the discovery of this relationship between mass and energy led directly to the age of atomic energy and the big bombs. Others feel that it only laid a theoretical groundwork by suggesting that vast amounts of energy are somehow available. The fact is that we have been making use of this relationship for a long time without knowing it. When you burn a log in the fireplace, you are converting a tiny fraction of the mass of the log into energy. The only reason we didn't know that a long time ago is that the fraction is far too tiny to measure by conventional means. The forms of atomic energy that we can now make use of by such nuclear reactions as fusion and fission still only convert a tiny amount of the available rest mass into energy, but the fraction is now large enough to measure. The only way we know of getting total mass-energy conversion is to allow anti-matter to combine with ordinary matter. However, no one has a sizable chunk of anti-matter, and no one knows where to find any either. For now, that is probably just as well. We had best first learn how to handle the tiny amounts of energy available from atomic physics.

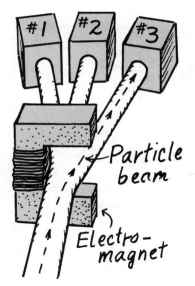

FIGURE 13-8 *The increased relativistic mass of particles coming from an accelerator makes them harder to deflect with a magnetic field.*

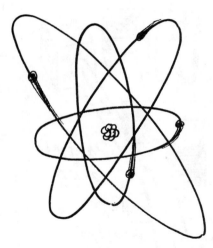

FIGURE 13-9 *A cartoonist's concept of an atom shows electrons going around the nucleus like little planets going around a tiny sun.*

▶ Atomic Physics

Having looked into the realm of the very fast, it is now time to look into the realm of the very tiny. Suppose you were to commit the most heinous of all crimes—falling asleep in class. Since the lecture was exciting, exhilarating, and dynamic, we must assume you were up all night (studying). Let us imagine that you shrank in size as you fell from your seat and became smaller and smaller as you approached the floor. Instead of falling onto the floor, you would find that you fell into it. As you approached microscopic dimensions, the relatively smooth floor would open up into jagged canyons, which would become even rougher the more you fell. The walls and bottom of the canyon you would fall into would begin to take on the structure of cells and organic fibers as you approached the level of microbiology. You would begin to see long chainlike hydrocarbon molecules as you shrank to the level of biochemistry. If you could get close enough to one of these molecules to see the atoms individually, you would see something very strange indeed. You would find that the picture most people form of what an atom should look like is a gross oversimplification of reality.

A bit of historical perspective is useful in sorting out the various mental models people use in dealing with atomic structure. The planetary model most people think of when they picture an atom, where the electrons orbit the nucleus like planets going around the sun, is actually an early model first proposed by Niels Bohr just after the birth of quantum theory. We still tend to think in terms of this simple picture, even though it has been replaced several times over by models that give progressively better results but become progressively more abstract.

Quantum theory got its start just after the turn of the century, when a man named Max Planck got the idea that energy comes in little chunks, which he called quanta, just like matter comes in little chunks called atoms. He showed how this assumption would explain the frequency distribution of radiation from hot radiating bodies. Nobody took Planck seriously until Einstein, a brash unknown at the time, showed how the same assumption would explain the energy of electrons excited by light falling on a vacuum tube photocell. This was the work, by the way, for which Einstein was later given the Nobel prize. It was an important piece of work, because it established Planck's quantum hypothesis as a significant physical theory describing the fundamental nature of matter and energy. Light, which had been thought of as a wave for centuries, was shown to be broken up into little quantized chunks called photons. The wave nature of light remained, but light was now understood to have particle properties as well.

Planck's quantum hypothesis having been lent strength by Einstein's use of it to explain the photoelectric effect, Niels Bohr then used it, together with new evidence that most of the mass of an atom and all of the positive charge is located at the center, to produce the planetary model of the atom. The electrons are shown traveling around the nucleus like planets going around the sun. This model really only lasted a couple of years, but we still use it because it was the last model that was easy to visualize. Bohr's planetary model is so beautiful and forms such a clear picture in our minds that we sometimes have to remind ourselves that the atom isn't really like that miniature solar system. The biggest problem that Bohr had with his planetary model was that he couldn't explain why the orbits were stable other than to say that the energy of the electron must be quantized somehow.

The next, more sophisticated, model of the atom was invented by a Frenchman named L. V. de Broglie. It occurred to de Broglie that if light waves had particle properties, maybe particles like electrons might also have wave properties. This stroke of genius immediately explained Bohr's quantized orbits in terms of standing waves. Only those orbits could exist that met the condition that the wave would fit in the sense that it would close on itself. The de Broglie wave hypothesis was an instant success. His picture of an atom with the electron spread around the orbit as a standing wave was so popular, as a matter of fact, that it lasted even less time than Bohr's planetary model. It attracted so much attention that very little time elapsed before other people figured out significant improvements.

Within a few years two different people had independently figured out the rules of physics that should apply to electrons treated as quantized waves. One man, Erwin Schrödinger, worked out the basic equation of wave mechanics that plays the same role in quantum mechanics as Newton's second law, $F = ma$, plays in classical mechanics. The Schrödinger equation describes the behavior of a de Broglie wave in terms of a partial differential equation, which cannot really be understood without the language of calculus. Even the picture of the atom described by the Schrödinger wave equation is a bit difficult to comprehend. As is frequently the case, the more sophisticated you make a model to better describe reality, the more difficult that model is to comprehend. This particular model of the atom has been very successful and has lasted for several decades, but it is so mathematically abstract that we can only paint a hazy picture of it here.

Instead of seeing an electron as located at a particular point in space, or even as being spread out around an orbit, the Schrödinger picture sees the electron's position as being described by a wave function, usually represented by the Greek letter psi, Ψ, which is spread out in three-dimensional space as a sort of cloud. This wave function cloud has greater or less density here and there, depending on the energy and angular momentum of the electron, but it does not have sharply defined edges. The physical meaning of this wave function, Ψ, is a bit abstract, as it happens to be a complex function having both real and imaginary parts, but the absolute square of the function, when evaluated at a particular point in space, turns out to be the probability of finding the electron at that point.

Different wave functions, Ψ, which form solutions to Schrodinger's wave equation for different energies and angular momenta of the electrons, can be represented by density clouds having different shapes. Some are round and centered on the nucleus with a finite probability of finding the electron in the center of the nucleus itself. Some are doughnut-shaped, with a hole in the middle. Some are clouds with two or more bulges.

A few people may still want to cling to the idea that there is such a thing as an electron in the form of a hard little particle, like a miniature BB, somewhere in that cloud of probability. This view is shot down by the fact that some wave functions have two density bulges separated by a null plane. There is equal probability of finding the electron in either part of the cloud above or below the plane, but there is exactly zero probability of finding the electron in the plane

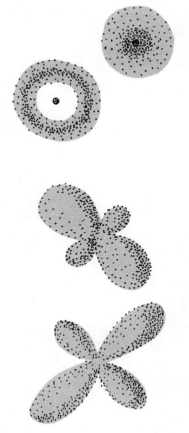

FIGURE 13-10 *Wave function solutions to Schrödinger's equation can be represented by clouds having different density shapes.*

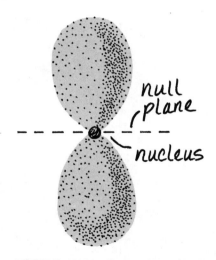

FIGURE 13-11 *Some wave functions have two density concentrations separated by a null plane.*

itself. A proponent of the pure particle theory of the electron would therefore have to figure out a way for the same electron to spend an equal amount of time traveling about above and below the plane without ever crossing over. It would seem that the electron just cannot be a simple particle.

Models such as particles and waves are pictures we form in our minds based on our everyday experiences with objects large enough to see and feel. These models may be dear to our hearts, but they simply fail to describe reality in atomic dimensions. Even the clouds that we have been using to describe the density of an electron's wave function are just models to help us picture what is going on. We should keep in mind the fact that electrons don't really look like that at all. What do they look like? The answer is that they don't look like anything you have ever seen. Electrons are so small that it doesn't even make sense to talk about looking at one in the conventional sense. To look at it would be to bounce several photons off it, but the very act of bouncing even just one photon off an electron will cause it to be long gone from the position in which it was "seen" by the photon. To ask what an electron looks like is something like asking what an earthquake would look like. You can see what it does and you can locate it in a general way, but you can't actually see it.

At the same time that Schrödinger was working out his wave mechanics, Werner Heisenberg was working out an equivalent set of quantum mechanical rules, starting from, of all things unanswerable questions. Heisenberg's so-called matrix mechanics are a set of rules that produce exactly the same results as Schrödinger's wave mechanics, but they are based on Heisenberg's **uncertainty principle.** Heisenberg pointed out that it is impossible to measure simultaneously the position and momentum of a particle to greater than a certain precision. Based on Planck's quantum hypothesis and de Broglie's wave-particle duality for an electron, Heisenberg figured out that the ultimate precision in measurement would in theory be when the uncertainty in the momentum, $\Delta(mv)$, times the uncertainty in the position, Δx, would be about equal to Planck's constant h.

$$\Delta(mv) \cdot \Delta x \cong h$$

Heisenberg also figured out that the same uncertainty between momentum and position would hold between energy and time. If you know the energy of a particle to within an uncertainty ΔE, then the smallest uncertainty of time, Δt, for which you know this energy is when the product of the uncertainties is approximately equal to Planck's constant.

$$\Delta E \cdot \Delta t \cong h$$

Heisenberg took these uncertainties as more than a limit of technology; he took them as a limit to knowledge itself. Starting from this theory of knowledge, he developed a calculus of observable quantities representing dynamic variables by matrices. The mathematics is just as much fun as that used by Schrödinger, and the result turns out to be the same. People who deal with particles all the time find it convenient to be conversant with both approaches.

▶ Nuclear Physics

One of the jolliest results of the quantum mechanics developed by Schrödinger and Heisenberg in dealing with atomic structure turned out to be that it worked nearly as well for nuclear structure. Nuclear structure, however, seems to be a great deal more complex than that of the outer atom, and we still haven't got it all figured out yet. The more we look at the nucleus, the more complex it seems to be. Still, there is a great deal that we do know and are able to put to practical use.

We know, for example, that there are two kinds of stable particles in the nucleus, and we know that they both have almost the same amount of mass. One kind has a positive charge and the other is neutral. The ones with a positive charge are called **protons;** the ones that are neutral are called **neutrons.** They both have a mass of about two-thousand times the mass of an electron, although the neutron has a tiny bit more mass than the proton. They both tend to stick to one another with about the same amount of force, sort of like candy-coated popcorn of two different colors in a popcorn ball. Because of their similarities, both kinds of particles are called **nucleons.**

The protons, with their positive charge, serve to hold the negatively charged electrons in their orbits. Each proton has the power to hold one electron in an electrically neutral atom, since protons have exactly the same size charge as an electron, but of opposite sign. The number of protons in the nucleus therefore determines the chemical properties of the atom by determining the possible structures of the electron orbits that can occur.

The number of neutrons in the nucleus has no direct effect on the chemistry of the atom. Neutrons serve, however, to help hold the nucleus together. Without neutrons, protons would have too much electrical charge to stick together. There is a strong force of attraction between nucleons but not strong enough to hold two naked protons together. You need a neutron or two around to lend attractive force and help stick two protons together. The more protons you have, the more neutrons you need to hold them together. For light elements, you need about equal numbers of neutrons and protons for a nucleus to stay together. For heavy elements, however, you have to throw in extra neutrons, because the strong nuclear force of attraction is very short-ranged. All the protons in a nucleus are electrically repelled by each other, even on opposite sides of the nucleus, but the force of attraction acts over such short distances that it is effective only between neighboring nucleons, whether they be neutrons or protons. The electrical repulsion builds up in heavy elements to the point where you need to mix in a disproportionately large number of neutrons to add their attractive force for stability. Lead, for example, has nearly one and a half times as many neutrons as protons. After a certain point, however, you just cannot add in enough neutrons to stabilize the nucleus.

One of the limiting factors in the size of a nucleus is the fact that neutrons are not stable by themselves. A naked neutron will spontaneously decay into a proton plus an electron. Out of a bunch of naked neutrons, about half of them will decay in 12 minutes. A neutron seems to need protons around to keep this from happening. After the size of a nucleus reaches a certain point, the neutrons so outnumber the protons that there are not enough protons in the mix

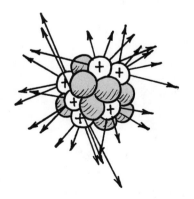

FIGURE 13-12 *The nuclear force of attraction holds a nucleus together, but it is so short-ranged that it only acts between neighboring nucleons, whereas all the protons push on each other.*

to prevent the neutrons from decaying. It is almost as if one of the neutrons in a big heavy nucleus doesn't see enough protons around and decides to "kick out" an electron and become a proton itself.

This ability of a neutron to change itself into a proton by "kicking out" an electron might make you think that maybe a neutron is really a proton with an electron inside. This is an attractive thought, but it is wrong. One of the many reasons why this cannot be true, or at least this simple, is that ordinary electrons are far too big to fit into a neutron. Only by adding a great deal of energy to an electron can you get its dimensions down to the point where it will even fit into the nucleus of an atom, let alone into a single nucleon. According to Heisenberg's uncertainty principle, the size of a particle, or the uncertainty in its position, is inversely proportional to its momentum. A neutron has a comparably large momentum because of its rest mass.

$$\Delta(mv) \cdot \Delta x \cong h$$

The smaller rest mass of an electron, nearly two thousand times smaller than a proton, means that the uncertainty in its position is nearly 2,000 times greater. It is therefore best to think of the neutron

$$\Delta(mv) \cdot \Delta x \cong h$$

as being a separate entity, distinct from a proton and an electron until it somehow makes the change. Someday, we hope to find some simple theory to explain what these different kinds of little boys and girls are made of, but we just don't understand it yet. What we do know is that it's not all as simple as sugar and spice.

▶ Alpha, Beta, and Gamma Rays

Early investigators found out that radioactive substances seemed to emit three distinct kinds of rays. They named them with the first three letters of the Greek alphabet in the order with which they could be stopped by progressively heavier shielding. The easiest to stop are alpha rays. They can be stopped by a reasonably heavy piece of paper. Beta rays go right through a piece of paper but are stopped by a fairly thin sheet of metal. Gamma rays are the hardest of all to stop and require lead or other heavy shielding to contain them.

Alpha rays are easy to stop because they are big, highly charged, particles. They are what amounts to helium nuclei that have "boiled off" the nucleus of a heavy unstable atom. A helium nucleus is a pair of protons held together with the help of a pair of neutrons. If the heavy atom has too many protons to be stabilized even by an extra number of neutrons, it is possible for excess energy to get into a pair of protons near the surface of the nucleus, and they will leave with a pair of neutrons like a molecule evaporating from a drop of hot liquid. It is always a pair of neutrons and protons, because this configuration is more stable and requires less energy to escape than a single proton leaving the nucleus by itself. The reason for this has to do with the wave nature of neutrons and protons, together with the fact that they act like little magnets. Their magnetic moments, or spins, cause them to pair up and stick together like magnets holding

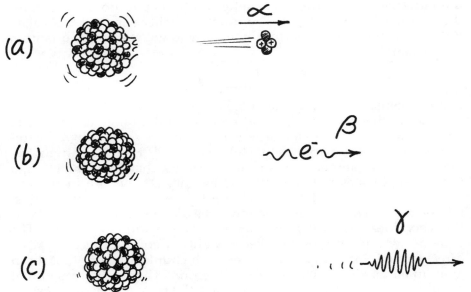

FIGURE 13-13 (a) *An alpha particle is a helium nucleus, two protons and two neutrons, given off by a heavy nucleus. (b) A beta particle is a high-speed electron given off when a neutron turns into a proton. (c) A gamma ray is an energetic photon of electromagnetic radiation.*

each other by the north and south poles. This pairing is not strong enough to hold two lonesome protons together, but a pair of neutrons added in makes a very stable configuration indeed. So when they come out, they all come out together.

An alpha particle is easy to stop, because it has two positive charges that produce such a strong electric field that it shakes apart just about every molecule it comes near, leaving a trail of positive and negative ions. This takes so much energy that the particle is soon brought to a stop. Even traveling through nothing but air, an alpha particle comes to a stop after only a few centimeters. Even a piece of paper will stop an alpha particle. After it stops, it grabs up a couple of electrons to become nothing more than a humble helium atom.

Beta particles are a little harder to stop, because they carry only a single charge. They are actually nothing more than high-speed electrons that have somehow been "kicked out" of the nucleus whenever a neutron changes into a proton. It makes up for its small mass by having a large velocity—which is another reason that it doesn't lose its energy in as short a distance as does an alpha particle. It doesn't stay around long enough to ionize many of the atoms that it comes near. It requires more of a direct hit before it gives up some of its energy.

Gamma rays are the hardest of the three to stop, because they don't have a charge of any kind. Like visible light, they are simply photons of electromagnetic radiation, but they are of much higher energy and frequency. Visible light is emitted when electrons jump from one orbit to another having a lower energy. Gamma rays are emitted when nucleons seem to do a similar sort of thing inside the nucleus. The energy states of the nucleus are widely separated, so that the photons of "light" emitted each carry a large amount of energy.

The relationship between the energy of a photon and its frequency forms the basis of Planck's original hypothesis, which started the whole quantum theory. It was Planck's idea that the energy, E, in a photon is just directly proportional to the frequency ν of the radiation (where the Greek letter nu, ν, is traditionally used for the frequency of radiation). The proportionality constant h is called **Planck's constant** in honor of the man who got the whole quantum theory started.

$$E = h\nu$$

Planck's constant has a value of 6.63×10^{-34} J·s.

The frequency of gamma radiation is in the range of 10^{20} Hz to 10^{25} Hz, which is five to ten orders of magnitude greater than the frequency range of visible light (10^{14} Hz to 10^{15} Hz). This high frequency gives gamma rays so much energy that they travel right through light materials just like X rays do. (X rays are in the frequency range of 10^{17} Hz to 10^{20} Hz.) All the energy in a photon must be given up in a single interaction or else none of it can be. That is the very idea of a quantum, an indivisible unit of energy. The gamma ray photon therefore keeps right on going until it makes a direct enough hit with a massive enough chunk of matter that it can give up all its energy in a single interaction. That is the reason that gamma rays are so hard to stop. It is also the reason why the most efficient shielding materials against gamma radiation are the heavy metals, whose large nuclei make good targets.

CHECK QUESTIONS

1. How much energy is in a visible light photon whose frequency is 5×10^{14} Hz?

    Answer: 3×10^{-19} J

2. What is the frequency of a gamma ray photon whose energy is 7×10^{-12} J?

    Answer: 1×10^{22} Hz

Alpha, beta, and gamma rays were only the first kinds of radiation identified by earlier investigators. After these three were identified and named for the order of difficulty in stopping them, other kinds of radiation were discovered that are even harder to stop. One of the most fantastic of these is made of tiny particles called **neutrinos,** which have zero rest mass and no charge. They have so little interaction with matter that a neutrino can go right into the center of the sun with a good chance that it will come out the other side without having been deflected. They are so difficult to detect that people looked for them for 26 years before they found hard evidence that they really exist. Fortunately, we are so transparent to them that they don't bother us much.

▶ Neutron Activation

Another form of radiation that is hard to stop, and which can be quite destructive, is made up of neutrons released by many nuclear reactions. Neutrons are difficult to shield against because they are electrically neutral. They don't even feel the strong electric fields that occupy most of the volume of an atom, but must bump right up against a nucleus if they are to have any interaction at all. Neutron radiation therefore passes right through heavy shielding material, making only occasional collisions with the tiny nuclei of the atoms in the material, and even these collisions are mostly glancing blows in which the neutrons keep on going.

The potential destructive nature of neutron radiation comes from the fact that neutrons are radioactive themselves and that they make the materials they pass through radioactive as well. We have mentioned the fact that naked neutrons tend to turn themselves into protons by "kicking out" an electron or beta particle. This beta decay of the neutrons forms only part of the problem, however. The other part of the radiation hazard comes from the tendency of neutrons to change the very nature of the material through which they pass. Since they are electrically neutral, they don't even need to be going particularly fast to bump right up against the nucleus of an atom and upset its balance in one of several ways. If the incoming neutron either sticks to or knocks out a proton, the whole character of the nucleus is changed.

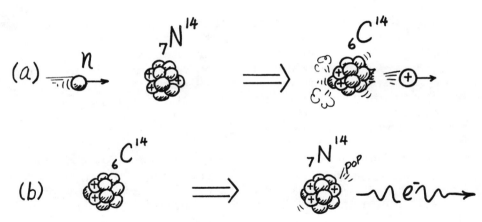

FIGURE 13-14 (a) Neutron activation of nitrogen 14 takes place when a neutron comes along and bumps out a proton, changing the nucleus to carbon 14. (b) Carbon 14 is radioactive and beta decays back into nitrogen with a half-life of 5,770 years.

Only certain combinations of neutrons and protons go together to form a stable nucleus. We have already mentioned the fact that the number of protons in a nucleus is the thing that determines the chemical character of the atom by determining the positive charge of the nucleus. The number of protons is therefore called the **atomic number** of the atom and dictates which element it is. Any atom having six protons in the nucleus will have the chemical properties of carbon, for example, while any atom having seven protons will be nitrogen. An atom can have different numbers of neutrons in the nucleus, however, and still remain the same element. The different number of neutrons will give the atoms of a certain element different amounts of mass but will have little effect on the chemical properties. The number of neutrons will determine whether the nucleus will be stable or not.

Atoms of the same element having different numbers of neutrons in the nucleus are called different **isotopes** of the element. Many elements have several stable isotopes as well as a few unstable ones. As we mentioned before, the isotopes of the lighter elements that tend to be most stable are the ones in which there are about the same number of neutrons as protons. The most common isotope of carbon, for example, has six neutrons and six protons. That is called carbon 12, because there are 12 nucleons in the nucleus all together. The total number of neutrons plus protons is called the **mass number** of the atom.

The usual way of identifying a particular isotope is to express its mass number as a superscript after the chemical symbol. For example, carbon 12 would be expressed with a 12 as a superscript:

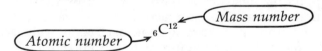

Sometimes it is convenient also to express the atomic number as a subscript before the chemical symbol, although it is redundant. Anything having six protons in the nucleus is carbon by definition.

The effect of neutron radiation can be to alter either the mass number or the atomic number of atoms in the path of the neutrons. If a neutron sticks in the nucleus of a target atom, the mass number increases by 1. If a neutron comes in and knocks out a proton, the mass number remains unchanged but the atomic number decreases by 1. Either way, a new isotope is formed, which may or may not be stable.

A natural example of the production of radioactive isotopes by free neutrons is the production of carbon 14, $_6C^{14}$, which has been going on in the upper reaches of our atmosphere since the birth of our planet. Cosmic radiation produces neutrons naturally in the upper atmosphere, and these neutrons bombard nitrogen nuclei to produce a certain low-level concentration of radioisotopes. The most abundant isotope of nitrogen is nitrogen 14, $_7N^{14}$, which has seven neutrons and seven protons. In the natural form of neutron activation, a neutron comes into the nucleus and "knocks out" a proton, leaving the nucleus richer by one neutron but poorer by one proton. This leaves the mass of the nucleus essentially unchanged, but it decreases its positive charge by one unit. The original nitrogen 14, with an atomic number of 7, is thereby changed into carbon 14, with an atomic number of 6.

$$_7N^{14} + {_0}n^1 \Rightarrow {_6}C^{14} + {_1}p^1$$

The neutron can be expressed as $_0n^1$, since it has zero charge and hence an atomic number of zero and a mass number of 1. The proton can be expressed as $_1p^1$, since it has one unit of charge and one unit of mass. Notice that the total charge as well as the total mass remain unaltered by the reaction. The superscripts and subscripts both add up to the same number on both sides of the reaction.

Carbon 14 produced in this way has two more neutrons than carbon 12, the most abundant isotope of carbon, but it behaves chemically the same. The two extra neutrons make carbon 14 slightly unstable; it beta decays slowly back into nitrogen 14. One of the neutrons turns into a proton by emitting an electron, or beta particle.

$$_6C^{14} \Rightarrow {_7}N^{14} + {_{-1}}e^0$$

There is also a neutrino emitted in this reaction, but we will not bother to show it in our expression because it has no effect on our simple calculations. Neutrinos have zero charge and zero rest mass. We do need to show the electron however. An electron is shown as $_{-1}e^0$ with an atomic number of -1, since it has one unit of minus charge, and a mass number of zero, since its loss does not change the total number of neutrons and protons in the nucleus. Here again, the sum of the superscripts is the same on both sides of the reaction, as is the sum of the subscripts. The total charge and the number of nucleons are both unchanged by the reaction.

In a natural sample of carbon, 98.89% is found to be $_6C^{12}$, the isotope with an equal number of neutrons and protons. Almost all the rest is $_6C^{13}$, an isotope with one more neutron, which is also stable. Only a trace amount of naturally occurring carbon turns out to be the radioactive isotope $_6C^{14}$, and all of it that is found was produced by natural neutron activation of nitrogen in the upper atmosphere by the cosmic radiation process. We know that none of the carbon 14 found today was present when the earth was formed, because we know the rate of the beta decay process is too rapid for $_6C^{14}$ to last that long. Any carbon 14 that was present when the earth was formed has long since turned into nitrogen 14.

The small amount of radioactive isotopes produced by cosmic radiation does not present much of a radiation hazard, but the huge amounts of neutron radiation from artificial nuclear reactions is one of the major problems of reactor design. The shielding material itself becomes radioactive and changes its chemical and structural properties as a result of the neutron bombardment. The working components of the reactor must be made of materials that can stand neutron bombardment for significant periods of time. Management of the radioactive waste that comes from the replacement of shielding and other parts of the reactor is still another problem that comes from working with neutrons.

Neutron activation can, on the other hand, be a useful tool. It is most frequently used for measuring even tiny amounts of certain elements in a sample. The radioactive decay of the isotopes in the sample after it has been bombarded with neutrons becomes a signature of the elements that were originally in the sample. For example, sodium has 11 protons in the nucleus and just a slightly larger number of neutrons. The only naturally occurring isotope of sodium is sodium 23, $_{11}Na^{23}$, with 12 neutrons in the nucleus. If a sample containing sodium is irradiated with neutrons, some of the sodium atoms will capture a neutron and become the unstable isotope sodium 24. A gamma ray, γ, is also produced in this reaction.

$$_0n^1 + {}_{11}Na^{23} \Rightarrow {}_{11}Na^{24} + \gamma$$

Sodium 24 decays with a half-life of 15 hours to yield gamma rays and beta particles of certain well-known energies.

$$_{11}Na^{24} \Rightarrow {}_{12}Mg^{24} + {}_{-1}e^0 + \gamma$$

Thus, the presence of simultaneous beta and gamma rays at a rate that decreases by half every 15 hours is a sure sign that sodium was present in the sample before it was subjected to neutrons. At least 70 elements can be activated in this way to produce radioactive isotopes that decay in a recognizable way.

► Half-Life and Carbon Dating

The existence of radioactive carbon 14 in the atmosphere means that you eat the stuff all the time. You can't get away from it. The

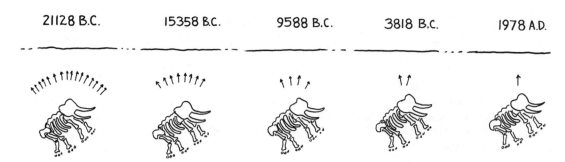

21128 B.C.   15358 B.C.   9588 B.C.   3818 B.C.   1978 A.D.

FIGURE 13-15 *The rate of decay of carbon 14 decreases by half every 5,770 years.*

carbon produced by neutron bombardment of nitrogen in the atmosphere turns into carbon dioxide and gets into growing plants. A certain amount of the oatmeal you eat in the morning is carbon 14. If you eat eggs instead, the chicken who laid the egg ate the oatmeal, and you still get the same ratio of carbon 14 to the other isotopes of carbon in everything you eat. The only way to avoid putting this radioactive substance into your body is to quit eating.

Looking at the bright side, you will eventually quit eating. The carbon 14 in your body will then eventually all decay, and you won't have anything more to worry about. It will take about 5,770 years for half the carbon 14 to go away. After another 5,770 years, half the remaining will decay. You will then be down to one-fourth of what you started with. After another 5,770 years, you'll be down to one-eighth. The amount present always goes down by a factor of two each time this period of time passes.

It is impossible to guess when any one radioactive nucleus will decay. The chances of its decay are independent of what has gone on in the past. Every instant of time is therefore a new ball game, so to speak. On the other hand, you can guess how long it will take for half of a whole bunch of radioactive nuclei to decay. You may not know which nucleus will be the next to go, but you do know that the rate of decay is proportional to the number that are left. When half are gone, the rate of decay will be half as great. The period of time for this to happen is called the **half-life** of the reaction. It is a known constant under most conditions for most nuclear decay reactions.

From the rate of decay of carbon 14 it is possible to tell when an animal had its last meal to within about 30 years. This fact has been useful to archaeologists and others interested in dating old organic remains. All that needs to be done is to figure out by how many factors of 2 the concentration of carbon 14 is smaller than the concentration in a fresh sample. The number of these factors of 2 is the number of half-lives of age of the sample.

► Nuclear Fission

There are two nuclear reactions which, given the state of our present technology, have the potential of giving us large amounts of power. One is nuclear fusion and the other is nuclear fission. It may seem remarkable that both reactions produce energy, since they are, in some ways, opposite kinds of reactions. Nuclear fusion is when the nuclei of two light elements are fused together to form a heavier nucleus; nuclear fission is when a heavy nucleus breaks apart to form nuclei of lighter elements. We will first consider these two reactions separately and then see how they can both produce energy.

The energy released in nuclear fission, when a heavy nucleus breaks up, is energy stored in the electrical forces with which the positively charged protons push on each other. As mentioned before, it is the build-up of the electrical repulsion between all the protons in a nucleus that makes it necessary to have more than the same number of neutrons as protons in a heavy nucleus to add in their short-ranged attractive force to help hold the nucleus together. After the number of protons gets up to a certain point, you just can't put enough neutrons into the nucleus to make it stable. Unstable heavy nuclei can reach a more stable configuration by alpha decay or beta decay or by breaking apart in nuclear fission.

The break-up of a heavy nucleus results in the release of neutrons as well as the lighter nuclei called fission fragments. The lighter nuclei don't need as many neutrons to balance the repulsive force between their protons, and so there are neutrons left over. These extra neutrons are important in the use of fusion to produce nuclear power.

Nuclear fission occurs spontaneously in heavy elements, but it takes place so slowly that the power release is at too low a level to be of much use. The neutrons released by the fission of one nucleus can be used, however, to artificially induce the fission of another unstable nucleus. As long as at least one neutron released by one fission induces the fission of another nucleus, a high rate of energy release will be propagated in what is called a chain reaction. If more than one neutron on the average induces a new fission, the rate of energy release will increase. An atomic bomb makes a fine example of such a runaway chain reaction. The trick in producing power in a nuclear reactor is to absorb just the right amount of neutrons, so that the chain reaction continues at a constant level, while keeping the neutron flux up to the point that the chain reaction can perpetuate.

Getting the neutron flux up to the point where a chain reaction is possible turns out to be quite a trick in itself. The neutrons need to be moving rather slowly or else you need a lot more of them. Most of the neutrons coming directly from a nuclear fission are moving so fast that they pass through nuclei they hit with only a small probability of an interaction. These are not hard little bullets that are sure to have an effect on everything they hit, but wave-particles, which have a smaller chance of interaction the shorter the time that they stay in any one place. This problem is solved in nuclear reactors by providing a moderator, a light element in which neutrons can bounce around and slow down without being absorbed. Slow neutrons that have lost enough kinetic energy so that they are in thermal equilibrium with their surroundings are called **thermal neutrons.**

Another problem encountered in getting a large enough neutron flux to sustain a chain reaction is that only the light isotope of uranium, $_{92}U^{235}$, fissions properly under neutron bombardment. The heavy isotope of uranium, $_{92}U^{238}$, absorbs thermal neutrons but doesn't fission. Thus, the heavy isotope soaks up the neutrons, but the only thing that happens to it is that it beta decays into plutonium $_{94}Pu^{239}$ after a while. This is a problem, because 99.3% of natural uranium is the heavy isotope. The light isotope, which will fission, only makes up 0.7% in a natural mix. Refining uranium to get a larger concentration of the fissionable light isotope is, fortunately, a difficult task. The isotopes cannot be separated by chemical means, because they are both chemically uranium. Gas diffusion plants, which separate by the slight difference in mass, are very difficult to build and operate.

This difficulty turns out to be fortunate because it is relatively easy to build an atomic bomb once you have enough of the enriched mixture of the light isotope, U^{235}. All you have to do is to get enough of it together in a small enough volume and hold it there while the neutron flux builds up. It is probably a good thing that this is as hard to do as it is.

There is another way to get around the neutron absorption of uranium U^{235}, and that is to use plutonium instead of uranium. While plutonium does not occur in nature, it does result from the operation of a uranium reactor. The heavy isotope U^{238} absorbs thermal neutrons and then beta decays into plutonium, and plutonium is fissionable in a chain reaction. What is more, plutonium can be chemically separated from uranium. It is produced as a by-product of refining spent reactor fuel.

Reactors can even be designed to optimize the production of plutonium from U^{238}. Such reactors are called **breeder reactors.** It happens that the most efficient operation of a breeder reactor requires the operation of the reactor with little or no moderator to slow down the neutrons. Such a reactor is called a fast breeder reactor, because it operates on vast quantities of fast neutrons rather than on smaller amounts of slow neutrons. Unfortunately, a fast neutron reactor is more unstable and difficult to control than a slow neutron reactor.

▶ Nuclear Fusion

Another nuclear reaction that can release a great deal of energy is the fusing together of light elements to make heavier ones. This is the reaction that is powering our sun. Hydrogen goes through several nuclear transformations to become helium. Several steps are necessary, because it is almost impossible to get four protons together and keep them there long enough for the fusion reaction to take place. In fact, it is a formidable task just to get two protons together. Two protons must be coming together at a tremendous velocity if they are to overcome their electrical repulsion to get close enough to interact. The temperature at which these kinds of velocities occur is on the order of many millions of degrees Kelvin. These temperatures occur in the stars, but they are very difficult for earthlings to produce.

Red stars as well as white stars that operate up to about the temperature of our sun all produce energy in just this way. Two protons come together long enough for one to turn itself into a neutron and emit a positron, which is an elementary particle much like an electron except that it has a positive charge. The resulting nucleus is deuterium $_1H^2$, a heavy isotope of hydrogen. A proton can then combine with this deuterium nucleus to form a helium 3 nucleus, $_2He^3$. Helium 3 is the light isotope of helium. The usual helium nucleus has two protons and two neutrons, but the helium 3 nucleus has two protons and only one neutron. It really wants another neutron. If two helium 3 nuclei come together, one of them will take both neutrons and the two protons in the other will then fly apart with a tremendous release of energy.

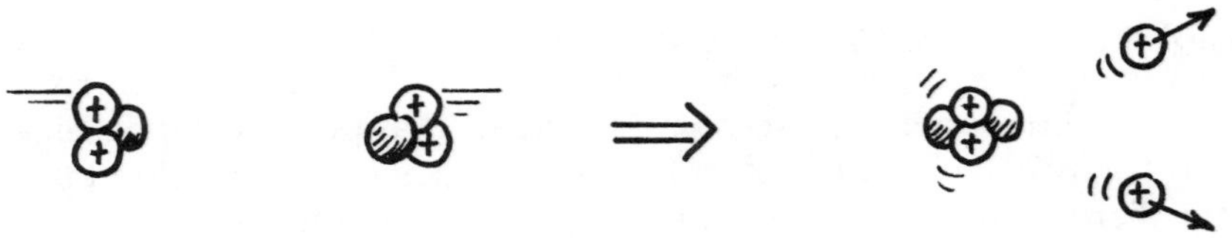

$$_2He^3 + {}_2He^3 \longrightarrow {}_2He^4 + 2{}_1H^1 + energy$$

FIGURE 13-16 *Two helium 3 nuclei come together and react to produce a helium 4 nucleus and two protons.*

Each step of the process releases energy, but the last step releases slightly more energy than all the other steps leading to two helium 3 nuclei taken together. This set of steps is collectively called the **proton-proton cycle.** It starts with ordinary hydrogen nuclei, or protons, and ends with a helium nucleus and two left-over protons.

Hot stars, hotter than our own sun, probably get their energy from a different set of steps called the **carbon cycle.** It involves protons interacting with a carbon nucleus as it changes back and forth between different isotopes of carbon and nitrogen to finally produce a helium nucleus and the original carbon. Carbon acts as a nuclear catalyst in this reaction.

Controlled nuclear fusion may be the energy source of the future if we can figure out a way to hold things together long enough at a high enough temperature to get out enough energy to make it pay. All of our efforts so far give less energy than we have had to put in to contain the reaction. It now looks like we will be able to get energy out, but not without a large number of problems still to be solved.

For one thing, the proton-proton cycle and the carbon cycle are still far beyond our capability. The only fusion reactions we are now working on are ones that generate vast numbers of neutrons as a by-product. As is the case with nuclear fission reactors, neutron activation of the working components and shielding material is a major problem of reactor design. Controlled nuclear fusion has many such problems to be solved before it can answer our energy needs of the future. For now, the only fusion power plant we can rely on is our sun.

You may wonder what will happen when the sun runs out of hydrogen to fuse into helium. This won't happen for a while; the sun has enough hydrogen to last for some tens of billions of years. Even after it runs out of hydrogen, however, it will keep on going. The fusion of hydrogen into helium is but one of many fusion reactions that can power a star.

After the sun runs out of hydrogen, the gravitational collapse of the helium that has been produced will raise its pressure and temperature at the center of the sun to the point where helium will start fusing into heavier elements. The heavier elements will then fuse into still heavier elements, and so on until the element iron is reached. Every fusion process liberates energy until we come to iron. Every fusion process after iron absorbs energy instead of giving it off. Therefore, after the sun develops an iron core, the next fusion process will be quite different. The energy produced by fusion reactions up to this point always tended to fight off the gravitational collapse, but fusion reactions past that point will actually favor gravitational collapse. Our sun will implode suddenly. The core will become immensely more dense; our sun will become a neutron star in a sudden event, called a supernova, and the mantle of various heavy elements will be blown away.

TABLE 13–1 A list of some isotopes shows that their masses are nearly, but not exactly, proportional to the number of nucleons.

Atomic Number	Isotope	Atomic Mass
1	H^1	1.007825
2	He^4	4.00260
3	Li^7	7.01600
4	Be^9	9.01218
5	B^{11}	11.00931
6	C^{12}	12.00000
7	N^{14}	14.00308
8	O^{16}	15.99491
9	F^{19}	18.99840
10	Ne^{20}	19.99244
11	Na^{23}	22.9898
12	Mg^{24}	23.98504
13	Al^{27}	26.98153
14	Si^{28}	27.97693
15	P^{31}	30.99376
16	S^{32}	31.9721
20	Ca^{40}	39.9626
25	Mn^{55}	54.9381
26	Fe^{56}	55.9349
27	Co^{59}	58.9332
28	Ni^{60}	59.9332
32	Ge^{74}	73.9219
37	Rb^{85}	84.9117
41	Nb^{93}	92.9060
47	Ag^{107}	106.9051
51	Sb^{121}	120.9038
53	I^{127}	126.9006
59	Pr^{141}	140.9074
66	Dy^{161}	160.9266
72	Hf^{180}	179.9468
76	Os^{190}	189.9586
82	Pb^{208}	207.9766
86	Rn^{215}	214.9987
86	Rn^{216}	216.0002
88	Ra^{220}	220.0110
92	U^{235}	235.0439
92	U^{238}	238.0508
94	Pu^{239}	239.0522
98	Cf^{250}	250.0766
100	Fm^{250}	250.0795
101	Md^{255}	255.0906

We believe that all the known elements that make up our planet were formed in the interior of an aging star and blown away in that cataclysmic event when the star imploded in a supernova. The elements up to iron were formed before the supernova; the elements past iron were formed in that cataclysmic event itself and just happened to get blown away with the rest of the mantle. You can therefore look upon yourself as literally being made of star dust. Those stars in your eyes are really stars in your eyes.

▶ Mass Defect

We are now in a position to answer two puzzles about nuclear reactions and nuclear structure. One is how you can get energy from both nuclear fission and nuclear fusion. It does seem strange that energy is released in an atomic bomb by allowing the nuclei of heavy elements to break apart into lighter elements and in a hydrogen bomb by allowing nuclei of light elements to combine into heavier elements. The other puzzle is one that bothered early investigators as well as present students who first study a chart of the elements. The atomic masses of the elements seem to bear some relationship to the atomic number, but the relationship doesn't seem to be exact. The relationship improves once we understand isotopes and the need for increasing numbers of neutrons to stabilize heavy nuclei, but the mass still doesn't seem to be exactly proportional to the number of nucleons in the nucleus. This defect, as it was called, in the mass relationship is caused by the same phenomenon that allows you to get energy both from fission and fusion.

Looking at Table 13-1, we see that the atomic mass is almost the same number as the mass number of the isotope, but not exactly. For atoms lighter than carbon, the number giving the atomic mass (in atomic mass units) is a tiny bit larger than the number of nucleons in the nucleus. At about that point, it dips below and becomes a tiny bit smaller. The mass stays below the mass number until about the element 86 where it crosses over and becomes larger again.

These tiny differences in the fifth, sixth, and seventh places turn out to be very important. They mean that neutrons and protons do not always have the same mass. The whole key to nuclear power resides in the fact that this tiny discrepancy in mass represents huge amounts of energy. The proportionality constant, in fact, is none other than the speed of light squared:

$$E = mc^2$$

We could make a graph of the numbers in the list to see what the relationship looks like, but what we are really interested in is the ratio of the atomic mass to the number of nucleons in each nucleus. This ratio gives us the average mass of neutrons and protons in that nucleus. A graph of this ratio (see Figure 13-17) shows a low point at element 26, iron, $_{26}Fe^{56}$, which has an atomic mass of 55.9349. Dividing this mass by 56, the number of neutrons and protons in this isotope of iron, we get a ratio of 0.998837 atomic mass units per nucleon. All other isotopes have more mass per nucleon.

Elements lighter than iron therefore lose mass when they combine by nuclear fusion to produce nuclei closer in atomic number to iron. Elements heavier than iron lose mass when they break apart or

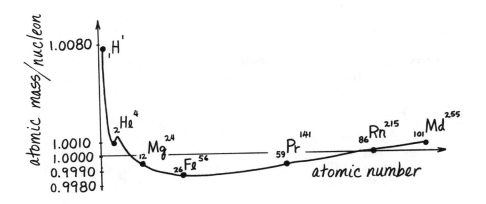

FIGURE 13-17 *A graph of the mass per nucleon shows that the neutrons and protons in the iron nucleus seem to have less mass than those in other elements.*

fission to approach iron. The decrease in atomic mass per nucleon in either process is accounted for by energy liberated in the reaction.

▶ Elementary Particles

We have been speaking of electrons, protons, and neutrons as though they are the only things that go into making up an atom. That is a useful model for understanding basic atomic and nuclear structure, but, as usual, there is more to it than that. All kinds of interesting particles seem to be in the nucleus. At least, all kinds of particles seem to come out of the nucleus when we hit it hard enough.

One of the most active fields of modern research is in trying to understand the particles that come out of a target when you send in a beam of high-energy particles. The number and diversity of particles that come out has become almost embarrassing as we have improved our techniques for sending particles in. Theoreticians have been hard-pressed to keep up with experimentalists in this field.

One result of particle study has been the discovery of anti-particles, which have symmetrical properties to ordinary particles. A positron, for example, has the same mass and spin as an electron but the opposite charge. Positrons don't last very long in the presence of ordinary matter. Their positive charge causes them to be attracted to an electron, whereupon the electron and positron combine to annihilate each other. Their rest masses are converted into energy in a huge burst of electromagnetic radiation. The same thing is true for the other anti-particles. They tend to annihilate ordinary particles. We have found anti-protons and anti-neutrons, which are symmetrical to protons and neutrons but which are annihilated and turned into energy when they find their opposite numbers.

We have seen very energetic photons of electromagnetic energy produce anti-particles and particles in pairs. It may be that the whole universe was created in this way. Maybe half the stars we see are made of anti-matter. Instead of protons and neutrons in the nucleus, there may be anti-protons and anti-neutrons. Instead of electrons in orbit about the nucleus, there may be positrons. Anti-matter would look just like ordinary matter. You wouldn't be able to tell that it was anti-matter without touching it. If you shook hands with an anti-matter person, however, it would be the last friendly gesture you would ever make before being literally annihilated.

High-energy particle research has also discovered strange particles that can do tricks for us that ordinary particles can't. A muon, for example, acts very much like an electron in terms of electric charge and everything else except that it is about 200 times as massive. It only lives for about 2.2×10^{-6} s, but that is plenty of time for it to do a rather remarkable thing. Replacing an electron in an atom, a muon will quickly decay into its lowest energy orbit. Because of its large mass, the inner orbits of a muon are actually smaller than the nucleus itself. We can therefore learn more about the charge distribution of the nucleus by studying the energy levels of a muon as it decays to orbits inside the nucleus.

It seems a bit strange, however, that a muon spends most of its short life in an orbit that is actually inside the nucleus without even bumping into the nucleons in the same way that they bump into each other. The muon doesn't even seem to feel the strong interaction force that acts between neutrons and protons to hold the nucleus together. Muons, like electrons, only interact electrically and by the weak interaction that governs most nuclear decay processes. Particles that interact by one kind of force can be transparent in some ways to particles that interact by another.

Just as the rules of behavior for things that move very fast differ from your ordinary experience with objects moving at ordinary velocities, the rules of behavior of these elementary particles differ markedly from what you might expect, based on your experience with ordinary-size objects. Modern physics is wonderful because it is so weird.

The special theory of relativity, applying to frames of reference moving at a constant velocity with no acceleration, is based on the fact that moving meter sticks must get short as their velocity approaches the velocity of light and that moving clocks must run slow as they approach the same speed. This **length contraction**

$$l' = l \sqrt{1 - \frac{v^2}{c^2}}$$

and **time dilation**

$$t' = \frac{t}{\sqrt{1 - \frac{v^2}{c^2}}}$$

are experimental facts resulting from the finite nature of the velocity of light,

$$c = 3 \times 10^8 \text{ m/s}$$

The special theory of relativity says that the laws of physics are the same in any frame of reference and that the velocity of light is also constant in any frame of reference. This implies that the length contraction and time dilation apply to the actual structure of space and time. Several results predicted by this theory have stood up to experimental test. Among these are the fact that Newton's second law, $F = \Delta(mv)/\Delta t$, remains valid even at speeds near the speed of light as long as the mass, m, is understood to be the **relativistic mass**

$$m = \frac{m_0}{\sqrt{1 - \frac{v^2}{c^2}}}$$

which increases as v approaches c. This relativistic mass results in the Einsteinian mass-energy relationship.

$$E = mc^2$$

Atomic and nuclear physics apply to dimensions so small that the wave nature of a particle puts limits on our knowledge of its momentum, (mv), and position, x, as well as its energy, E, and time, t. Heisenberg's uncertainty principle states that the product of our uncertainties in our knowledge of these variables is limited by the wave nature of the particle to about the size of Planck's constant:

$$\Delta(mv) \cdot \Delta x \cong h$$
$$\Delta E \cdot \Delta t \cong h$$

Planck's constant, h, is the proportionality constant between energy, E, and frequency, v, of electromagnetic radiation that got the whole theory of quantum mechanics started.

$$E = h\nu$$

The particle nature of waves, suggested by Einstein when he proposed that energy of electromagnetic radiation comes in little quantum bundles called photons whose energy is proportional to the frequency of the radiation, was what eventually suggested the idea that particles also have a wave nature. The wave-particle nature of both waves and particles is not apparent until you get to small dimensions, because Planck's constant is a very small number ($h = 6.63 \times 10^{-34}$ J·s).

A nucleus can decay by emitting an **alpha particle,** which is a helium nucleus consisting of two neutrons and two protons; a **beta particle,** which is a high-speed electron; or a **gamma ray,** which is an energetic photon. Alpha decay decreases the mass number by 4 and the atomic number by 2. Beta decay changes a neutron into a proton, leaving the mass number unchanged but increasing the atomic number by 1. Gamma decay leaves both mass number and atomic number unchanged.

$$_{a}X^{b} \Rightarrow {}_{a-2}Y^{b-4} + {}_{2}\alpha^{4}$$

$$_{a}X^{b} \Rightarrow {}_{a+1}Y^{b} + {}_{-1}\beta^{0}$$

13-1 You are watching a large man-made earth satellite zip across the night sky in a low circular orbit and start wondering how fast it would have to be going to relativistically contract a measurable fraction of its original length.

Given: The satellite is 30.0 m long and is traveling at 8×10^3 m/s, but you imagine that it somehow speeds up to the point where it will be only 29.9 m long.

Find: (a) How fast would the satellite have to travel to contract to the length that you imagine?

(b) How much time would a clock on the satellite lose in an hour of earth time because of traveling at this imaginary high velocity?

(c) By how much does the length of the satellite actually contract?

Your Solution

Discussion

This is an exercise in using the length contraction and time dilation. It gives you an opportunity to calculate the length contraction for the fastest large object you are likely to see near the surface of the earth.

The calculation in part (c) involves taking the square root of a number that is so close to unity (1) that a slide rule is of little use. If you don't have a calculator that takes square roots of this many digits, try a hand calculation squaring a number which is very close to 1. Then try another number that gives a result even closer to the number whose square root you want. This is called the method of successive approximations.

Sample Solution 13-1

(a) $\ell' = \ell \sqrt{1 - \dfrac{v^2}{c^2}}$

$\left(\dfrac{\ell'}{\ell}\right)^2 = 1 - \dfrac{v^2}{c^2}$

$\dfrac{v^2}{c^2} = 1 - \left(\dfrac{\ell'}{\ell}\right)^2$

$V = c\sqrt{1 - \left(\dfrac{\ell'}{\ell}\right)^2}$

$\quad = c\sqrt{1 - \left(\dfrac{29.9\,m}{30.0\,m}\right)^2}$

$\quad = c\sqrt{1 - (0.9967)^2}$

$\quad = c\,(0.08158)$

$\quad = \boxed{2.45 \times 10^7\,m/s}$

(b) $t' = \dfrac{t}{\sqrt{1 - v^2/c^2}}$

$\quad = \dfrac{1.0\,hr}{\sqrt{1 - \dfrac{(2.45 \times 10^7\,m/s)^2}{(3.0 \times 10^8\,m/s)^2}}}$

$\quad = 1.0033\,hr$

$\Delta t = t' - t = 0.0033\,hr = \boxed{12\,s}$

(c) $\ell' = \ell\sqrt{1 - v^2/c^2}$

$\quad = (30m)\sqrt{1 - \dfrac{(8 \times 10^3\,m/s)^2}{(3 \times 10^8\,m/s)^2}}$

$\quad = (30m)\sqrt{0.9999999993}$

$\quad = (30m)(0.9999999996)$

$\Delta\ell = \ell - \ell'$

$\quad = (30m) - (29.99999999\,m)$

$\quad = \boxed{1.1 \times 10^{-8}\,m}$

Discussion

The velocity found in part (a) is far greater than the velocity the satellite would actually have for a circular orbit. In fact, it is about 2,000 times the escape velocity of a space vehicle.

The fractional change in time, namely 0.9967, turns out to be the same as the fractional change in length, namely 29.9 m divided by 30.0 m, because a dilation of one is associated with a contraction of the other by the same rule of relativity.

The change in length found in part (c) for the velocity of an actual orbiting satellite would be about 2% as long as the wavelength of yellow light.

13-2   A science fiction spacecraft is somehow able to combine a small amount of anti-matter with an equal amount of ordinary matter and convert all the energy into its own kinetic energy.

Given: The rest mass of the spacecraft is 5×10^6 kg. It converts a total mass of 2.0 kg into its own kinetic energy.

Find: (a) How much energy, in joules, is produced?
(b) If that energy all goes into the kinetic energy of the space-craft, how fast does the spacecraft go?
(c) By how much does the relativistic mass of the spacecraft increase as a result of its increase in kinetic energy?

Your Solution

Discussion

This problem illustrates the concept of relativistic mass and the Einsteinian mass-energy relation. You will also need to remember that kinetic energy is one-half the mass times the square of the velocity.

We know of no way that a spacecraft could propel itself forward by giving itself all the kinetic energy expended. In rocket propulsion, most of the energy goes into the exhaust gas. Here we assume that science fiction technology has overcome this problem.

Sample Solution 13-2

(a) $E = mc^2$

$= (2.0 \, kg)(3.0 \times 10^8 \, m/s)^2$

$= 2.0 \times 3.0^2 \times 10^{2 \times 8} \, kg \, m^2/s^2 \left(\frac{1 \, Joule}{1 \, N \cdot m}\right)\left(\frac{1 \, N}{1 \, kg \, m/s^2}\right)$

$= \boxed{1.8 \times 10^{17} \, Joule}$

(b) $E_k = \frac{1}{2} m v^2$

$v = \sqrt{\frac{2 E_k}{m}}$

$= \sqrt{\frac{2(1.8 \times 10^{17} \, Joule)}{(5 \times 10^6 \, kg)} \left(\frac{kg \, m^2/s^2}{Joule}\right)} = 2.683281573 \times 10^5 \, m/s$

$\cong \boxed{2.7 \times 10^5 \, m/s}$

(c) $m = \frac{m_o}{\sqrt{1 - v^2/c^2}}$

$= \frac{(5 \times 10^6 \, kg)}{\sqrt{1 - \frac{(2.7 \times 10^5 \, m/s)^2}{(3.0 \times 10^8 \, m/s)^2}}} = 5.000002 \times 10^6 \, kg$

$\Delta m = m - m_o = (5.000002 \times 10^6 \, kg) - (5.000000 \times 10^6 \, kg) = \boxed{2 \, kg}$

Discussion

You can see that $E = mc^2$ at least has the correct units, since a joule of energy is a newton of force times a meter, while a newton is defined as that amount of force that accelerates 1 kg at 1 m/s².

Even the huge amount of energy that would be liberated by combining a kilogram of anti-matter with the same amount of ordinary matter would, according to part (b), only speed a 5,000-metric-ton spacecraft up to about a thousandth the velocity of light.

The increase in relativistic mass of the spacecraft found in part (c) is exactly equal to the mass converted into energy found in part (a). The mass-energy relationship works both ways.

13-3   Looking through a microscope at a 10-micron animal swimming about in a drop of water, you wonder if these dimensions are small enough to bring the uncertainty principle into play. Planck's constant is equal to 6.63×10^{-34} J·s.

Given: The microscopic animal has a mass of about 10^{-12} kg and a velocity of about 10^{-4} m/s. You estimate the uncertainty in these measurements to be 10^{-13} kg and 10^{-5} m/s.

Find: (a) What is the uncertainty in the animal's momentum from your estimates?
(b) What uncertainty in position is predicted by the uncertainty principle?
(c) What uncertainty in time is associated with the uncertainty in the animal's kinetic energy?

Your Solution

Discussion

This exercise provides an opportunity to see if Heisenberg's uncertainty principle has much of an effect on observations made with an optical microscope. Remember that momentum is mass times velocity, and kinetic energy is one-half mass times the square of velocity.

Sample Solution 13-3

(a) Momentum including uncertainty

$$= (m + \Delta m)(v + \Delta v)$$

$$= 10^{-12-4} kg\ m/s \pm (10^{-12-5} \pm 10^{-13-4}) kg\ m/s \pm \underbrace{10^{-13-5} kg\ m/s}$$

$$\qquad\qquad\qquad\qquad\qquad\qquad\qquad\qquad\qquad\quad \text{small enough to ignore}$$

$$= 10^{-16} kg\ m/s \pm (2 \times 10^{-17} kg\ m/s)$$

where $\Delta(mv) = \boxed{2 \times 10^{-17} kg\ m/s}$

(b) $\Delta(mv)\Delta x \cong h$

$$\Delta x \cong \frac{h}{\Delta(mv)} \qquad\qquad \text{conversion factor}$$

$$= \frac{(6.63 \times 10^{-34} Joule \cdot s)}{(2 \times 10^{-17} kg\ m/s)} \left(\frac{kg\ m^2/s^2}{Joule}\right)$$

$$= \boxed{3 \times 10^{-17} m}$$

(c) Energy including uncertainty

$$= \tfrac{1}{2}(m \pm \Delta m)(v \pm \Delta v)^2$$

$$= \tfrac{1}{2}(10^{-12} kg \pm 10^{-13} kg)(10^{-4} m/s \pm 10^{-5} m/s)^2$$

$$= \tfrac{1}{2} 10^{-12} kg(10^{-4} m/s)^2 \pm \tfrac{1}{2}(10^{-12} kg)\, 2(10^{-4} m/s)(10^{-5} m/s)$$

$$\qquad \pm \tfrac{1}{2}(10^{-13} kg)(10^{-4} m/s)^2 \pm (\text{insignificantly small terms})$$

$$= \tfrac{1}{2} \cdot 10^{-20} kg\ m^2/s^2 \pm 10^{-21} kg\ m^2/s^2 \pm \tfrac{1}{2} \cdot 10^{-21} kg\ m^2/s^2$$

$$= E \pm \Delta E$$

where ΔE = "plus or minus" terms above

$$= 1.5 \times 10^{-21} kg\ m^2/s^2 = 1.5 \times 10^{-21} Joule$$

$$\Delta E\, \Delta t \cong h$$

$$\Delta t \cong \frac{h}{\Delta E} \cong \frac{(6.63 \times 10^{-34} Joule \cdot s)}{(1.5 \times 10^{-21} Joule)} \cong \boxed{4 \times 10^{-13} s}$$

Discussion

The method of finding the error, done as a special case in part (a), is done in general in calculus as the "chain rule." The result is as follows:

$$\Delta(mv) = m(\Delta v) + (\Delta m)v$$
$$\Delta(\tfrac{1}{2}mv^2) = \tfrac{1}{2}(\Delta m)v^2 + \tfrac{1}{2}m(2v\ \Delta v)$$

Those students who haven't had calculus can do part (c) by finding the uncertainty in the energy as a special case, as shown in part (a).

The resulting uncertainties in position and time found in parts (b) and (c) are far too small to be observed. The uncertainty principle therefore comes into play only at dimensions much smaller than are observable with an optical microscope. Quantum effects are therefore not noticeable on these dimensions.

13-4   A low-power laser produces a very bright spot of light, the power content of which is measured in milliwatts. You decide to estimate the number of photons per second in the laser beam. You know that Planck's constant $h = 6.6 \times 10^{-34}$ J·s.

Given: The laser beam is rated at a quarter of a milliwatt, 2.5×10^{-4} W, for light having a frequency of 4×10^{14} Hz.

Find: (a) How much energy does one photon of this light have?
(b) How many photons of this energy come from the laser in 1 s to produce the rated power?

Your Solution

Discussion

This is an exercise in using Planck's constant and Einstein's quantum theory of radiation. Remember that power is defined as energy per unit time. This exercise gives you an opportunity to see how many photons go into making a bright spot of light.

Sample Solution 13-4

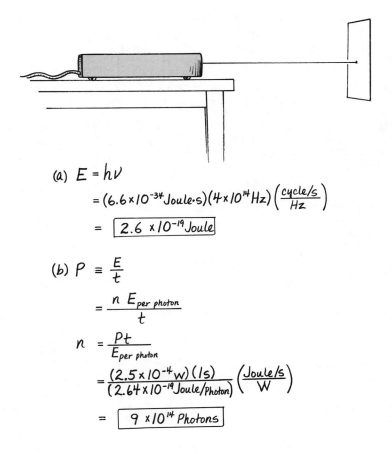

(a) $E = h\nu$

$= (6.6 \times 10^{-34} \text{Joule} \cdot \text{s})(4 \times 10^{14} \text{Hz})\left(\frac{\text{cycle/s}}{\text{Hz}}\right)$

$= \boxed{2.6 \times 10^{-19} \text{Joule}}$

(b) $P \equiv \dfrac{E}{t}$

$= \dfrac{n\, E_{\text{per photon}}}{t}$

$n = \dfrac{Pt}{E_{\text{per photon}}}$

$= \dfrac{(2.5 \times 10^{-4} \text{w})(1\text{s})}{(2.64 \times 10^{-19} \text{Joule/Photon})}\left(\dfrac{\text{Joule/s}}{\text{W}}\right)$

$= \boxed{9 \times 10^{14} \text{Photons}}$

Discussion

The energy in part (a) is found from Planck's hypothesis that the energy, E, per photon in electromagnetic radiation is proportional to the frequency, ν, with Planck's constant being the proportionality constant. The unit for frequency, the hertz, can be converted to cycles per second, with the cycles being merely understood in the final answer. The result is understood to be the energy per photon.

The result of part (a) is the energy used in part (b) with the units joule per photon being explicitly stated when inserted into the general solution. This general solution is obtained from the definition of power where the total energy is assumed to be the number of photons, n, times the energy per photon, as found in part (a).

13-5   In the thorium series of natural radioactivity, thorium 232 undergoes a series of alpha and beta decays to become eventually a stable isotope of lead.

Given: Thorium, $_{90}Th^{232}$, has an atomic number of 90 and a mass number of 232. It alpha decays with a half-life of about 10^{10} years into an isotope of radium. The radium then beta decays, with a half-life of 6.7 years into actinium, which again beta decays into another isotope of thorium. This second isotope of thorium is much less stable than thorium 232 and alpha decays with a half-life of 1.9 years.

Find: (a) What is the mass number of radium, $_{88}Ra^?$, into which $_{90}Th^{232}$ decays?
  (b) What are the mass numbers of actinium (Ac) and thorium resulting from beta decay?
  (c) What is the next isotope in the chain resulting from alpha decay of thorium?

Your Solution

Discussion

This problem in nuclear transformation involves remembering the difference between an alpha particle and a beta particle. The isotope resulting in part (c) then alpha decays again into radon, which alpha decays once more into polonium. Two more alpha decays and two more beta decays can then be seen to result in lead 208, $_{82}Pb^{208}$.

Sample Solution 13-5

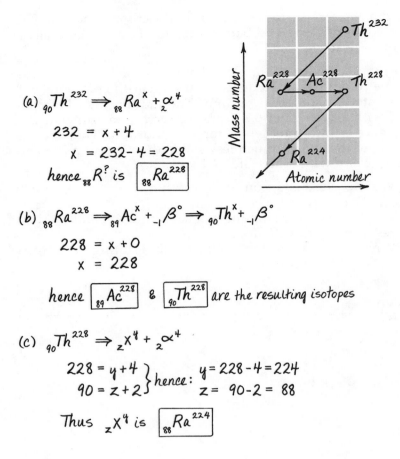

(a) $_{90}Th^{232} \Rightarrow \,_{88}Ra^{x} + \,_{2}\alpha^{4}$

$232 = x + 4$

$x = 232 - 4 = 228$

hence $_{88}R^{?}$ is $\boxed{_{88}Ra^{228}}$

(b) $_{88}Ra^{228} \Rightarrow \,_{89}Ac^{x} + \,_{-1}\beta^{0} \Rightarrow \,_{90}Th^{x} + \,_{-1}\beta^{0}$

$228 = x + 0$

$x = 228$

hence $\boxed{_{89}Ac^{228}}$ & $\boxed{_{90}Th^{228}}$ are the resulting isotopes

(c) $_{90}Th^{228} \Rightarrow \,_{z}X^{y} + \,_{2}\alpha^{4}$

$\left.\begin{array}{l} 228 = y + 4 \\ 90 = z + 2 \end{array}\right\}$ hence: $\begin{array}{l} y = 228 - 4 = 224 \\ z = 90 - 2 = 88 \end{array}$

Thus $_{z}X^{y}$ is $\boxed{_{88}Ra^{224}}$

Discussion

Both the mass numbers and atomic numbers add up on both sides of the reaction in part (a). The atomic number of radium is 88, two less than that of thorium, because the alpha particle carries away two protons. This mass number, 228, does not change in beta decay, since a beta particle has no mass number. Thorium again alpha decays into radium in part (c), since we already know from part (a) that the element having atomic number 88 is called radium. The rest of the thorium series is

$$_{88}Ra^{224} \Rightarrow \,_{86}Rn^{220} + \,_{2}\alpha^{4} \Rightarrow \,_{84}Po^{216} + \,_{2}\alpha^{4}$$

$$\Rightarrow \,_{82}Pb^{212} + \,_{2}\alpha^{4} \Rightarrow \,_{83}Bi^{212} + \,_{-1}\beta^{0} \begin{array}{c} \nearrow \,_{81}Tl^{208} + \,_{2}\alpha^{4} \searrow \\ \searrow \,_{84}Po^{212} + \,_{-1}\beta \nearrow \end{array} \,_{82}Pb^{208} + \begin{array}{c} \alpha \\ \beta \end{array}$$

ESSENTIAL PROBLEMS

© *DC Comics, Inc.*

13-6   Superman is a big six foot two husky fellow when at rest, but he is considerably shorter when flying at near the speed of light.

Given: Superman is 1.88 m tall when at rest, but loses some of that when traveling at $0.5c$, half the speed of light relative to you. When he goes still faster, his height is reduced to 1.21 m (4 ft tall).

Find: (a) What is Superman's height when traveling at half the speed of light?

(b) How much does his watch lose in an hour of your time at this speed?

(c) How fast is he going when his height is reduced still more?

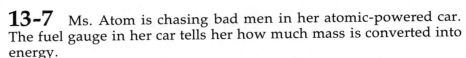

13-7   Ms. Atom is chasing bad men in her atomic-powered car. The fuel gauge in her car tells her how much mass is converted into energy.

Given: Ms. Atom's car has a mass of 1.2×10^3 kg and converts 1.0 g of matter into kinetic energy.

Find: (a) How much energy, in joules, is produced?

(b) How fast is Ms. Atom's car going if, starting from rest, all that energy is converted into the kinetic energy of the atomic car?

(c) By how much does the relativistic mass of the car increase as a result of this increase in velocity?

13-8   The molecules in a flask of oxygen at 0 °C under atmospheric pressure have an average velocity of about 460 m/s. The atoms in the molecule have an equilibrium separation of 1.12×10^{-10} m and a combined mass of 5×10^{-26} kg. ($h = 6.63 \times 10^{-34}$ J·s)

Given: A molecule has a momentum of 2.4×10^{-23} kg-m/s, give or take about 10%, and an average kinetic energy of 5×10^{-21} J, with about the same uncertainty.

Find: (a) What is the uncertainty in the molecule's momentum?

(b) What uncertainty in position is predicted by the uncertainty principle.

(c) What uncertainty in time is associated with the molecule's uncertainty in kinetic energy?

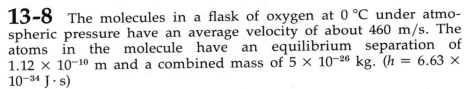

13-9   A mobile ham radio transmitter is tuned into the 10-m band and is sending out photons of energy in every direction. ($h = 6.63 \times 10^{-34}$ J-s)

Given: The radio transmitter is tuned to a frequency of 3.3×10^8 Hz and is operating at a power of 100 W.

Find: (a) How much energy does 1 photon of this frequency radio wave have?

(b) How many photons of this energy are radiated in 1 s?

13-10   In the uranium series of natural radioactivity, uranium 238 undergoes a series of alpha and beta decays to eventually become a stable isotope of lead, $_{82}Pb^{206}$.

Given: Uranium $_{92}U^{238}$ has an atomic number of 92 and a mass number of 238. It alpha decays with a half-life of about 10^9 years into an isotope of thorium. The thorium (Th) then beta decays with a half-life of about 24 days into an isotope of protactinium (Pa). The protactinium then beta decays into another isotope of uranium, which again alpha decays.

Find: (a) What is the mass number of the thorium isotope, $_{90}Th^?$, into which the uranium 238 decays?
(b) What are the isotopes of the Pa and U atoms resulting from beta decay?
(c) What is the next isotope in the chain resulting from alpha decay?

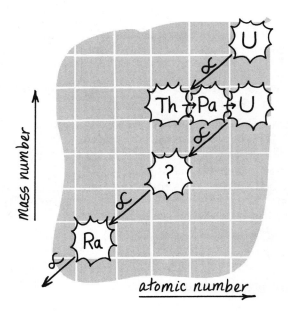

MORE INTERESTING PROBLEMS

13-11   Besides the thorium series and uranium series, there is only one other series of natural radioactivity. It is called the actinium series and starts with the less abundant isotope of Uranium, $_{92}U^{235}$. Like the other natural radioactive series, it ends in a stable isotope of lead.

Given: Uranium 235 alpha decays with a half-life of 10^8 years into an isotope of thorium. The thorium beta decays into an isotope of protactinium, which then alpha decays into actinium.

Find: (a) What isotope of thorium results from alpha decay of $_{92}U^{235}$?
(b) What isotope of actinium results from alpha decay of Pa?

13-12 You are on a spaceship going at near the speed of light traveling from one star to another several light-years distant. To you it seems that the distance between these stars is foreshortened, because you see them moving relative to you at near the speed of light.

Given: You are traveling at eight-tenths the speed of light, $0.8c$. The distance between the stars is 2.5 light-years in a frame of reference in which they are stationary, which you decide to call the star frame.

Find: (a) How much star frame time does it take you to make the journey in the star frame?
(b) Your clock runs slow because you are traveling so fast. How much time does your clock measure for the journey?
(c) How far does it seem to be between stars as measured in your frame of reference?
(d) How long does it seem to you for the distance between the stars to zip past going at the velocity you see the stars moving at?

13-13 A 5-W flashlight bulb emits a large number of photons of visible light and even a larger amount of heat energy photons. ($h = 6.63 \times 10^{-34}$ J·s)

Given: Thirty per cent of the energy going into a 5-W flashlight bulb is radiated as visible light with an average wavelength of 5×10^{-7} m. The flashlight is left on for 1 min.

Find: (a) How much energy does 1 photon of the average wavelength have?
(b) If all the visible light were of this wavelength, how many photons would be emitted in 1 min?
(c) What is the mass equivalent of this many photons of this energy?

13-14 The size of individual nuclear power plants is expected to increase from the 200-megawatt experimental plants to something between 3,000 and 3,500 megawatts by the mid-1980s.

Given: A 3×10^9-W nuclear power plant operates for 180 days on one set of fuel rods. Each uranium nucleus liberates about 1.5×10^{-13} J of energy when it fissions.

Find: (a) How much energy is produced on one set of fuel rods?
(b) What is the mass equivalence of this much energy?
(c) How many uranium nuclei must fission to produce this much energy?

13-15 The energy liberated in a chemical explosion produces a tiny change in mass, in accordance with the Einsteinian mass-energy relationship.

Given: One kilogram of dynamite liberates 5.4×10^6 J of energy when it explodes.

Find: (a) What is the mass equivalence of this much energy?
(b) What percentage of the total mass is converted to energy?

ANSWERS

13-6   (a) 1.63 m; (b) 0.13 hours = 8 min; (c) $.765c = 2.3 \times 10^8$ m/s.

13-7   (a) 9×10^{13} J; (b) approx 3.9×10^5 m/s; (c) 1×10^{-3} kg = 1 g.

13-8   (a) 2.4×10^{-24} kg-m/s; (b) $\sim 3 \times 10^{-10}$ m; (c) $\sim 1.3 \times 10^{-12}$ s.

13-9   (a) 2.19×10^{-25} J; (b) 4.6×10^{26} photons.

13-10   (a) 234; (b) $_{91}Pa^{234}$ and $_{92}U^{234}$; (c) $_{90}Th^{230}$.

13-11   (a) $_{90}Th^{231}$; (b) $_{89}Ac^{227}$.

13-12   (a) 3.1 years; (b) 1.9 years; (c) 1.5 light-years; (d) 1.9 years.

13-13   (a) 4×10^{-19} J/photon; (b) 4×10^{18} photons; (c) 2×10^{-17} kg.

13-14   (a) 4.7×10^{16} J; (b) 0.5 kg; (c) 3×10^{29} nuclei.

13-15   (a) 6.0×10^{-11} kg; (b) 6×10^{-9} %.

Algebra Review

Algebra is the next level of abstraction beyond arithmetic. In arithmetic you learn certain rules of logic regarding a set of things called real numbers. If you take a 2 and add it to a 7, you always get a 9:

$$2 + 7 = 9$$

If you take a 2, however, and multiply it by a 7, you always get a 14:

$$2 \times 7 = 14$$

The usefulness of these rules is that the real numbers to which they apply can stand for anything. Once you know that two apples added to seven apples always gives you nine apples, you also know that two dollars added to seven dollars will give you nine dollars.

In algebra, we let certain symbols, such as x, y, z, a, or t, stand for real numbers that are either unknown or of no importance at the moment. The great secret of algebra is that you can let a symbol stand for anything and still the rules and operations of arithmetic hold the same for them as for the real numbers. If you don't know what something is, you just fake it. You put down an x or y and pretend that you know what you are talking about. This is a valuable trick, and if you follow the correct rules of logic you will come out all right in the end.

For example, if someone gave you two dollars and you later count your money to find that you have nine dollars all together, it would be easy to figure out how much money you had to start with. The number 2, added to whatever it was, let's call it x, is 9:

$$2 + x = 9$$

If you subtract 2 from both sides of the equation, you see that x must have been equal to 7 if the equality was true in the first place:

$$x = 9 - 2 = 7$$

Getting the unknown isolated on one side of the equation so that you know what it is in terms of other symbols and numbers is called **solving** the equation. Certain operations are permissible in solving an equation, because they do not change the equality. One of these Great Truths of algebra is that you can move an additive term from one side of the equation to the other by changing its sign. This First Great Truth is an almost direct application of the more general rule that you can do almost anything you like to an equation as long as you do the same thing to both sides. In the example, we subtract 2 from both sides of the equation:

$$2 + x = 9$$
$$-2 = -2$$
$$x = 9 - 2 = 7$$

Most people would say that that is all there is to it; it seems obvious to most people that subtracting 2, or adding -2, to $(2 + x)$ gives x as an answer. People who know about forms of algebra that are more sophisticated than the simple kind that we will use in this course have to think about associative and distributive laws to persuade themselves that this is the case, but we will not worry about such things. We should note, however, that the addition and subtraction rule works as well for algebraic symbols as it does for the real numbers they represent.

EXERCISES

Solve the following equations for x. (Cover answers and then compare.)

		Answers
1. $5 + x = 10$	$x =$	5
2. $x + 7 = y$	$x =$	$y - 7$
3. $x - 3 = 7$	$x =$	10
4. $-4 + a = x + y$	$x =$	$-4 + a - y$
5. $5 + x = a - 7$	$x =$	$a - 12$
6. $a + x = 2a - y$	$x =$	$a - y$
7. $5a + y = 12a + x + y$	$x =$	$-7a$
8. $5 + 3x = 7 + 2x$	$x =$	2
9. $12x + y = 7x$	$x =$	$\frac{-y}{5}$
10. $x + 3y = 4x + 3a$	$x =$	$y - a$

▶ Multiplication and Division Rule

The preceding Exercises 9 and 10 involve dividing both sides of the equation by the same number to get x all by itself on one side of the equation. In Exercise 9, for example, subtracting $7x$ and y from both sides of the equation gives $5x$ equal to minus y. This expression can be solved for x by dividing both sides of the equation by 5;

$$\frac{5x}{5} = \frac{-y}{5}$$

Since 5 divides into itself leaving unity (1), the answer is $-y$ divided by 5.

$$x = \frac{-y}{5}$$

The Second Great Truth of algebra is that you can take something out of the numerator on one side of the equation and put it into the denominator on the other side. The reverse works equally well. You can also move something from the denominator on one side to the numerator on the other. If we were solving for y in the above expression, for example, we could go backwards to the preceding expression to get $-y$ equal to $5x$. There is one exception. Division by zero is undefined in the sense that the reciprocal of an infinitesimally small number is an infinitely large number.

Exercise 10 illustrates a point of caution that must be kept in mind when using the Second Great Truth. Subtracting $4x$ and $3y$ from both sides of the equation, and then changing the signs of both sides, gives

$$3x = 3y - 3a$$

Dividing both sides by 3 gives

$$\frac{3x}{3} = \frac{3y - 3a}{3}$$
$$x = y - a$$

One Infernal Pit which must be avoided is to divide the 3 into only one of the two terms on the right side of the equation instead of into both terms. Whenever we perform an operation on an equation, we must perform that operation on *every* member of the equation. Many people who would know better if only they would stop to think about it have stepped into this First Infernal Pit of algebra, and it has pulled them right down.

EXERCISES

Solve the following equations for x. (Cover answers and then compare.)

Answers

1. $12x = 24y$ $\qquad x =$ $\qquad 2y$

2. $ax = 2y$ $\qquad x =$ $\qquad \dfrac{2y}{a}$

3. $\dfrac{x}{3} = 3$ $\qquad x =$ $\qquad 9$

4. $\dfrac{x}{2} = y + 3$ $\qquad x =$ $\qquad 2y + 6$

5. $2y + 3x = 6a$ $\qquad x =$ $\qquad 2a - \dfrac{2y}{3}$

6. $4x + 3y = 3x + 5a$ $\qquad x =$ $\qquad 5a - 3y$

7. $7(x + y) = a$ $x =$ $\dfrac{a - 7y}{7} = \dfrac{a}{7} - y$

8. $\dfrac{1}{4}x + a = x + 6y$ $x =$ $\dfrac{4a}{3} - 8y$

9. $\dfrac{1}{x} = y + \dfrac{1}{a}$ $x =$ $\dfrac{a}{ay + 1}$

10. $\dfrac{1}{a} = \dfrac{1}{x} + \dfrac{1}{y}$ $x =$ $\dfrac{ay}{y - a}$

▶ Fractions

In several of the preceding exercises, the unknown was part of an algebraic fraction. The $\frac{1}{4}x$ in Exercise 8, for example, is subtracted from x to give $\frac{3}{4}x$ after rearranging the expression to get the unknown on one side:

$$x - \tfrac{1}{4}x = a - 6y$$

As in adding any fractions, a common denominator must be found. In this case, the terms containing x have 4 as a common denominator.

$$\frac{4x}{4} - \frac{x}{4} = a - 6y$$

$$\frac{3x}{4} = a - 6y$$

The unknown, x, is then isolated by multiplying both sides of the equation by the reciprocal of $\frac{3}{4}$, namely $\frac{4}{3}$:

$$x = \frac{4}{3}(a - 6y) = \frac{4a}{3} - 8y$$

Exercise 9 also requires finding a common denominator. A common denominator must be found in the first step of this exercise when it comes to adding y and $1/a$. In this case, a itself is the common denominator:

$$\frac{1}{x} = \frac{ay}{a} + \frac{1}{a}$$

The two terms on the right can then be added, since they are fractions with the same denominator:

$$\frac{1}{x} = \frac{ay + 1}{a}$$

Some people would be tempted to let the a in the denominator cancel the a in the numerator but they must put down such temptation and recognize it as the same First Infernal Pit of algebra discussed in the last section

$$\frac{\cancel{a}y + 1}{\cancel{a}} \qquad \text{WRONG!}$$

If you divide a into one term of the numerator, you must divide it into both, and that would put you back at the original statement of the problem in Exercise 9.

People who mistakenly believe that you can cancel the a in the denominator have probably forgotten the difference between an *additive term* and a *multiplicative factor* in an algebraic expression. Additive components of an algebraic expression connected with a plus or minus sign are called **terms,** while multiplicative components are called **factors.** Thus, in an expression such as $4x^2 + 3ay + 2axy$, there are three terms, which are $4x^2$, $3ay$, and $2axy$ all added together. In the last term, 2 and the symbols a, x, and y are all factors. It is okay to divide something into one factor of an expression, but not one term of an expression. Thus,

$$\frac{2axy}{a} = 2xy \qquad \text{but} \qquad \frac{4x^2 + 2axy}{a} \neq 4x^2 + 2xy$$

The proper final step in solving Exercise 9 is to take the reciprocal of both sides of the equation, just as in Exercise 10. This gets the unknown out of the denominator of the fraction. The first step in Exercise 10, however, is to get the term containing the unknown, x, alone on one side of the equation even though x is in the denominator.

$$\frac{1}{x} = \frac{1}{a} - \frac{1}{y}$$

Then, after adding the two terms on the right with the same common denominator, we are ready to take the reciprocal.

$$\frac{1}{x} = \frac{y}{ay} - \frac{a}{ay} = \frac{y - a}{ay}$$

Taking the reciprocal of both sides is perfectly legal, as it may be looked upon as a multiple application of the Second Great Truth of algebra, whereby things are moved from the denominator on one side to the numerator on the other.

$$x = \frac{ay}{y - a}$$

Taking the reciprocal of both sides may, on the other hand, be looked upon as an application of a Third Great Truth of algebra, which is that both sides of an equation can be raised to the same exponential power. In this case, the exponential power is -1.

▶ Exponents

By an exponential power, we mean the number of times something is multiplied by itself. Thus x squared is x to the second power; x cubed is x to the third power, and so on.

$$x \cdot x = x^2 \qquad \text{and} \qquad x \cdot x \cdot x = x^3$$

In this case, x is called the **base** and 2 or 3 is called the **exponent.** The rule for multiplying two things together, if they have the same base, is to add their exponents. Thus, x squared times x cubed is equal to x to the fifth power.

$$x^2 \cdot x^3 = (x \cdot x)(x \cdot x \cdot x) = x^{2+3} = x^5$$

Application of this rule to division leads to the concept of a negative exponent. Dividing by x cubed is the same as subtracting its exponent.

$$\frac{x^2}{x^3} = \frac{x \cdot x}{x \cdot x \cdot x} = x^{2-3} = x^{-1}$$

By x^{-1} we mean the reciprocal of x, or $1/x$.

Another application of this rule to division leads to the fact that anything raised to the zero power equals 1. That is what happens when you divide something by itself. For example, x^2 divided by x^2 equals unity (1).

$$\frac{x^2}{x^2} = x^{2-2} = x^0 = 1$$

One very useful application of exponents is in expressing very large or very small numbers in terms of powers of 10. A million, for example, is 10 to the sixth power:

$$1,000,000 = 10^6$$

The speed of light, which is two hundred ninety-nine million seven hundred ninety-three thousand meters per second, can be expressed in power-of-ten notation as

$$c = 299,793,000 \text{ m/s} = 2.99793 \times 10^8 \text{ m/s}$$

This power-of-ten notation is therefore useful for placing the decimal point without using zeros, which may or may not be significant in terms of information content. In the speed of light, for example, the last three zeros are not significant. To nine significant figures, the speed of light is 299,793,456 m/s.

Another use of exponents is in expressing square and cube roots when solving algebraic equations for an unknown that is raised to an exponential power. The exponent of the root, in this case, turns out to be a fraction rather than a whole number.

$$\sqrt{x} = x^{1/2}$$

The logic of this notation comes from the rule for multiplying. Squaring both sides of the equation gives x to the first power just as it should:

$$(\sqrt{x})^2 = (\sqrt{x})(\sqrt{x}) = x^{1/2}x^{1/2} = x^{1/2+1/2} = x^1 = x$$

As an example of the fractional exponent notation, we can solve the Pythagorean theorem for one of the sides. The Pythagorean theorem, you may recall, states that the square of the hypotenuse, h, of a right triangle is equal to the sum of the squares of the sides x and y.

$$h^2 = x^2 + y^2$$

Solving this expression for x^2,

$$x^2 = h^2 - y^2$$

and then taking the square root of both sides of the equation gives the solution in terms of the side x.

$$x = (h^2 - y^2)^{1/2}$$

This brings us to the edge of the Second Infernal Pit of algebra, into which the unwary sometimes fall. Some people might be tempted to let the square root of the difference $(h^2 - y^2)$ be equal to the difference of the square roots. While it is true that the exponential power of a product is equal to the product of the exponential powers, it is *not* true that the same rule holds for sums and differences. People fall into this Second Infernal Pit for the same reason that they fall into the first. They fail to distinguish additive terms and multiplicative factors.

$$(A \cdot B)^2 = A^2 \cdot B^2 \qquad \text{but} \qquad (A + B)^2 \neq A^2 + B^2$$

The rule that does hold for the square of a sum of two terms will be discussed in the next section.

EXERCISES

Solve for x. (Cover answers and then compare.)

Answers

1. $\dfrac{x}{m} + 3m = 4m$ $\qquad x =$ $\qquad\qquad m^2$

2. $\dfrac{x}{r} - 5 = r + 2$ $\qquad x =$ $\qquad\qquad r^2 + 7r$

3. $x^2 + y^2 = 5y^2$ $\qquad x =$ $\qquad\qquad 2y$

4. $\dfrac{x^2 - a^2}{y} = \dfrac{2y}{a}$ $\qquad x =$ $\qquad\qquad \left(\dfrac{2y^2}{a} + a^2\right)^{1/2}$

5. $5x^4 = 3x^4 + 4a^3$ $\qquad x =$ $\qquad\qquad (2a^3)^{1/4}$

Evaluate.

6. $x^2 x^5 =$ $\qquad\qquad\qquad\qquad\qquad x^7$

7. $\dfrac{x^2}{x^5} =$ $\qquad\qquad\qquad\qquad\qquad x^{-3}$

8. $x^4(xa)^{-2} =$ $\qquad\qquad\qquad\qquad x^2 a^{-2}$

9. $\dfrac{(6c^2 d)^2 (a^4 d)}{(2acd)} =$ $\qquad\qquad\qquad 18a^3 c^3 d^2$

10. $x^2 + x^5 =$ $\qquad\qquad\qquad\qquad x^2 + x^5$

▶ Operations with Polynomials

Exercise 10 in the preceding section might be thought of as a trick question in that many people will try to add the exponents in the same way as in Exercise 6. If you stand back and look at it, you will see that this is just another edge of that First Infernal Pit of algebra that people keep sliding into. We again need to draw a distinction between multiplying two things together, as in Exercise 6, and adding them, as in Exercise 10. The two factors x^2 and x^5 are part of the same term in Exercise 6, since they are multiplied together. They are separate terms in Exercise 10, since they are added.

An algebraic expression having only one term is frequently called a **monomial** to distinguish it from an expression having two or more terms, which is called a **polynomial.** A polynomial with two terms is called a **binomial,** and one with three terms is called a **trinomial,** and so on, but all polynomials have more than one term.

Polynomials differ from monomials in the way in which they are multiplied or divided by things. Consider multiplying the polynomial $(x^2 + y)$ by the number 2, for example:

$$2(x^2 + y) = 2x^2 + 2y$$

Two times anything is just two of those things added together. Two times the quantity x^2 plus y is just $2x^2$ plus $2y$. The same would be true of any number times the polynomial.

$$n(x^2 + y) = nx^2 + ny$$

Thus, a monomial times a polynomial is the monomial times each term of the polynomial.

The equality works both ways. The polynomial in Exercise 10, for example, could be produced by multiplying x^2 times $(1 + x^3)$.

$$x^2(1 + x^3) = x^2 + x^5$$

In fact, $x^2(1 + x^3)$ could be used as an answer to Exercise 10, but it is not particularly better or simpler than the original statement $x^2 + x^5$. There are some applications, however, in which it might be desirable to "factor out" common members of a polynomial and express it as the product of that factor times a simpler polynomial. Consider the following expression, for example:

$$6xy^2 + 2xy^3 = 2xy^2(3 + y)$$

Whether the factored form of the expression, $2xy^2(3 + y)$, or the unfactored form, $6xy^2 + 2xy^3$, is better depends on the problem. When you see a polynomial having common factors, you only need to be aware of the fact that you can factor them out if need be.

We can extend the rule for multiplying a monomial by a polynomial to the case of multiplying polynomials together by looking again at the above development. There we considered multiplying 2 times a polynomial and then multiplying n times that same polynomial. We might consider multiplying the polynomial $(2 + n)$ times that polynomial. Since $(2 + n)$ means two of anything added to n of anything, we can just add the two above expressions to get

$$(2 + n)(x^2 + y) = 2x^2 + 2y + nx^2 + ny$$

Thus, the rule for multiplying polynomials together is to multiply each term of one polynomial by each term of the other and add the results.

Here again, the equality runs both ways; there are many cases in which an algebraic expression is best expressed in a factored form rather than the "multiplied out" form. It all depends on the particular problem. Consider, for example, the following expression:

$$\frac{2x^2 + 2y + nx^2 + ny}{x^3 + xy}$$

This expression can be simplified by factoring out common terms in both numerator and denominator:

$$\frac{2(x^2 + y) + n(x^2 + y)}{x(x^2 + y)}$$

Dividing the $(x^2 + y)$ in the denominator into both terms of the numerator gives a much simpler expression:

$$\frac{2 + n}{x}$$

There are two polynomials that you should be able to recognize as expressible in factored form. They are the polynomials resulting from the square of a sum and the product of the sum and difference of two terms.

$$(x + y)^2 = x^2 + 2xy + y^2$$
$$(x + y)(x - y) = x^2 - y^2$$

When you see the polynomials on the right, you should know that they can be expressed in terms of the products on the left.

EXERCISES

Multiply out the following expressions. (Cover answers and then compare.)

		Answers
1.	$a\,(x + y) =$	$ax + ay$
2.	$x^2(1 + x^3) =$	$x^2 + x^5$
3.	$\frac{5y^2}{x^2}\,(3y + x^3) =$	$15y^3x^{-2} + 5y^2x$
4.	$(x + 1)(a + b) =$	$xa + xb + a + b$
5.	$(x + 1)(x^2 + x) =$	$x^3 + 2x^2 + x$

Factor the following polynomials.

6.	$ax + ay =$	$a(x + y)$
7.	$4y^2 + y^4 =$	$y^2(4 + y^2)$
8.	$mx + my + 4x + 4y =$	$(m + 4)(x + y)$
9.	$x^2 + 6x + 9 =$	$(x + 3)^2$
10.	$y^2 - 4 =$	$(y + 2)(y - 2)$

▶ Solving Equations Having Polynomial Fractions

Our review of algebra has led us to consider three Great Truths of algebra, or rules of logic that allow us to move things around in an equation to isolate one unknown in terms of the others. The First Great Truth allows us to move a whole additive or subtractive term from one side to another by changing the sign. The Second Great Truth allows us to move a factor from one side to the other by changing it between numerator and denominator. The Third Great Truth allows us to raise both sides of an equation to the same exponential power. We have also considered two Infernal Pits, or mistakes of reasoning which some people occasionally fall into. These errors in reasoning are based on a failure to distinguish between terms and factors in an algebraic expression. With three rules of reason to guide us and only two pitfalls to avoid, we can do all the algebra needed in this course. We can even do algebra more complex than we need at present.

One example of these more difficult problems is an algebraic equation having both monomials and polynomials as parts of a fraction. It is useful to work problems of this nature at this point because they summarize the algebraic operations that you will need to master in the first part of the course. Consider, for example, the following equation to be solved for x:

$$\frac{x^2 + y^2}{2xy} = \frac{x}{y^2}$$

We see that the unknown we are solving for is on both sides of the equation and that it is in both the numerator and denominator on the left. The best first step is frequently to get the unknown out of the denominator. Multiplying both sides by the denominator on the left, or using the Second Great Truth of algebra,

$$x^2 + y^2 = \frac{x(2xy)}{y^2}$$

which is the same as

$$x^2 + y^2 = \frac{2x^2}{y}$$

Using the First Great Truth of algebra to get terms containing the unknown on one side of the equation,

$$x^2 - \frac{2x^2}{y} = -y^2$$

Factoring out x^2,

$$x^2 \left(1 - \frac{2}{y}\right) = -y^2$$

Or, using negative exponent notation, we can write the same thing as follows:

$$x^2(1 - 2y^{-1}) = -y^2$$

The Second Great Truth of algebra can then be used to solve for x^2:

$$x^2 = \frac{-y^2}{1 - 2y^{-1}}$$

We could just take the square root of both sides at this point and be done with it, but it is nice to try to simplify the fraction on the right. Multiplying both numerator and denominator by $-y$ gets rid of the negative exponent in the denominator.

$$x^2 = \frac{-y^2(-y)}{-y + 2}$$

$$= \frac{y^3}{2 - y}$$

Then taking the square root of both sides, we get the solution. By the Third Great Truth of algebra,

$$x = \frac{y^{3/2}}{(2 - y)^{1/2}}$$

Thus, this one problem makes use of all three algebra rules we are reviewing and confronts us with both pitfalls that people should learn to avoid. You might find it instructive to go back over this example to see if you can find where the pitfalls lie.

EXERCISES

Solve for x. (Cover answers and then compare.)

Answers

1. $\dfrac{x^2 + y^2}{2z} = \dfrac{z}{y}$ $x =$ $\left(\dfrac{2z^2}{y} - y^2\right)^{1/2}$

2. $\dfrac{x + 2y + a}{3x} = a$ $x =$ $\dfrac{2y + a}{3a - 1}$

3. $a = \dfrac{x + y}{x - y}$ $x =$ $y\,\dfrac{a + 1}{a - 1}$

4. $\dfrac{x^2 + 4x + 4}{x + 2} = 4$ $x =$ 2

5. $\dfrac{(x + 1)(x^2 - 1)}{(x - 1)(x + 1)} = 1$ $x =$ 0

Trigonometry Review

Trigonometry is the study of angles. There are several ways of measuring an angle. One is to draw a circle around the apex and then divide the circle up into some number of parts and call them **degrees.**

The number of parts can be any arbitrary number. We obtained the convention of using 360° from the ancient Mesopotamians. It seems likely that they got that number by noting that there are about 360 days in the year. No one knows why they did not use $364\frac{1}{4}$. It probably has to do with the fact that they had a numbering system based on 60.

Another approach would be to measure an angle by taking the ratio of the arc length to the radius of the circle.

$$\angle\,\theta = \frac{C}{r}$$

This ratio is called the radian measure of the angle. Since the arc length of a complete circle is 2π times the radius, one complete circle, or 360°, is equal to 2π radians.

$$360° = \frac{2\pi r}{r} = 2\pi \text{ radians}$$

We do not have to draw a circle in order to measure an angle. Instead we can drop a perpendicular from one side of the angle to the other, forming a right triangle.

Right triangles have two legs and a hypotenuse. The hypotenuse, labeled h, is the side opposite the right angle. One of the two remaining sides is **adjacent** to the angle θ and the other is **opposite.** Let us label them a and o, respectively. The angle can now be measured in terms of the ratios of the sides of the triangle. The ratio of the opposite over the hypotenuse depends only on the size of the angle, not on the size of the triangle we draw. A larger triangle would have a larger leg opposite the angle, but it would also have a proportionally larger hypotenuse. The ratio of opposite to hypotenuse would be the same in the large triangle as in the small one.

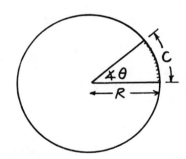

FIGURE II-1 *An angle can be measured with a circle divided into parts.*

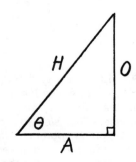

FIGURE II-2 *An angle can also be measured using a right triangle. One leg will be opposite the angle and the other adjacent.*

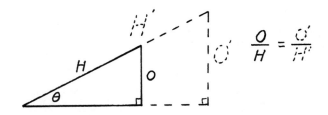

FIGURE II-3 *The ratio of the leg opposite the angle to the hypotenuse is independent of the size of the triangle. It depends only on the angle.*

The ratio of the opposite to the hypotenuse is a *signature* of the angle θ, called the **sine** of the angle.

$$\sin \theta \equiv \frac{o}{h}$$

The ratio of the adjacent leg to the hypotenuse is also independent of the size of the triangle. It is the *co-signature*, called the **cosine,** of the angle.

$$\cos \theta \equiv \frac{a}{h}$$

The only other ratio left is that of the opposite over the adjacent leg. This ratio is called the **tangent** of the angle.

$$\tan \theta \equiv \frac{o}{a}$$

(Its name comes from the fact that this ratio is used in finding the slope of a line that is tangent to a curve at a particular point.)

A good way to remember the sine and the cosine is to visualize that part of the triangle with which they are associated. We will often be given the hypotenuse and wish to know the lengths of the legs. Multiplying the definitions of sine and cosine by the hypotenuse, we find that the hypotenuse times the sine of the angle; and the adjacent side equals the hypotenuse times the cosine of the angle.

Eventually you will think sine when you see the side of a right triangle opposite an angle, and you will think cosine when you see the side adjacent. The tangent is then the ratio of the sine to the cosine.

The sum of all three angles in any triangle is 180°. Since one of the angles is a right triangle, equal to 90°, the other two add up to the remaining 90°. The other angle in the triangle, unmentioned until now, is called the **complement** of θ.

FIGURE II-4 *The basic trigonometric definitions give the legs of a right triangle in terms of the hypotenuse. The side opposite the angle is the hypotenuse times the sine, and the side adjacent is the hypotenuse times the cosine. The tangent is the ratio of the sine to the cosine, which is the ratio of the legs they are associated with.*

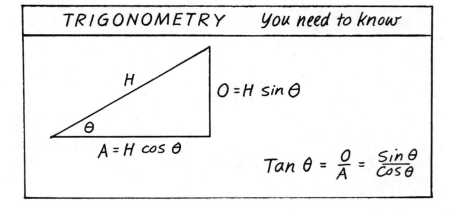

The side of the triangle opposite θ is adjacent to that other angle. This fact is the basis of the trigonometric identity that says, **The sine of an angle is the cosine of its complement.**

$$\sin \theta = \cos(90° - \theta)$$

When you look at the side of a right triangle, look at it in relation to an angle. The side is the hypotenuse times the sine of one of the angles. It is also the hypotenuse times the cosine of the other angle.

APPENDIX 3

Slide Rule Use

Often people taking their first physics course are inclined simply to do all computations by hand. This is an inefficient use of their time. Arithmetic is the least important and most tedious part of a physics problem, and yet it can be the most time-consuming.

Because the models we use most are of a proportional nature, most of our computations will be multiplication and division, the most difficult arithmetic to do by hand. This course, therefore, calls for the use of a calculator. The least expensive calculator for this course is a slide rule, three sticks which have logarithmic scales printed on them. Multiplication and division are done by adding and subtracting distances on these movable logarithmic scales.

Although at first a slide rule requires a bit of study and practice, it is actually better than an inexpensive digital calculator for some purposes. The rules of significant figures are easier to see in terms of the slide rule scales. Besides, slide rules are kind of fun to play with if you have the time and patience to learn how they work. What is more, an inexpensive slide rule that is fancy enough to handle trigonometric functions has little or no market value. You can put it down, turn your head for a moment, and there will still be a finite probability that it will be there when you look for it. If you want a cheap alternative to hand arithmetic that can be left on during an exam without running its batteries down or if you happen to have an old slide rule around, you may well wish to learn how a slide rule works.

You don't need to know anything about logarithms to work a slide rule. All that is necessary is to notice that someone has been very clever about the spacing of the numbers on the scales so that everything stays in proportion. If you place the 1 on the B scale opposite the 2 on the A scale, all of the other numbers on the A and B scales are in the proportion of 1 to 2. The 2 on the B scale is opposite the 4 on the A scale. The 3 on the B scale is opposite the 6 on the A scale, and so on. The slide rule is set up to multiply anything times 2.

Note that it tells that 2 times 4 is equal to 8, and then verify that 2 times 5 is equal to 10. Although the 10 on the A scale opposite the 5 on the B scale looks like a 1, we will call it 10, since it follows 9 on the

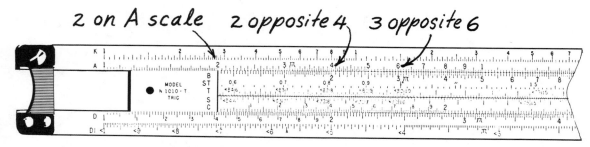

Slide rule set up to multiply 2 times any number

FIGURE III-1 *Slide rule set up to multiply 2 times anything.*

A scale. The second 2 on the A scale can also stand for 20, and all of those scratches between the 1 and 2 are then numbers between 10 and 20. Convince yourself that the scratch on the A scale opposite the 6 on the B scale could be called 12.

The rule for multiplying two numbers on the A and B scales is place the 1 on the B scale opposite one of the numbers on the A scale and find the other number on the B scale. The answer then is found opposite that number on the A scale. Suppose you are multiplying 3 times 4. Put the 1 on the B scale opposite the 3 on the A scale. Then you look to find the 4 on the B scale. The answer is 12 on the A scale. We know that 3 times 5 is 15. The number 15 on the A scale is the large mark halfway between the 10 on the A scale and the 20. Three times 6 is found to be two medium marks short of 20.

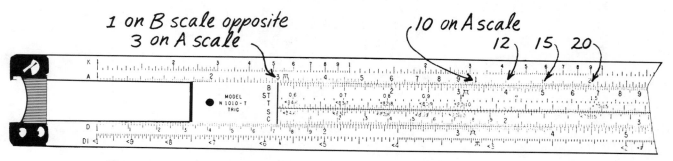

Slide rule set up to multiply 3 times any number

FIGURE III-2 *Slide rule set up to multiply 3 times anything.*

There are different numbers of subdivisions between numbers on different parts of the scale, because the spacing between the numbers changes. Looking to find the value of π on the B scale, notice that there is room for 10 subdivisions between the 3 and the 4. Pi (π) is indicated between the first and second marks, counting from the 3. Thus, π is between 3.1 and 3.2. Estimating the distance between subdivisions to be about halfway, we might read π as 3.15. Of course, the value of π is really 3.14159265 . . . ; but three significant figures is about all most slide rules can handle. Fortunately, that's plenty for this course.

Turn your attention to the C and D scales for a moment. We find that they are the same as the A and B scales except that they are twice as long. The value of π is easily read to three significant figures on the C scale. The π scratch on the C scale is clearly short of the midway mark between 3.1 and 3.2. You can easily estimate that

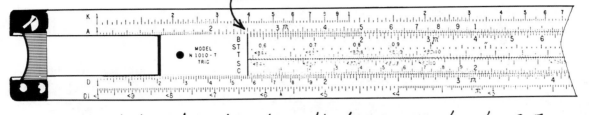

1 on B scale opposite 2.5 on A scale

Slide rule set up to multiply any number by 2.5

$\pi = 3.14$ on this scale. Once you master the A and B scales, you do the same calculations with more accuracy on the C and D scales. However, two significant figures will be adequate for most of the problems in this course.

Let us return to the A and B scales to multiply 2.5 times 3. By moving the 1 on the B scale opposite the 2.5 on the A scale, we set the slide rule to multiply 2.5 times anything. Looking at the 3 on the B scale, we find that it is halfway between the 7 and the 8 on the A scale. The answer is 7.5. We can use the same setting of the rule to multiply 2.5 times 4 to get 10.

We can even use the same setting of the rule to multiply 25 times 4 to get 100. If we call the first 2 a 20 and the first 3 a 30, the halfway mark between the 2 and the 3 becomes 25. Multiply that by 4 to get 100. It is necessary to keep track of the decimal place in your head. With a little practice, you will find it easy to deal with numbers between a thousandth and a hundred thousand. One trick in placing the decimal point is to choose nearby numbers that we can deal with in our heads. Before multiplying 25 times 620, for example, we might figure that 20 times 500 is 10,000. The mark on the A scale above the 6.2 mark on the B scale would then be read as 15,500, since we know it will be between 10,000 and 20,000. When numbers get either too large or too small to deal with in our heads, they can always be handled using the power-of-ten notation explained in chapter 1.

Division is just the reverse of multiplication. The slide rule scales, set up to multiply 2.5 times 4 to get 10, can be used to divide 10 by 4 to get 2.5. The rule for division is to find the numerator on the A scale, place the denominator below it on the B scale, and find the answer on the A scale opposite the 1 on the B scale. This operation may be thought of as finding that number by which you would have to multiply the denominator to get the numerator. But, of course, that is not the way you will think of it after you get used to the slide rule. You just know that you start with the 1 on the movable scale when you multiply and end up with the 1 on the movable scale when you divide.

Square roots are found on the slide rule by going from the A scale to the D scale. Looking at the 2 on the D scale of any of the figures, notice that 4 is right above it on the A scale. The 9 is above the 3 on the D scale. All numbers on the A scale are squares of the numbers on the D scale. All the numbers on the D scale are square roots of the numbers on the A scale. Looking at the slide rule in Figure III–1, notice that when the 1 on the B scale is opposite the 2 on the A scale, the 1 on the C scale is opposite the square root of 2 on the D scale.

$$\sqrt{2} = 1.41$$

FIGURE III-3 *Slide rule set up to multiply anything by 2.5.*

The slide rule in Figure III-2 is also set up to take the square root of 3. The 1 on the C scale is opposite $\sqrt{3}$.

$$\sqrt{3} = 1.73$$

The slide rule in Figure III-3 shows that

$$\sqrt{2.5} = 1.58$$

You need to be careful which end of the A scale you use. The left half of the A scale is for numbers between 1 and 10, while the right half is for numbers between 10 and 100. The square roots of numbers between 100 and 1,000 is found by using the left half of the A scale, while the square roots of numbers from 1,000 to 10,000 are found using the right half.

Alternate powers of ten use alternate ends of the A scale. Here again, a useful trick is to estimate in your head with nearby even numbers and then to read the exact number on the scale.

▶ Trigonometric Functions on a Slide Rule

The sine of an angle is found on the S scale of a slide rule. This scale is always read in angles, but the corresponding value of the sine is found on the C scale on some rules and the B scale on others. It is useful to know the sine and cosine of three common angles so that you can verify that you are using the proper scales for any particular rule that you happen to pick up. These numbers come up frequently enough so that it is worthwhile remembering them. They are

.500
.707
.866

You can identify these as the sines and cosines of 30°, 60°, and 45° by remembering the shapes of two common right triangles. The 45°–45° triangle is isosceles, while the 30°–60° is not.

If both of these triangles are drawn so that the hypotenuse is one unit of length, the smallest of the three values you have memorized is associated with the short side of the 30°–60° triangle. The value .500 is the sine of 30° and the cosine of 60°, since it is the side opposite the little angle and adjacent the large angle. Similarly, .866 can be visualized as the cosine of 30° and the sine of 60°, since it is adjacent and opposite those angles. The remaining value, .707, is represented by the length of the two sides of the isosceles triangle and is the sine and cosine of 45°.

Using the angles for which we know the sines and cosines, we can figure out how to use the S scale of any slide rule. Suppose we wanted to know the sine of 50°. We know that the result must be between the sines of 45° and 60°; that is, between .707 and .866. Look at the slide rule in Figure III-5. The numbers on the S scale to the right of the scratches are the angles for which the sine is indicated on the C scale. The number 60, to the right of the mark on the S scale, is located at a point on the C scale that is two-thirds of the way from the 8 to the 9. We read this as .866, since we know that the sine of an angle is always less than 1. Also note that 45°, the midway mark between 40° and 50°, is located at .707 on the C scale. Moving the hairline to 50°, we see that its sine is .766.

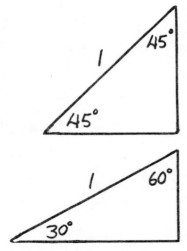

FIGURE III-4   *A 30°–60° and a 45°–45° triangle have the sines and cosines of 30°, 60°, and 45° as sides if the hypotenuses of both triangles are equal to unity.*

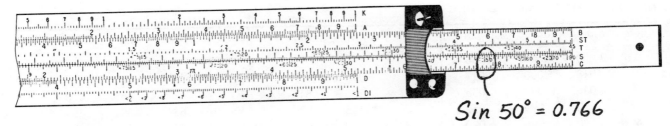

Sin 50° = 0.766

FIGURE III-5 *A slide rule set up to take the sine of 50°.*

The numbers on the S scale to the left of the scratches are the angles for which the cosine of the angle is given on the C scale. These numbers are all complements of the numbers to the right of the scratches. Some slide rules do not have numbers to the left and to the right of the scratches. To find the cosine of an angle on such a rule, subtract the angle from 90° and look up the sine of the complement of the angle.

The tangent of an angle is found using the T scale. The numbers on the T scale refer to angles. The values of the tangent of an angle less than 45° are found on the C scale. Angles greater than 45° have tangents greater than 1.00. Their value is found on the C1 scale or the reciprocal C scale.

NATURAL TRIGONOMETRIC FUNCTIONS

Angle	Sine	Cosine	Tangent	Angle	Sine	Cosine	Tangent	Angle	Sine	Cosine	Tangent
0°	0.000	1.000	0.000								
1°	.018	1.000	.017	31°	.515	.857	.601	61°	.875	.485	1.804
2°	.035	0.999	.035	32°	.530	.848	.625	62°	.883	.470	1.881
3°	.052	.999	.052	33°	.545	.839	.649	63°	.891	.454	1.963
4°	.070	.998	.070	34°	.559	.829	.675	64°	.899	.438	2.050
5°	.087	.996	.088	35°	.574	.819	.700	65°	.906	.423	2.145
6°	.105	.995	.105	36°	.588	.809	.727	66°	.914	.407	2.246
7°	.122	.993	.123	37°	.602	.799	.754	67°	.921	.391	2.356
8°	.139	.990	.141	38°	.616	.788	.781	68°	.927	.375	2.475
9°	.156	.988	.158	39°	.629	.777	.810	69°	.934	.358	2.605
10°	.174	.985	.176	40°	.643	.766	.839	70°	.940	.342	2.747
11°	.191	.982	.194	41°	.656	.755	.869	71°	.946	.326	2.904
12°	.208	.978	.213	42°	.669	.743	.900	72°	.951	.309	3.078
13°	.225	.974	.231	43°	.682	.731	.933	73°	.956	.292	3.271
14°	.242	.970	.249	44°	.695	.719	.966	74°	.961	.276	3.487
15°	.259	.966	.268	45°	.707	.707	1.000	75°	.966	.259	3.732
16°	.276	.961	.287	46°	.719	.695	1.036	76°	.970	.242	4.011
17°	.292	.956	.306	47°	.731	.682	1.072	77°	.974	.225	4.331
18°	.309	.951	.325	48°	.743	.669	1.111	78°	.978	.208	4.705
19°	.326	.946	.344	49°	.755	.656	1.150	79°	.982	.191	5.145
20°	.342	.940	.364	50°	.766	.643	1.192	80°	.985	.174	5.671
21°	.358	.934	.384	51°	.777	.629	1.235	81°	.988	.156	6.314
22°	.375	.927	.404	52°	.788	.616	1.280	82°	.990	.139	7.115
23°	.391	.921	.424	53°	.799	.602	1.327	83°	.993	.122	8.144
24°	.407	.914	.445	54°	.809	.588	1.376	84°	.995	.105	9.514
25°	.423	.906	.466	55°	.819	.574	1.428	85°	.996	.087	11.43
26°	.438	.899	.488	56°	.829	.559	1.483	86°	.998	.070	14.30
27°	.454	.891	.510	57°	.839	.545	1.540	87°	.999	.052	19.08
28°	.470	.883	.532	58°	.848	.530	1.600	88°	.999	.035	28.64
29°	.485	.875	.554	59°	.857	.515	1.664	89°	1.000	.018	57.29
30°	.500	.866	.577	60°	.866	.500	1.732	90°	1.000	.000	∞

Index